This book is due for return not later than the
last date stamped below, unless recalled sooner.

Urban Environments
and Wildlife Law

Urban Environments and Wildlife Law

A MANUAL FOR SUSTAINABLE DEVELOPMENT

Paul A. Rees

School of Environment & Life Sciences
University of Salford

Blackwell
Science

With natural history illustrations by
Alan J. Woodward
BSc (Hons) MPhil PhD

© 2002 by Blackwell Science Ltd,
a Blackwell Publishing Company
Editorial Offices:
Osney Mead, Oxford OX2 0EL, UK
   Tel: +44 (0)1865 206206
Blackwell Science, Inc., 350 Main Street,
Malden, MA 02148-5018, USA
   Tel: +1 781 388 8250
Iowa State Press, a Blackwell Publishing
Company, 2121 State Avenue, Ames, Iowa
50014-8300, USA
   Tel: +1 515 292 0140
Blackwell Publishing Asia Pty, 550
Swanston Street, Carlton, Victoria 3053,
Australia
   Tel: +61 (0)3 9347 0300
Blackwell Wissenschafts Verlag,
Kurfürstendamm 57, 10707 Berlin,
Germany
   Tel: +49 (0)30 32 79 060

First published 2002 by Blackwell Science
Ltd

Library of Congress
Cataloging-in-Publication Data is available

ISBN 0-632-05743-2

A catalogue record for this title is available
from the British Library

Set in 10/12$\frac{1}{2}$pt Palatino
by DP Photosetting, Aylesbury, Bucks
Printed and bound in Great Britain by
MPG Books, Bodmin, Cornwall

For further information on
Blackwell Science, visit our website:
www.blackwellpublishing.com

For Katy, Clara, Mum and Dad

In Europe, 38% of bird species and 45% of all butterflies are threatened. In North and Western Europe, some 60% of wetlands have been lost. Some two-thirds of trees in the European Union are under stress. . . . Some fish stocks are under risk of collapse and some marine life other than commercial fish have been decimated. . . . International trade in wildlife is recognised as a threat to some 30,000 species.

. . . The diversity, distribution, composition in terms of size and age and abundance of different species are indicators of the well-being of the natural systems of the Earth that society depends upon. We must take action before it is too late to preserve the irreplaceable resources of nature and biodiversity.

Environment 2010: Our future, Our choice
The Sixth Environment Action Programme of the European Community

Each Contracting Party shall . . .:

(a) Develop national strategies, plans or programmes for the conservation and sustainable use of biological diversity. . .

United Nations Convention on Biological Diversity, 1992, Article 6

You can't build a tortoise.

Gerald Durrell

A dead Nature aims at nothing. It is the essence of life that it exists for its own sake, as the intrinsic reaping of value.

A.N. Whitehead

Contents

Preface   xv
Acknowledgements   xvii
Introduction   xix

Part I   Law, Ecology and Organisations   1

1   An introduction to the law   3
1.1   The English legal system and devolution   3
   1.1.1   Introduction   3
   1.1.2   Criminal and civil law   4
   1.1.3   Types of legislation   4
   1.1.4   Law reports   8
   1.1.5   Judicial review and *ultra vires*   9
1.2   Where to find the law   9
   1.2.1   Titles of legal cases and where to find
       law reports   9
   1.2.2   How to find English legislation   10
1.3   Lawyers and the court system   13
   1.3.1   The legal profession   13
   1.3.2   The courts   14
   1.3.3   Planning inquiries   15
   1.3.4   Legal personality and *locus standi*: who has a
       right to bring an action to the courts?   15
1.4   European Union law   16
   1.4.1   The history of the European Union   16
   1.4.2   The institutions of the European Union   17
   1.4.3   Primary legislation   18
   1.4.4   Secondary legislation   19
   1.4.5   The supremacy of European law   20
   1.4.6   Where to find EC law   22
   1.4.7   The EU and international law   22
1.5   International law   23
   1.5.1   The nature of international treaties   23
   1.5.2   Sources of international law   24
Reference   24

2   The principles of classification and ecology   25
  2.1   Animal and plant names and the law   25
       2.1.1   Introduction   25
       2.1.2   The binomial system   26
       2.1.3   Scientific names and the law   29
       2.1.4   The major groups of living things   29
       2.1.5   Species identification   29
  2.2   Basic concepts in ecology   33
       2.2.1   Introduction   33
       2.2.2   Energy flow   34
       2.2.3   Nutrient cycling   35
       2.2.4   Species diversity   36
       2.2.5   Population dynamics   37
       2.2.6   Ecological succession   40
       2.2.7   Soil   42
       2.2.8   Ecological surveys   44
       2.2.9   The distribution of species   48
  2.3   Nature conservation   48
       2.3.1   Why bother to conserve nature?   48
       2.3.2   Pollution and biological indicators of
                environmental quality   50
       2.3.3   Conservation management   53
       2.3.4   Design of nature reserves   55
       2.3.5   Urban ecology   56
  References   57
      Identification guides   58

**3   The organisation and administration of nature
    conservation in the UK   59**
  3.1   Central government and its agencies   59
       3.1.1   Department for Environment, Food and
                Rural Affairs   59
       3.1.2   The statutory nature conservation agencies   62
       3.1.3   National Parks Authorities   65
       3.1.4   Wildlife Inspectorate   65
       3.1.5   Heritage   66
       3.1.6   Other government agencies   67
       3.1.7   Local government and police   74
  3.2   Non-statutory conservation organisations   76
       3.2.1   Species-based organisations   76
       3.2.2   Habitat-based organisations   78
       3.2.3   Environmental pressure groups   80
  3.3   An outline of nature conservation law   81
       3.3.1   A brief history of the law in the UK   81

	3.3.2	European law	83
	3.3.3	International law	87
3.4	UK government policy		87
	3.4.1	The Urban and Rural White Papers	88
	3.4.2	Countryside and Rights of Way Act 2000	89
References			90

Part II Species and Habitat Protection   91

4 Species protection under UK law   93

4.1	The protection of birds		93
	4.1.1	Introduction	93
	4.1.2	Classification and natural history	98
	4.1.3	Birds and the law	103
	4.1.4	Mitigation measures	109
	4.1.5	Case studies	110
4.2	The protection of animals		113
	4.2.1	Introduction	113
	4.2.2	Classification and natural history	113
	4.2.3	Animals and the law	125
	4.2.4	Mitigation measures	130
	4.2.5	Case studies	132
4.3	The protection of bats		132
	4.3.1	Introduction	132
	4.3.2	Natural history	133
	4.3.3	Bats and the law	135
	4.3.4	Mitigation measures	138
	4.3.5	Case studies	140
4.4	The protection of badgers		141
	4.4.1	Introduction	141
	4.4.2	Natural history	141
	4.4.3	Badgers and the law	144
	4.4.4	Mitigation measures	148
	4.4.5	Case studies	150
4.5	The protection of deer		152
	4.5.1	Introduction	152
	4.5.2	Natural history	152
	4.5.3	Deer and the law	154
	4.5.4	Mitigation measures	157
4.6	The protection of seals		157
	4.6.1	Introduction	157
	4.6.2	Natural history	157
	4.6.3	Seals and the law	158
4.7	Protection of wild plants		161

4.7.1   Introduction   161
4.7.2   Natural history   162
4.7.3   Plants and the law   163
4.7.4   Mitigation measures   164
4.7.5   Case study   164
4.8   Introduction of new species   165
4.8.1   Introduction   165
4.8.2   Introduced species and the law   167
4.8.3   A legal obligation to restore wildlife
and habitats   169
4.8.4   Case studies   170
4.9   Fish, game and quarry species   173
4.9.1   Fish, game and quarry species and the law   173
4.9.2   Foxes   174
4.10   Access to land for conservation purposes   176
4.10.1   Case study   176
4.11   Cruelty to wildlife   177
4.11.1   Introduction   177
4.11.2   The law and cruelty to wild mammals   177
4.11.3   Cruelty to other species   179
4.11.4   Illegal poisoning of wildlife   179
References   179

5   **Habitat and landscape protection under UK law   183**
5.1   Introduction   183
5.2   Multiple designation of sites   184
5.3   Landscape designations   185
5.3.1   National Parks   185
5.3.2   Areas of Outstanding Natural Beauty
(AONBs), National Scenic Areas (NSAs)
and Natural Heritage Areas (NHAs)   187
5.3.3   Heritage Coasts and coastal erosion   189
5.3.4   Natural Areas   191
5.3.5   National Trails   193
5.4   Sites of Special Scientific Interest   193
5.4.1   Introduction   193
5.4.2   Definition, selection and notification of SSSIs   194
5.4.3   Public bodies   197
5.4.4   Management schemes   198
5.4.5   Management agreements and compulsory
purchase   199
5.4.6   Offences   200
5.4.7   Entry onto land   202
5.4.8   Case law   202
5.4.9   Case study   205

5.5	National Nature Reserves		206
	5.5.1	Case study	206
5.6	Forests		207
	5.6.1	Forest Nature Reserves	207
	5.6.2	Community forests and the National Forest	208
5.7	Local wildlife sites		208
	5.7.1	Local Nature Reserves	208
	5.7.2	Local wildlife sites and the Wildlife Trusts	210
	5.7.3	Country parks	211
	5.7.4	Privately owned nature reserves	212
5.8	Protection of important geological sites		212
	5.8.1	Limestone pavements	213
	5.8.2	Regionally Important Geological and Geomorphological Sites (RIGS)	215
	5.8.3	Protection of caves	216
5.9	Geological structures and landscape features as ancient monuments		217
	5.9.1	Case study	217
5.10	Protection of freshwater environments		218
	5.10.1	The ecological importance of freshwater habitats	218
5.11	Protection of marine wildlife and habitats		220
	5.11.1	Marine Nature Reserves	222
5.12	Farming, wildlife and access to the countryside		223
	5.12.1	Agri-environment designations	224
	5.12.2	Farmyard manure and wildlife	225
	5.12.3	Destruction of wildlife for disease control	225
	5.12.4	Restriction of access to the countryside for the purposes of conservation	227
	5.12.5	Genetically modified organisms and the threat to the countryside	230
5.13	Heritage conservation		230
	5.13.1	Statutory heritage designations	231
	5.13.2	Non-statutory heritage designations	232
References			232
6	**The protection of trees and hedgerows**		**234**
6.1	The protection of trees		234
	6.1.1	Introduction	234
	6.1.2	Natural history	235
	6.1.3	Trees and the law	238
	6.1.4	Mitigation measures	246
	6.1.5	Case study	246
6.2	The protection of hedgerows		247
	6.2.1	Introduction	247

	6.2.2	Natural history	247
	6.2.3	Hedgerows and the law	248
	6.2.4	Planting new hedges	253
	References		254

7 European and international wildlife law   255

7.1		European law	255
	7.1.1	Wild Birds Directive	255
	7.1.2	Habitats Directive	259
	7.1.3	The Conservation (Natural Habitats etc.) Regulations 1994	266
	7.1.4	The Water Directives	272
	7.1.5	Trade in endangered species regulation in the EU	275
	7.1.6	Environmental impact assessment in the EU	275
7.2		International protection of wildlife and habitats	275
	7.2.1	Biosphere Reserves	275
	7.2.2	Biogenetic Reserves	277
	7.2.3	Ramsar Convention	277
	7.2.4	World Heritage Convention	282
	7.2.5	Bonn Convention	284
	7.2.6	Berne Convention	286
	7.2.7	International Whaling Convention	288
	7.2.8	Kyoto Protocol	289
	7.2.9	CITES Convention	290
	7.2.10	Biodiversity Convention	293
	References		307

Part III Planning, Urban Environments and Environmental Impact Assessment   309

8 Planning, highways and wildlife   311

8.1		Planning and development	311
	8.1.1	Planning, development control and conservation	311
	8.1.2	The Planning Inspectorate	312
	8.1.3	Planning Policy Guidance Notes	312
	8.1.4	The National Land Use Database	313
8.2		Highways and wildlife	314
	8.2.1	The Highways Agency and conservation	314
	8.2.2	Preventing roadkills	315
	8.2.3	The Highways Agency's *Design Manual for Roads and Bridges*	319

8.2.4   Trunk road Biodiversity Action Plan
for Scotland   319
8.2.5   Managing roadside verges   320
8.2.6   Case studies   321
8.2.7   Translocations   322
8.3   Case study: Planning and biodiversity in
Cambridgeshire   324
8.3.1   Cambridgeshire's Biodiversity Partnership   324
8.3.2   Good practice examples from
Cambridgeshire   328
8.4   Case study: The *Life* Econet Project, Cheshire   328
8.5   Case study: Dibden Terminal Inquiry   329
8.5.1   Introduction   330
8.5.2   The proposal   330
8.5.3   The inquiry   330
8.5.4   English Nature's objections   332
8.5.5   Mitigation measures proposed by
Associated British Ports   333
8.5.6   Objections to the mitigation   333
8.6   Building design and wildlife   334
References   336

9   Environmental impact assessment   337
9.1   Introduction   337
9.1.1   The EIA Directive and its implementation   337
9.1.2   Town and Country Planning (Environmental
Impact Assessment) (England and Wales)
Regulations 1999   338
9.1.3   Types of project that require EIA   341
9.1.4   Exemptions of certain projects from EIA   341
9.1.5   EIA of projects not approved under the
planning system   342
9.1.6   EIA of projects in Scotland and
Northern Ireland   344
9.1.7   What impacts must be assessed?   345
9.2   Nature conservation and EIA   349
9.2.1   Identifying impacts   349
9.2.2   Mitigation of ecological damage   351
9.3   Case studies   351
9.3.1   The second runway at Manchester
International Airport   351
9.3.2   The Samlesbury to Helmshore Natural Gas
Pipe-line   353
References   355

10   The future   **356**

Appendix 1   History of wildlife and conservation law in
the UK   361
Appendix 2   Selected legislation, consultation and policy
documents affecting nature conservation   364
Appendix 3   Egg-laying times and typical nest locations for
some British birds found in urban areas   366
Appendix 4   Schedules to the Wildlife and Countryside Act
1981 and European listed species   368
Appendix 5   Annex I projects, EIA Directive   382
Appendix 6   Annex II projects, EIA Directive   384
Appendix 7   Habitats Directive, Annex I   386
Appendix 8   Vegetation classification systems   389

Glossary and acronyms   391
    Legal terms   391
    Ecological terms   392
    Some acronyms used in the text   393

Information sources   395

Table of laws   401
    International law   401
    European Community law   402
        Treaties establishing the European Community   402
        Directives   402
        Regulations   403
        European Community case law 403
    UK law   403
        Acts of Parliament   403
        Statutory Instruments   408
        UK case law   410

Index   411

Preface

The purpose of this book is to provide an outline of wildlife and nature conservation law in the United Kingdom. It does not contain details of licensing or planning procedures, or of penalties or other fine legal details. It is not an academic discussion of the merits of particular laws or legal problems. It is merely a simple guide to the law, and certainly no substitute for the law itself. By its very nature, law changes with time. In particular, nature conservation law has changed a great deal in recent years because of the influence of international and European law and also as a result of a change in government.

The aim of this book is to provide information about wildlife law for students and professionals whose primary interest is environmental management, planning or construction, rather than the law itself. I hope it will also be used by ecologists, conservationists and other environmental scientists.

I have adopted the definition of nature conservation used in s. 131(6) of the Environmental Protection Act 1990, namely the conservation of flora, fauna or geological or physiographical features. I have used the term 'wildlife' to include all living things, following the definition used by Richard and Maisie Fitter in their *Penguin Dictionary of British Natural History* which was first published in 1967 – 'a general term covering all living things, both flora and fauna' (Fitter and Fitter, 1967).

The title of this work includes the term 'urban environments' and I have attempted to concentrate on those aspects of wildlife law which are designed to protect nature in its various forms from the effects of urbanisation. These include threats from construction projects, industrial pollution and agriculture, all of which have effects upon individual species and habitats. Game and fisheries law in the UK is extensive. I have not attempted to consider this in detail here as it is primarily concerned with the control of hunting and fishing.

Most non-lawyers have little understanding of the law so I have attempted to explain the way the law works before discussing particular laws. I have also attempted to explain how ecosystems work and why certain species and certain habitats need legal protection.

Some brief notes on the natural history of selected species will, I hope, give the reader a better understanding of their environmental requirements. Where possible I have included cases studies in order to demonstrate how the law has functioned in particular instances and how individuals and companies have been dealt with by the law or have adapted in order to fulfil their legal obligations.

It has been a difficult, but interesting, time to attempt to summarise the law and policy relating to wildlife in the UK. Since I began working on this book, there have been considerable changes to the law and government departments concerned with the protection of the environment. In particular, the passing of the Countryside and Rights of Way Act 2000, the demise of the Ministry of Agriculture, Fisheries and Food (MAFF), and the Department of the Environment, Transport and the Regions (DETR), and the creation of the Department for Environment, Food and Rural Affairs (DEFRA) and the Department for Transport. Further complications have been created by the devolution of certain powers to the Welsh Assembly, the Scottish Parliament and the Northern Ireland Assembly. Most of the legislation detailed here refers to English law, although I have attempted to draw attention to differences in other parts of the UK where they occur.

Following the Interpretation Act 1978, in any Act, unless the contrary is clearly intended, references to the masculine gender include the feminine and vice versa, and words in the singular include the plural and vice versa.

It is essential that any reader who wishes to determine the law in a particular area in detail consults original and up-to-date legal documents, or better still, a specialist lawyer. I hope this book will provide a summary of the most important areas of wildlife and nature conservation law and signposts to sources of more detailed information.

This work is based upon the author's understanding of the law at 1 June 2002.

REFERENCE

Fitter, R. and Fitter, M. (1967) *The Penguin Dictionary of British Natural History.* Penguin Books Ltd, Harmondsworth, Middlesex.

Paul A. Rees
BSc (Hons) LLM PhD Cert Ed
School of Environment & Life Sciences
University of Salford

June 2002

Acknowledgements

I should like to record my thanks to the many people who have been involved in the production of this book.

The idea for this book evolved from a single lecture given to Construction Management students at the University of Salford in 1999 and I am indebted to Chris March, formerly Dean of the Faculty of Environment, for inviting me to give this lecture. He was instrumental in the formation of the original concept for the book and I am grateful to him for his encouragement in the early stages.

My friend and ex-colleague Dr Alan Woodward produced the drawings of wildlife which, I hope, will make reading about the law less tedious than it would otherwise have been. At the University of Salford, Dr Philip James introduced me to the *Life* Econet Project and Mary Carruthers and Michael Carrier (Academic Information Services) kindly assisted me by locating various law sources.

Madeleine Metcalfe, Senior Commissioning Editor at Blackwell Publishing, encouraged me to write this book and saw it through from the original concept to the final production. Madeleine must take at least fifty per cent of the credit for the original idea and I am extremely grateful for her support and enthusiasm. I am also grateful to Robert Chaundy of Bookstyle who was responsible for the overall production of the book. His careful design has succeeded in making the final product look much more attractive than I had anticipated.

Deric Newman, Director of *Civic Trees*, kindly supplied the photograph of a tree translocation used on the front cover.

Crown Copyright material is reproduced with the permission of the Controller of Her Majesty's Stationery Office and the Queen's Printer for Scotland. The versions of legislation reproduced here have no official status and are not endorsed by any part of government.

Finally, my wife Katy and my daughter Clara kindly helped to check various parts of the manuscript; however, any errors that remain are my own.

Introduction

Charles Elton, the famous British ecologist, once said that the 'balance of nature' does not exist, and perhaps has never existed (Elton, 1930). His study of animal populations convinced him that the balance of nature was a fallacy, and as evidence of this, he noted that the numbers of animals vary constantly and with irregular patterns.

Elton was right. Nature, left to itself, is not in balance, and since it is not, perhaps man cannot disturb it. There is no single grand plan for nature. It varies with time and place. If we seek to keep ecosystems just as they are now, we miss the point. However, man has now squeezed natural systems into such small spaces that they often require sensitive ecological management to prevent them from disappearing altogether. The trick is to achieve relative ecological stability without the need for excessive human intervention.

Many parts of the world are man-made urban environments that largely exclude natural habitats. The traditional environmental view would claim that we should return to a simpler lifestyle without towns or cities; that we should return to a former existence, living off the land in harmony with nature. If the urban populations dispersed to the land we would consume much of what is left of the natural habitats of the Earth. Urban living and intensive agriculture are here to stay and, although they do great damage to the immediate surroundings, they protect the wider environment from the human control that would accompany a wider dispersion of the population.

In the UK, much of our landscape is artificial. We subsidise the maintenance of an intensive agricultural system in an artificially created countryside. Our islands support a relatively low diversity of animal and plants species that has been influenced by ice ages and by the ravages of man. We have caused the wholesale destruction of many of our native species while introducing exotic animals and plants from Europe, America, the Far East and elsewhere. We have lost the lynx, the wolf and the bear, and gained the sika deer and the North American bullfrog.

International pressures, along with pressure from the European Union and national concerns, have forced governments, at last, to recognise the importance of protecting biodiversity. This cannot be achieved without protecting and restoring habitats. The law has been slow to reflect the concerns of ecologists for the loss of global biodiversity, but in recent years we have seen significant advances at the international, European and national level. However, global change can only be achieved by a change in attitude at a local level. Global deforestation will not be reversed unless individual people plant trees. The loss of biodiversity will not be arrested unless individual planners and politicians take a stand and make sound, sustainable decisions about land use. Breeding populations of rare songbirds will increase only if we stop poisoning their habitats and start putting up nest boxes.

Governments around the world are slowly coming to accept that change in biodiversity is an indicator of the condition of the environments in which people live. This acceptance of the importance of biodiversity needs to be coupled with the recognition that we must actively protect and maintain Earth's ecosystems because we cannot live anywhere else. Ecologists and environmental lawyers both have an important part to play in this protection.

REFERENCE

Elton, C. (1930) *Animal Ecology and Evolution*. Oxford University Press, Oxford, p. 17.

PART I

Law, Ecology and Organisations

Part I begins with a description of the workings of the law affecting the UK, with the main emphasis on English law, but with an account of how this is affected by European and international law. It continues with an explanation of ecological principles and concludes with a consideration of the structure and organisation of nature conservation in the UK.

1   An introduction to the law   3

2   The principles of classification and
    ecology   25

3   The organisation and administration of
    nature conservation in the UK   59

1
An introduction to the law

It is impossible to understand the interrelationships between different aspects of wildlife and nature conservation law without a brief consideration of the structure and workings of the English legal system and its relationship with European and international law.

1.1 THE ENGLISH LEGAL SYSTEM AND DEVOLUTION

The basic principles of English law are described below as they apply to England and Wales. The legal system of Scotland is different, and with devolution of power to the Scottish Executive, the National Assembly for Wales and the Northern Ireland Assembly, both the details of the law and the mechanisms of law-making will undoubtedly diverge as time progresses.

The devolution of certain legislative and executive powers to Wales, Scotland and Northern Ireland has been made possible by:

- The Government of Wales Act 1998
- The Scotland Act 1998, and
- The Northern Ireland (Elections) Act 1998

Devolved issues generally include planning, agriculture, natural and built heritage, forestry, fishing, local government and some aspects of transport.

Readers should refer to specialist texts for more information on the legal systems of Scotland, Wales and Northern Ireland. Readers who wish to gain a more detailed insight into the workings of English law could do no better than to read Glanville Williams' *Learning the Law* (Williams, 1993) which was first published in 1945.

1.1.1 Introduction

English law is based on both legislation and common law. The legislation consists of Acts of Parliament and Statutory Instruments produced by the government. The common law is based upon the decisions of the courts as recorded in law reports. These reports represent a complex record of precedent decisions that are used to interpret the law and to predict the likely outcome of similar cases.

A great deal of wildlife and nature conservation law exists in the form of Acts of Parliament and Statutory Instruments. However, the interpretation of this law must be considered in the light of court decisions. For example, the protection of Sites of Special Scientific Interest (SSSIs) depends in part upon the legal definition of the term 'occupier', as some offences may only be committed by 'owners or occupiers'. The term is not defined in the legislation (Wildlife and Countryside Act 1981) so it has fallen to the courts to decide who is and who is not an 'occupier' in particular circumstances (see Section 5.4).

Acts of Parliament are primary legislation. Statutory Instruments are an example of secondary or delegated legislation. Often an Act of Parliament will delegate the power to change the law to a Secretary of State. He or she may achieve this by producing a Statutory Instrument. This mechanism allows the details of the law to be changed without the need for Parliament to produce new Acts each time a change is required.

1.1.2   Criminal and civil law

It is difficult to make a clear distinction between criminal offences and civil offences. In general the criminal law is concerned with behaviour of which the state disapproves while the civil law is concerned with disputes between individuals. However, the difference is in the consequences that follow the wrongful behaviour rather than in the behaviour itself. The criminal law is primarily concerned with punishing wrongdoers while the civil law is concerned with providing a remedy to those who have been wronged. This remedy is often compensation, but it may take the form of an injunction that restricts the future actions of the defendant. Some acts may give rise to both criminal and civil proceedings. Sometimes criminal courts make compensatory awards to victims of crime, while civil courts may 'punish' wrongdoers with heavy fines as a deterrent to others.

1.1.3   Types of legislation

Green Papers and White Papers

When the government proposes new legislation the process starts with a period of consultation during which a Green Paper and then a White Paper may be produced. A Green Paper is essentially a broad statement of the intention of the proposed changes to the law. Its purpose is to stimulate discussion and to attract comment from interested parties. A White Paper is a more detailed statement of government policy. This is not itself law but may eventually be implemented by legislation.

Acts of Parliament

An Act of Parliament begins its life as a 'Bill'. However, the process of producing new legislation begins long before this. Initially someone must have an idea for a change in the law. This idea may come from the government of the day (resulting in a 'Government Bill') or it may come from an individual Member of Parliament (resulting in a Private Member's Bill). Government Bills are produced by the government to push through its own legislative programme. A Private Member's Bill is legislation proposed by an individual Member of Parliament. Such legislation is usually concerned with areas of special interest to the member concerned, for example fox hunting. New legislation is drafted by lawyers and then presented to Parliament as a Bill. The proceedings of Parliament are recorded in *Hansard*, including debates and amendments made to Bills as they pass through the various procedural stages.

Each Act of Parliament is assigned a chapter number. This number and the date that the Act received Royal Assent identifies the Act, for example 2000 chapter 37, is the Countryside and Rights of Way Act 2000. In some legal documents only the year and the chapter number are used to identify Acts.

The main body of an Act of Parliament is divided into numbered 'sections'. These are subdivided into subsections. In very large Acts the sections may be grouped together into parts which are divided into chapters (unrelated to the chapter number of the Act itself), so that the hierarchy of divisions is:

- Part
- Chapter
- Section
- Subsection

Schedules at the end of the Act may contain information necessary for the interpretation of the law. For example, they may contain lists of species that receive various levels of protection, as in the Wildlife and Countryside Act 1981 (WCA 1981). Without the information contained in the schedules many sections of the Act cannot be interpreted. Some schedules are used to list amendments made to other legislation.

Acts of Parliament are primary legislation and may be supplemented by delegated legislation known as Statutory Instruments. Northern Ireland Orders in Council are UK Statutory Instruments and equate to primary legislation.

PANEL 1.1   GENERALISED STRUCTURE OF AN ACT OF PARLIAMENT

The structure of an Act varies, depending upon its length and purpose. It may be a single page long with a simple structure, or a complex, multi-part document of considerable length. The following is a description of the general form of the more complex Acts.

Short title

Arrangement of sections

Short title
Official citation (year and chapter)
Long title
Date of Royal Assent

Enacting formula

Main provisions
(divided into parts, chapters, sections etc.)

Miscellaneous and supplementary provisions
(divided into sections etc.)

Repeals

Commencement

Interpretation, short title and extent

Schedules
(including consequential amendments and tables of repeals)

Statutory Instruments

Statutory Instruments may be used to add detail or make small changes to existing legislation. In many cases, a statutory instrument must be read in conjunction with the appropriate Act of Parliament (the 'parent Act') if its meaning is to be clear. Statutory Instruments are divided into numbered 'regulations', which are subdivided into 'paragraphs'.

The government produces explanatory notes to accompany legislation. These notes explain the main changes to the law in relatively simple language, but they do not form part of the law itself.

PANEL 1.2   INTRODUCTION TO THE WILD MAMMALS (PROTECTION) ACT 1996

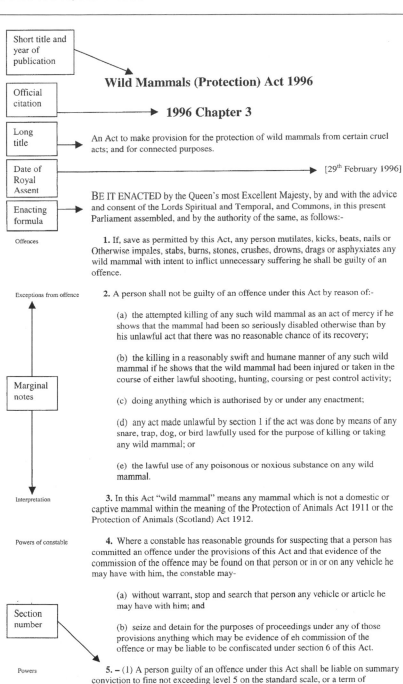

Short title and year of publication

Wild Mammals (Protection) Act 1996

Official citation

1996 Chapter 3

Long title

An Act to make provision for the protection of wild mammals from certain cruel acts; and for connected purposes.

Date of Royal Assent

[29th February 1996]

Enacting formula

BE IT ENACTED by the Queen's most Excellent Majesty, by and with the advice and consent of the Lords Spiritual and Temporal, and Commons, in this present Parliament assembled, and by the authority of the same, as follows:-

Offences

1. If, save as permitted by this Act, any person mutilates, kicks, beats, nails or Otherwise impales, stabs, burns, stones, crushes, drowns, drags or asphyxiates any wild mammal with intent to inflict unnecessary suffering he shall be guilty of an offence.

Exceptions from offence

2. A person shall not be guilty of an offence under this Act by reason of:-

(a) the attempted killing of any such wild mammal as an act of mercy if he shows that the mammal had been so seriously disabled otherwise than by his unlawful act that there was no reasonable chance of its recovery;

(b) the killing in a reasonably swift and humane manner of any such wild mammal if he shows that the wild mammal had been injured or taken in the course of either lawful shooting, hunting, coursing or pest control activity;

Marginal notes

(c) doing anything which is authorised by or under any enactment;

(d) any act made unlawful by section 1 if the act was done by means of any snare, trap, dog, or bird lawfully used for the purpose of killing or taking any wild mammal; or

(e) the lawful use of any poisonous or noxious substance on any wild mammal.

Interpretation

3. In this Act "wild mammal" means any mammal which is not a domestic or captive mammal within the meaning of the Protection of Animals Act 1911 or the Protection of Animals (Scotland) Act 1912.

Powers of constable

4. Where a constable has reasonable grounds for suspecting that a person has committed an offence under the provisions of this Act and that evidence of the commission of the offence may be found on that person or in or on any vehicle he may have with him, the constable may-

(a) without warrant, stop and search that person any vehicle or article he may have with him; and

Section number

(b) seize and detain for the purposes of proceedings under any of those provisions anything which may be evidence of eh commission of the offence or may be liable to be confiscated under section 6 of this Act.

Powers

5. – (1) A person guilty of an offence under this Act shall be liable on summary conviction to fine not exceeding level 5 on the standard scale, or a term of imprisonment not exceeding six months, or both.

PANEL 1.2   CONTINUED

---

(2) Provided that where the offence was committed in respect of more than one wild mammal, the maximum fine which may be imposed shall be determined as if the person had been convicted of a separate offence in respect of each such wild mammal.

Court powers of confiscation &c.
6. – (1) The court before whom any person is convicted under this Act may, in addition to any punishment, order the confiscation of any vehicle or equipment used in the commission of the offence.

(2) The Secretary of State may, by regulations made by statutory instrument and subject to annulment in pursuance of a resolution of either House of Parliament, make provision for the disposal or destruction in prescribed circumstances of any vehicle or equipment confiscated under this section.

Citation, commencement and extent.
7. – (1) This Act may be cited as the Wild Mammals (Protection) Act 1996.

(2) This Act shall come into force with the expiration of the period of two months beginning with its passing.

(3) This Act shall not apply in Northern Ireland.

(4) Section 6 of this Act shall not apply to Scotland, and so much of section 4 as refers to that section shall also not apply there.

1.1.4   Law reports

A law report is a version of the proceedings of a particular case heard in a court. It contains the 'facts of the case' along with the judgment of the court and comments made by the judge. Many cases are not reported at all, while others may appear in more than one series of law reports. Since there may be more than one version of the proceedings in a particular case it is important to appreciate that some law reports are more authoritative than others. The series of reports known as *The Law Reports* contain versions of the court proceedings that have been authorised by the presiding judge. Other commercially produced reports are also available.

The judgment contains two distinct parts. The most important is called the *ratio decidendi*. This is the fundamental legal principle that has been used to decide the outcome of the case, based upon the facts. This principle may be important in acting as a precedent in future cases. The second element of the judgment is called the *obiter dicta*. These are things 'said in passing' by the judge. They are a statement of law which is not based on the facts of the case, and as such do not form part of the decision. For example, the judge may consider what the law might have been if the facts of the case had been slightly different. Such comments will not generally form part of a legal precedent since they do not relate to the facts of the case that was under consideration.

When does the decision of a court act as a precedent? If all court decisions were to act as precedents for all other courts, the result would be chaos. The system of precedent acts within a strict hierarchy so that, in general, the 'lower' courts are bound by the decisions of the 'higher' courts, and all courts are bound by the decisions of the House of Lords, the highest court in the land. The House of Lords, however, is not bound by its own previous decisions.

Judges may choose to broaden or narrow the application of a precedent by their interpretation of the *ratio decidendi*. For example, it may be inappropriate to widen the application of a previous court decision if this would lead to an unacceptably high number of claims for compensation. Alternatively, a court may extend a decision where it believes that this is in the interests of justice.

1.1.5 Judicial review and *ultra vires*

The process of judicial review is designed to allow a re-examination of a judicial or administrative decision. It is concerned only with the decision-making process, and not the decision itself. Even if a judicial review finds that a particular case has not been conducted properly it does not follow that the decision itself will be changed. A judicial review will only be granted in certain quite specific circumstances, for example if there has been some procedural impropriety or if an official has acted *ultra vires* (beyond his or her powers). In March 2001, Kindersley (the publisher and organic farmer) was given leave to apply for judicial review of the Ministry of Agriculture's decision to cull healthy animals in an attempt to control an outbreak of foot-and-mouth disease.

1.2 WHERE TO FIND THE LAW

Law books are a useful starting point for anyone researching the law. However, it is inevitable that law books tend to date very quickly. Libraries are often filled with out-of-date law texts and great care should be taken to check the publication date of any texts that appear to be authoritative. Some areas of law change faster than others. Unfortunately, many aspects of environmental law have changed significantly in recent years and there is every reason to believe that frequent change will be a permanent feature of this area of the law.

1.2.1 Titles of legal cases and where to find law reports

Law reports are generally published in volumes which are eventually bound in chronological order at the end of each year. Some law reports appear in newspapers, for example *The Times*. Universities that teach law, particularly those that have a law school, will devote part of their library to collections of law reports and may employ a specialist law librarian. Case law is also accessible online through systems such as:

- Lawtel   *www.lawtel.co.uk*
- Lawdirect   *www.lawdirect.co.uk*
- Lexis   *www.lexis.com*

Legal cases are identified by the names of the parties concerned and the date. They are usually written in italics:

Smith v. *Jones* (1966)

The full reference to a case includes its location in the law reports:

Hasley v. *Esso Petroleum Company* [1961] 1 WLR 683

This case is to be found in the first volume of *Weekly Law Reports* at page 683. It is the convention to use round brackets for the date when the case was pronounced, and square brackets for the date when the case appears in the law reports, which may be later.

A list of the abbreviated titles of law reports should be available in a good law library. Commonly cited reports and journals include:

- All ER   *All England Law Reports*
- EnvLR   *Environmental Law Reports*
- HL   *House of Lords Appeals*
- HL Cas   *House of Lords Cases*
- JEL   *Journal of Environmental Law*
- JP   *Justice of the Peace & Local Government Review*
- JPL   *Journal of Planning Law*
- LR   *Law Reports*
- P&CR   *Planning & Compensation Reports*
- QB (or KB)   *Law Reports, Queen's or King's Bench Division*

1.2.2   How to find English legislation

There are a number of publications that are designed continually to update the law. Some of these publications update legislation while others record changes to case law.

The law is clearly not static. If Acts of Parliament simply replaced each other over time, life would be relatively straightforward. Unfortunately, although some Acts are superseded in their entirety by others, most are altered in a piecemeal fashion by subsequent legislation. Sometimes this may mean the addition or deletion of a section of

---

PANEL 1.3   CITATION OF LEGAL DOCUMENTS

The usual form of citation of legislation and case law is given here for UK, EC and international law.

UK national legislation

Act of Parliament
> Wildlife and Countryside Act 1981 c.69

Statutory Instrument
> Hedgerow Regulations 1997 (SI 1997/1160)

Case law
> *R* v. *Secretary of State for Trade and Industry, ex parte Greenpeace* (No. 2) [2000] Env LR 221

EC secondary legislation

Directive
> EC Directive on the Conservation of Natural Habitats and of Wild Fauna and Flora, 92/43/EEC, OJ, L206, 22 July 1992

Regulation
> EC Regulation on Implementation of CITES, 3636/82/EEC, OJ, L384, 31 December 1982

ECJ case law
> *Commission* v. *Germany* (case C-57/89) [1991] ECR I-883 ('Leybucht Dykes')

International treaty
> Convention on Biological Diversity (Rio de Janeiro) 1992, UNEP/Bio.Div.Conf./L.2

---

an Act. However, on occasion a single word may be added or deleted, thereby completely changing the meaning of a sentence. For example, the Countryside and Rights of Way Act 2000 made the following changes to the Conservation of Seals Act 1970:

Chapter	Short title	Extent of repeal
1970 c. 30	The Conservation of Seals Act	Section 10(4)(c) and the following word 'or'

The WCA 1981 has undergone a number of revisions, including those made by the Wildlife and Countryside (Amendment) Act 1985 and the

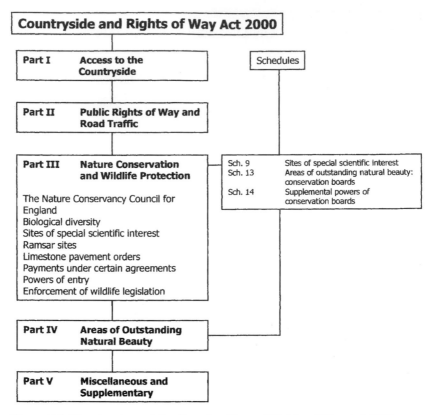

Figure 1.1 The structure of the Countryside and Rights of Way Act 2000 (abridged).

Countryside and Rights of Way Act 2000 (Figure 1.1). There have also been various amendments to the schedules of the WCA 1981. It is therefore impossible to determine the law without tracing all of the subsequent amendments.

Each Statutory Instrument (SI) can be identified by the year it was issued and a consecutive number allocated chronologically. Statutory Instruments are bound chronologically and are therefore relatively easy to locate in a law library. However, some SIs are very short, often making very small changes to existing legislation, so the total number produced in any year is very large.

Changes in legislation can be tracked using:

- *Chronological Table of Statutes*
- *Halsbury's Laws of England*
- *Halsbury's Statutory Instruments*
- *Current Law Statute Citator*
- *Current Law Legislation Citator*
- *Current Law Statutes Annotated*

■ *Current Law Monthly Digest*

Case law can be updated by reference to:

■ *Current Law Case Citator*
■ *Current Law Year Books*

Halsbury's Laws of England provides a concise statement of the law in a particular area with reference to the relevant statutes and case law.

Green Papers and White Papers all have Command Numbers which may be used to locate them in a library. These documents are available from The Stationery Office and also often appear on the websites of the relevant government department.

Occasionally, consolidated versions of important areas of law may be published in book form. These can be a useful starting point for determining the law, but it should be appreciated that some of these publications provide incomplete versions of the law (with certain sections omitted) and, in any event, they are not official versions. Such publications contain updated versions of legislation sometimes showing the original provisions along with any amendments and repeals. However, it is inevitable that these books will quickly become dated and the reader should note the date of publication or any reference to the date at which the author considered the law to be accurate.

The Stationery Office publishes a large number of legal documents on the internet at *www.hmso.gov.uk*, including:

■ Acts
■ Explanatory notes to Acts
■ Statutory Instruments
■ Draft Statutory Instruments

Other documents, including *Hansard*, are available on the UK Parliament website at: *www.parliament.the-stationery-office.co.uk*.

1.3   LAWYERS AND THE COURT SYSTEM

1.3.1   The legal profession

The legal profession in England is made up principally of:

■ Solicitors
■ Barristers
■ Queen's Counsels
■ Judges
■ The Director of Public Prosecutions
■ The Attorney-General
■ The Lord Chancellor

A solicitor of the Supreme Court may conduct certain legal proceedings or give advice on legal problems. He may also instruct a barrister to represent a client in legal proceedings. Legal executives are employed to assist solicitors, for example as managing clerks.

A barrister's function is primarily as an advocate who has exclusive right of audience in certain types of judicial proceedings. Barristers are generally self-employed and work in small groups from the same offices (chambers), employing a clerk to distribute written instructions from solicitors (briefs) between them. A barrister has a duty to represent his client's interests but his primary duty is to the court. A Queen's Counsel (QC) is a senior barrister (also called a 'silk') who has been appointed by the Lord Chancellor. He may appear in any case, for or against the Crown, and may be assisted by junior barristers. Recent changes to the law have allowed some solicitors to become QCs.

Judges are generally appointed from practising barristers and have the power to decide disputes and determine appropriate penalties. They include judges of the High Court, circuit judges, recorders, Lords of Appeal in Ordinary, the Lord Chief Justice and the Master of the Rolls.

The Director of Public Prosecutions (DPP) works under the supervision of the Attorney-General and heads the Crown Prosecution Service (CPS). The Attorney-General is the chief law officer of the Crown and head of the English Bar. The Lord Chancellor is the principal legal dignitary. He is a member of the cabinet and the government's legal advisor.

Much of the work of the court system is performed by lay magistrates, who are part-time justices of the peace (with a small number of stipendary magistrates) in some 400 Magistrates' Courts.

1.3.2   The courts

A detailed description of the structure and functioning of the courts is beyond the scope of this book. However, it is useful to consider the role of the courts that are likely to hear wildlife law cases.

Minor infringements of the law are referred to the Magistrates' Court. However, this court may also decide whether or not there is sufficient evidence to refer more serious cases to the Crown Court.

A Magistrates' Court usually consists of a small number of lay magistrates (between three and seven) assisted by a 'clerk to the justices' who is legally qualified and advises the magistrates on points of law. In addition, most large conurbations have stipendary magistrates who are professional lawyers. The magistrates act as both judge and jury in minor (summary) offences.

More serious (indictable) offences may be referred to the Crown Court by the magistrates. The Crown Court may hand down more severe penalties than those available to the magistrates. In the Crown Court the defendant is represented by a barrister and tried by a jury. It may hear appeals from the Magistrates' Courts and may sentence persons committed for sentencing by those courts. The main civil court is the County Court. Appeals lie to the Court of Appeal or in a few cases, the High Court. The High Court of Justice has unlimited civil jurisdiction and appellate jurisdiction in both criminal and civil proceedings.

The Court of Appeal has a Criminal Division and a Civil Division. The Criminal Division hears, for example, appeals by persons convicted on indictment, or against sentence from the Crown Court. The Civil Division hears, for example, appeals from the High Court, the County Courts and various tribunals. Appeal from either the Criminal or Civil Divisions is to the House of Lords.

The House of Lords hears appeals in civil and criminal cases from the Court of Appeal. The judges in the Lords are Lords of Appeal in Ordinary and peers who hold or have held high judicial office. At least three designated Lords of Appeal must hear an appeal.

1.3.3   Planning inquiries

Planning inquiries may be set up to deal with objections to development proposals. During a planning inquiry, all parties are allowed a fair chance to put forward their view in writing or at a public hearing. Where a particular site is involved, it will normally be visited by an Inspector or another appointed officer. Government policy will be considered in any decision. In most cases, the decision is taken by the Inspector, but in some, the Inspector will prepare a report and make a recommendation to the appropriate Secretary of State (see Chapter 8).

Inquiries into trunk road and motorway proposals are held by a separate group of Inspectors called the Lord Chancellor's Panel. These are independent Inspectors who are nominated by the Lord Chancellor.

1.3.4   Legal personality and *locus standi*: who has a right to bring an action to the courts?

If anyone could bring an action in any court about anything, the court system could very quickly become overwhelmed by frivolous cases. In order to prevent this, a potential party in a case must be able to demonstrate to the court that it has *locus standi*: a right to be heard, or sufficient interest in the case.

The first requirement is that the party should possess a 'legal personality'. There are essentially two types of legal person: natural persons and artificial persons. A natural person is an individual human being or a group of people. An artificial person is an organisation like a company or an institution, where the organisation has a separate legal identity to that of the individuals of which it is composed. Natural persons and artificial persons may be granted *locus standi* in a case provided they can demonstrate sufficient interest in that case.

Unfortunately in English law there are no generally recognised 'environmental rights' available to the public. Decisions as to who has sufficient interest in a particular environmental case are made by the court considering the legal and factual context. Some interest groups have been successful in establishing *locus standi* in cases involving planning issues where there is a clear public interest in relation to a particular locality, but this is more difficult in purely environmental matters.

In recent years courts have recognised the environmental pressure group Greenpeace as having sufficient interest in cases concerned with marine pollution and pollution from radioactive materials to grant it *locus standi* in these cases. Greenpeace has a very large membership amongst the general public and has been recognised by the courts as representing the interests of the public in matters of general environmental concern.

In some cases it has been necessary for voluntary organisations to work together in order to bring an action in the courts. In *R* v. *Poole BC, ex parte Beebee* [1991] the judge considered whether various environmental groups had sufficient standing to challenge a decision by Poole Borough Council to grant itself planning permission for housing to be built on a site of special scientific interest. The site was important because of the presence of populations of rare reptiles and amphibians. The judge ruled that the World Wide Fund for Nature did not have standing, but the British Herpetological Society was held to have standing because of its long association with the site. The WWF was involved mainly to fund the action. The judge also made it clear that the Nature Conservancy Council (now English Nature) would also have standing in such a case because of its official responsibilities for nature conservation.

In English law, the Attorney-General always has *locus standi* as the guardian of the public interest. Where an applicant does not have 'sufficient interest' in a case he could ask the Attorney-General to bring an action on his behalf.

1.4   EUROPEAN UNION LAW

1.4.1   The history of the European Union

The European Union (EU) came into existence as a result of the signing of the Maastricht Treaty by the member states in 1992 although it has a history stretching back to the 1950s (Panel 1.4). European law has a significant influence on nature conservation law in the UK and has led to a number of changes in national law.

In order for the EU to promulgate new laws there must first be a relevant provision within a treaty to which the member states have already agreed. For example, the EU could not make new laws to protect the environment of individuals who reside within the EU unless the member states had previously agreed to this as an objective and had signed a treaty containing this provision. Such a provision may be found in Article 174 of the Treaty Establishing the European Community (previously Art. 130r before renumbering by the Treaty of Amsterdam).

1.4.2   The institutions of the European Union

The institutions of the EU include:

- The Council of the European Union (formerly the Council of Ministers)
- The Commission
- The Parliament
- The European Court of Justice (ECJ)

PANEL 1.4   THE FOUNDATION OF THE EUROPEAN COMMUNITY (EC)

Treaty Establishing the European Coal and Steel Community (Paris) 1951

Treaty Establishing the European Economic Community (Rome) 1957

Treaty Establishing the European Atomic Energy Authority (Euratom) (Rome) 1957

Single European Act 1986
  Amended the Treaty of Rome and introduced the first references to the environment

Treaty on European Union (Maastricht) 1992
  Created the European Union and altered the name of the EEC Treaty to the EC Treaty

Treaty of Amsterdam 1997
  Amended the EC Treaty, including a renumbering of the Treaty

> ## Treaty Establishing the European Community Art. 174 (ex Art. 130r)
>
> Community policy on the environment shall contribute to the pursuit of the following objectives:
>
> - preserving, protecting and improving the quality of the environment;
> - protecting human health;
> - prudent and rational utilisation of natural resoures;
> - promoting measures at international level to deal with regional or worldwide environmental problems.

The Council of the European Union is the supreme legislative body of the Community. It is made up of the ministers of the member states and it votes on new laws. The Commission is the executive of the Community and is made up of 20 Commissioners appointed by the member states. It formulates policy, proposes new laws and institutes proceedings in the ECJ where it believes that there has been a breach of EC law. The Commission is divided into 24 directorates-general. For example, DG XI is concerned with Environment, Nuclear Safety and Civil Protection. The ECJ interprets and enforces EC law. Finally, the European Parliament consists of 626 members (MEPs) whose powers include the control of the EU's budget and the power to dismiss the Commission (see Panel 1.5).

1.4.3   Primary legislation

The primary law of the EC is made up of the treaties created by the member states (Panel 1.4). These treaties set out the general areas in which the EC is to have competence in the creation of new legislation by laying down a framework. Article 249 of the Treaty of Rome gives the EC the power to create secondary legislation.

> ## Treaty Establishing the European Community Art. 249 (ex Art. 189)
>
> In order to carry out their task and in accordance with the provisions of this Treaty, the European Parliament acting jointly with the Council, the Council and the Commission shall make regulations and issue directives, take decisions, make recommendations or deliver opinions.

PANEL 1.5 THE ROLES OF THE INSTITUTIONS OF THE EUROPEAN UNION

Council of the European Union
- Consists of government ministers or their representatives from all the member states
- Members of the Council represent their national interests
- The Council has legislative powers
- Acts on the recommendation of the Commission and can amend Commission proposals
- Meets in Brussels

Commission
- Acts as the executive of the EU
- Commissioners are drawn from the member states
- Responsible to the Parliament
- Initiates most EU proposals and policies
- Implements EU legislation
- Advocates EU interest in the Council
- Located in Brussels

European Parliament
- The Parliament has no legislative powers
- Council must consult the Parliament before legislating
- Controls EU budget
- May dismiss the Commission
- May take other EU institutions to the ECJ for infringements of the Treaty
- Located in Luxembourg

European Court of Justice (ECJ)
- Judges are appointed by the member states
- Supervises the uniform application of EU law within the Union
- Interprets EU treaties and legislation
- Makes rulings on points of law referred to it by domestic courts of member states
- May give preliminary rulings which bind all the courts of all member states
- ECJ has no machinery for enforcement of its decisions and rulings; this is left to the national court system
- Located in Luxembourg

1.4.4 Secondary legislation

The secondary legislation of the EC takes the form of:

- Regulations

- Directives
- Decisions

European Regulations are generally applicable and binding in their entirety. They are directly applicable in all member states without the need for individual states to make any changes to their domestic law. Since it requires no action by the member states, a regulation can bring about swift and immediate reform of the law. However, this mechanism is rarely used in the sphere of environmental law, except to fulfil obligations under international treaties, e.g. Regulation 338/97 on Trade in Endangered Species, which fulfils commitments under the Convention on International Trade in Endangered Species of Fauna and Flora.

European Directives are addressed to one or more member states and require that they change their national law. Directives are binding in policy but leave the choice of form and method of implementation to the member states. In the UK this is done by Act of Parliament (for example, Directive 85/374/EEC on product liability was implemented by the Consumer Protection Act 1987) or by delegated legislation in the form of, for example, a Statutory Instrument. The European Communities Act 1972 s. 2(2) makes provision for EC law to be transposed into UK law even if no parent statute authorises it. Directives have the advantage of allowing for the gradual harmonisation of laws within the Community, but there is a time limit within which any Directive must be incorporated into national law.

Decisions may be addressed to a state, a company or a person and are binding on them in their entirety. For example, the Commission may order a company to discontinue any anti-competitive practice. The purpose of a decision is to bring a state, corporation or individual into line with EC policy and law.

The Council and the Commission may also make recommendations, give opinions and issue notices, but these are not legally binding.

1.4.5   The supremacy of European law

Where a conflict arises between UK national law and EC law the court should follow EC law (s. 2 European Communities Act 1972). Where an individual member state has failed properly to incorporate EC law into its national legislation EC law may still have effect through two doctrines established by the ECJ: 'direct effect' and 'indirect effect'.

Direct effect

In 1963 the ECJ created the doctrine of direct effect in the *Van Gend en Loos* v. *Nederlandse Administratie der Belastingen* case. This established that an individual could rely on EC law even if it has not been implemented by the member state. In this case the court decided that EC law could have *direct effect* if:

- the law is clear and unambiguous;
- it is unconditional and not dependent on the exercise of discretion;
- it is not dependent upon further action by the EC or a member state which would seem to prevent Directives from having direct effect.

Two types of direct effect may be distinguished:

- vertical direct effect;
- horizontal direct effect.

Vertical direct effect gives rights against governments or the institutions of government under EC law. Horizontal direct effect gives individuals rights against each other under EC law.

Treaties and Regulations can be directly relied upon by individuals in their national courts. They have both *vertical* and *horizontal* direct effect. Decisions can be relied upon by the addressee (i.e. the state, company or individual) and are immediately binding.

Directives have vertical effect so individuals can use a directive against a government institution. This was established in *Marshall* v. *Southampton & South West Hampshire Area Health Authority (Teaching)* [1986]. However, Directives cannot have horizontal effect, so an individual cannot sue a private company using a Directive as a source of law (*Duke* v. *GEC Reliance* [1988]). Through the mechanism of vertical direct effect the government, and its institutions, can be held accountable for their unlawful actions.

Why would it be unjust to allow horizontal direct effect? If an individual has a grievance against his employer and claims that a European Directive, which has not been implemented by the government, has horizontal direct effect, he would have to sue his employer for the unlawful action of the government. It is clearly unjust to hold a company, which has behaved lawfully (i.e. complied with UK law) liable for the failure of the government.

The ECJ has established that a person who is caused loss by failure of the state to implement a Directive can sue the state for damages (*Francovich and Others* v. *Italian Republic* [1991]). This principle has come to be known as 'Francovich liability'.

Indirect effect

In some cases, a party may bring an action in a domestic court using the provisions of an EC Directive even if the member state has not implemented that Directive. The principle of *indirect effect* was established in *Marleasing SA* v. *La Comercial Internacional de Alimentación SA*

[1990] and *Von Colson and Kamann* v. *Land Nordrhein-Westfalen* [1984]. This principle states that national laws should be interpreted so as to comply with EC laws and allows horizontal actions. This imposes an unfair burden on private bodies where they have relied upon British law, for the reasons discussed above. The impact of *indirect effect* depends upon:

■ the willingness of British judges to adopt the EC interpretation of the law; and
■ the existence of some national law that can plausibly be 'interpreted' in an appropriate manner.

1.4.6   Where to find EC law

EC legislation is published in the *Official Journal of the European Communities* (OJ). This is divided into two series:

■ the L series (Legislation) contains the texts of EC legislation; and
■ the C series (Information and Notices) contains key extracts from judgments of the European Court of Justice and the Court of First Instance (i.e. one of the national courts of a member state), and Preparatory Acts (proposed legislation).

Other useful sources of European case law include:

■ *Reports of Cases before the Court*
■ *European Court Reports* (ECR)
■ *Common Market Law Reports* (CMLR)

European legislation can be obtained from a number of European documentation centres located in libraries around the UK. Salford University houses one such centre.

Individual EC Directives and Regulations change with time, as does national law. It is, therefore, essential to consult up-to-date versions of EC documents which incorporate any amendments to the law.

An excellent source of European law is the European Union website at *www.europa.eu.int*. The site contains consolidated versions of the law, i.e. they incorporate all of the changes that have been made to the law since it first came into force. This is particularly useful since it removes the need for the tedious task of tracing all of these amendments in obscure legal documents.

1.4.7   The EU and international law

In addition to creating its own laws, the EU has the power to enter into treaties with other (non-EU) countries on behalf of the member states. Treaties signed by the EU include the Convention on

International Trade in Endangered Species of Wild Fauna and Flora 1973 (CITES).

1.5 INTERNATIONAL LAW

International law is increasingly concerning itself with wildlife and the conservation of the natural environment. Although much of this law is regional in its application some of it has global implications and has led to changes in both European and UK wildlife and nature conservation law.

Added protection has been achieved by creating additional designations for protected areas, banning trade in some wildlife species, and by the further listing of protected species beyond what was previously the case under UK law.

1.5.1 The nature of international treaties

The international community has no legislature capable of formulating laws that are binding on individual states. When states wish to cooperate in areas of mutual concern they may voluntarily enter into mutually binding legal obligations. Such obligations take the form of treaties or conventions. However, this system relies entirely upon the cooperation of individual states for its effectiveness, since no state can be bound by international law without its consent. States are free to join and leave treaties as they please. Even if a state joins a treaty it may still ignore its provisions, although international political pressure may be applied in an attempt to achieve compliance.

International treaties are divided up into sections called 'articles', and often contain commitments made in very general terms. For example, some treaties require parties to exchange knowledge or to cooperate in research. Others commit parties to the establishment of protected areas aimed at the conservation of particular species or habitats, or they seek to control the emission of particular pollutants into the environment. Such commitments will usually only have legal effect when they are written into the domestic law of the parties to the treaty. In the UK this may require a new Act of Parliament or Statutory Instrument. In some cases our international commitments may be incorporated into European law in the form of a Regulation (requiring no further UK legislation).

Parties to an international treaty may meet regularly at events known as the 'Conference of the Parties' where they may agree changes to the treaty. Treaties usually establish a secretariat which arranges and services these conferences, and has various administrative functions in relation to the operation of the treaty. The Convention on International

Trade in Endangered Species of Wild Fauna and Flora (CITES) asked the United Nations Environment Programme (UNEP) to undertake the functions of the secretariat established by the treaty in 1973. The UNEP contracted the task to the International Union for the Conservation of Nature and Natural Resources (IUCN) which is now known as the World Conservation Union. The secretariat's office is located in Gland, near Geneva.

1.5.2   Sources of international law

There are a number of useful sources of international law available on the internet, for example:

- The Center for International Environmental Law:
  www.ciel.org
- The Library of Congress – Law reading room, Global Legal Network Information:
  www.loc.gov/law/guide/index.html
- International Union for the Conservation of Nature and Natural Resources (World Conservation Union):
  www.iucn.org
- United Nations Environment Programme (UNEP):
  www.unep.org

International treaties are published in the UK as Command Papers and are available from the Stationery Office.

REFERENCE

Williams, G. (1993) *Learning the Law*. Stevens & Sons Ltd, London.

2
The principles of classification and ecology

Ecology is the scientific study of living things and their relationship with each other and the environment. Biologists classify organisms into groups for convenience. Taxonomists use groups that contain close relatives and reflect the process of evolution, grouping genetically similar forms together, for example mammals or spiders. Ecologists group organisms together on the basis of their associations within a particular ecological unit such as wetland species or forest species.

In order to specify which species or habitats we wish to protect under the law it is essential that we have a means of clearly defining them. The law must rely on scientists to provide these definitions but sometimes, for convenience, lawyers use artificial groupings that have no scientific basis. The law sometimes also refers to particular types of ecosystem or habitat, such as limestone pavements or wetlands, and may provide legal definitions of these.

The purpose of this chapter is to provide a basic understanding of the way species are named, the fundamental characteristics of ecosystems, the methods used in ecological surveys and the principles of nature conservation.

2.1 ANIMAL AND PLANT NAMES AND THE LAW

2.1.1 Introduction

Often the biological description of animal and plant species does not correspond with the legal description. The use of collective legal terms to encompass a wide range of biological forms can lead to considerable confusion. Such terms are generally defined within legal instruments and it is important that these definitions are consulted if the law is to be interpreted correctly. For example, the biological term 'animal' is

---

Wildlife and Countryside Act 1981 s. 27(1)

'wild animal' means any animal (other than a bird) which is or (before it was killed or taken) was living wild;

'wild bird' means any bird of a kind which is ordinarily resident in or is a visitor to Great Britain in a wild state...

---

used by scientists to mean all animal species, but in the law it is often not intended to include birds. In the Wildlife and Countryside Act 1981 there are separate sections referring to the 'protection of birds' and the 'protection of other animals'.

The definition of 'fish' has also given rise to some difficulty. In *Caygill* v. *Thwaite* (1885) the court held that crayfish (which are crustaceans) were 'fish' in the absence of any indication to the contrary, and that a statute that made it unlawful to take 'fish' applied equally to crayfish. The term 'sea fish' is defined under s. 20(1) of the Sea Fisheries Regulation Act 1966 to mean fish of any kind found in the sea, and includes shellfish (defined as crustaceans and molluscs of any kind) but excludes salmon and migratory trout. Zoologically this is nonsense since it excludes some fish species from the definition of 'fish' while, at the same time, extending the definition to include crustaceans and molluscs which belong to quite separate phyla.

2.1.2   The binomial system

The law only recognises the scientific name of a species for the purposes of identification. In theory the scientific name of an organism is the same in every country in the world but the common or vernacular names of animals vary from country to country. For example, the animal known as a caribou in North America is usually called a reindeer in Europe. To avoid the confusion caused by vernacular names each species is assigned a scientific or Latin name. This system was devised in the eighteenth century by a Swedish physician called Carolus Linnaeus (the Latinised name of Carl Linné) and is known as the binomial system. Each name is written in italics and consists of two parts: the generic name and the specific name. The scientific name of the blackbird is:

Turdus	*merula*
(genus)	(species)

Similar species belong to the same genus, for example:

Mistle thrush	*Turdus viscivorus*
Song thrush	*Turdus philomelos*
Redwing	*Turdus torquatus*

Scientists often abbreviate the scientific names of organisms by using only the initial letter of the genus once it has been written out in full, for example, *T. merula*. Sometimes the generic name alone may be used in order to refer to all members of the genus, for example, *Turdus* or *Turdus* spp. This avoids the need to list all of the known species. Some authorities recognise subspecies and where these occur a third name is added. The range of the great tit (*Parus major*) extends from the British Isles and northern Europe across Asia to the Pacific. A number of subspecies have been identified based on geographical differences in morphology for example:

Subspecies	Region
P. m. major	From Brittany to Okhotsk
P. m. newtoni	British Archipelago
P. m. excelsus	Spain, Portugal, North Africa
P. m. aphrodite	Balearics, Cyprus, Crete, Greece
P. m. corsus	Corsica, Sardinia

The binomial system is hierarchical in nature, with similar species being collected together into larger and larger groups (Figure 2.1). Related species are grouped into genera; similar genera are grouped into families; similar families are grouped into orders and so on. The main divisions (or taxa) of this hierarchical system are:

Kingdom	
Phylum	(plural phyla)
Class	
Order	
Family	
Genus	(plural genera)
Species	(plural species)

The complete system is much more complex than this and includes groupings such as suborders, superfamilies and infraclasses which fall between those listed above.

It is sometimes convenient to use the names of higher taxa, instead of listing each species individually. The order Cetacea includes all whale, dolphin and porpoise species. The family Falconidae includes all species of falcons. The abbreviation 'spp.' is generally used to denote all species of a higher taxon. For example, *Panthera* spp. means all species of the genus *Panthera*, which includes many of the big cats.

Although in theory scientific names are the same all over the world and remain the same for ever, in practice animal and plant groups are

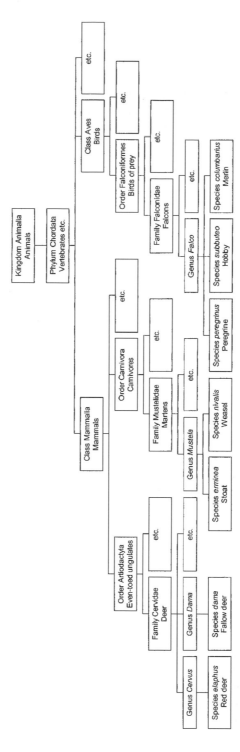

Figure 2.1 The binomial system of classification: a simplified classification of some mammal and bird species. The diagram is intended to illustrate the principles of classification only and is not a complete classification of the taxa shown.

occasionally reclassified. Recently it was discovered that the pipistrelle bat (*Pipistrellus pipistrellus*) is two distinct species that communicate on different wavelengths. In such circumstances scientists must allocate a new name to the more recently discovered form (see p. 133).

2.1.3   Scientific names and the law

It is normal practice in legal documents to list protected species using their scientific names. These are the names by which species are known in any dispute or legal proceedings. Such lists may be found, for example, in the schedules to the Wildlife and Countryside Act 1981, the Habitats Directive and the Appendices to the Convention on International Trade in Endangered Species of Wild Fauna and Flora (CITES). It is generally more useful to list species in schedules or appendices to legislation rather than in the main body of the documents themselves, as they may be more easily amended when the law is changed.

Environmental assessments often contain lists of species recorded on development sites. Where these lists consist of scientific names without accompanying common names it is quite impossible for planners or other non-specialists to identify most if not all of these species. Such an approach makes it difficult for the lay person to determine whether a particular species is a tree or a small herbaceous flowering plant.

2.1.4   The major groups of living things

The term vertebrate refers to those animals that possess a backbone (essentially, mammals, birds, reptiles, amphibians and fish). The term invertebrate refers to animals that do not possess a backbone (essentially all other animal types). This simplistic classification is not very useful to biologists but, nevertheless, these terms are widely used by non-biologists.

The most widely used classification of living things is known as the 'Five Kingdom Classification'. This system divides all organisms into animals (Panel 2.1), plants (Panel 2.2), fungi, single-celled organisms and bacteria and blue-green algae (Panel 2.3).

2.1.5   Species identification

Most of the species found in the UK are relatively small and unfamiliar to the majority of people. The identification of species is best undertaken by specialists but many amateur naturalists have expertise in particular groups of animals or plants. Naturalists groups are also likely to be aware of the location of particularly rare species and may even actively monitor specific sites. Specialist biologists are employed in museums, zoos and university departments and may be able to offer advice on the

PANEL 2.1   THE MAIN GROUPS OF ANIMALS

Kingdom Animalia

Phylum Porifera
The sponges. These organisms are extremely simple and mostly marine.

Phylum Cnidaria
Simple aquatic organisms including coral-forming species and jellyfish.

Phylum Platyhelminthes
Free-living and parasitic flatworms, e.g. tapeworm.

Phylum Nematoda
Free-living and parasitic roundworms. Extremely numerous but rarely seen because most are either very small and live in the soil or they are parasites inside plants or other animals.

Phylum Annelida
Segmented worms found in soil and aquatic environments, e.g. earthworm, ragworm.

Phylum Arthropoda
Segmented animals possessing a hard outer skeleton with paired, jointed limbs, e.g. insects, crabs, woodlice, spiders.

Phylum Mollusca
Soft-bodied animals with a muscular 'foot' and often possessing a shell, e.g. snails, slugs, octopuses, squid.

Phylum Echinodermata
Marine animals with a radially symmetrical structure, e.g. starfish, brittlestars, sea cucumbers, sea urchins.

Phylum Chordata
This group includes the vertebrates: mammals, birds, reptiles, amphibians and fish.

PANEL 2.2   THE MAIN GROUPS OF PLANTS

Kingdom Plantae

Phylum Bryophyta
Small, green terrestrial plants, found in permanently moist environments. Mosses and liverworts.

Phylum Filicinophyta
Terrestrial green plants consisting of stem, leaf (frond) and root. Ferns.

Phylum Lycopodophyta
Small, green spore-bearing, herbaceous plants with moss-like leaves. Club mosses.

Phylum Coniferophyta
Cone-bearing plants that produce seeds. Most are large trees with needle-like leaves. Conifers.

Phylum Angiospermophyta
Flowering plants. Occur as trees, shrubs, herbaceous plants, shrubs and epiphytes. Dominant plant forms in almost all ecosystems.

Note: Details of plant classifications vary between authorities. Some plant classifications use the term division instead of phylum.

PANEL 2.3   THE MAIN GROUPS OF MICROORGANISMS

Kingdom Monera

Phylum Cyanobacteria
Single-celled and filamentous, photosynthetic organisms. Commonly known as blue-green algae.

Phylum Bacteria
Parasitic (disease-causing) and free-living bacteria.

Kingdom Protoctista

Subkingdom Protozoa
Single-celled animal-like forms, e.g. amoebae, ciliates.

Subkingdom Algae
Single-celled and filamentous algae, e.g. diatoms, algae.

Kingdom Fungi

Phylum Zygomycota
Fungi found on decaying organic material, e.g. molds.

Phylum Ascomycota
Fungi usually with a cup-shaped fruiting body, e.g. yeasts, truffles, powdery mildews.

Phylum Basidiomycota
Contains many of the more familiar fungi, e.g. mushrooms, toadstools, puffballs, rusts.

Phylum Mycophycophyta
Lichens: an association between an alga and a fungus. Low-growing forms that are the first to colonise bare rock.

identification and location of rare species. Local museums with a natural history department are likely to employ biologists and they may also be the focus for the local natural history society. Local amateur naturalists often possess a wealth of information about the species and habitats in their area. Museums and natural history societies may publish checklists of local species, particularly birds, but some also maintain lists of local insects, flowering plants and even fungi.

The Natural History Museum in London has a wide range of expertise. For example, the Department of Entomology offers insect identification services on a commercial basis. It can also identify and advise on other arthropod species, such as spiders, scorpions, mites, ticks and millipedes.

Identification guides

The long-established interest in natural history in the UK has led to the publication of a wide range of identification guides. A short list of publications covering a wide range of species, mostly written for the serious amateur, appears at the end of this chapter.

Figure 2.2 Frond of a fern plant showing sori used to produce reproductive spores.

2.2   BASIC CONCEPTS IN ECOLOGY

2.2.1   Introduction

Ecology is the scientific study of the relationships between living things and their environment. Many of the terms used by ecologists appear in environmental legislation so it is useful to consider some basic ecological concepts.

Ecologists find it convenient to study units of the natural world called ecosystems. Ecosystems have three main characteristics:

- energy flow;
- nutrient cycles; and
- species diversity (biodiversity).

An ecosystem may be of any size from the smallest pond to the largest tropical forest. All ecosystems consist of a biological community and a physical environment. The biological community is made up of all of the animals, plants and microorganisms living in a particular place. The physical environment is the air, water, soil and other non-living components of the system.

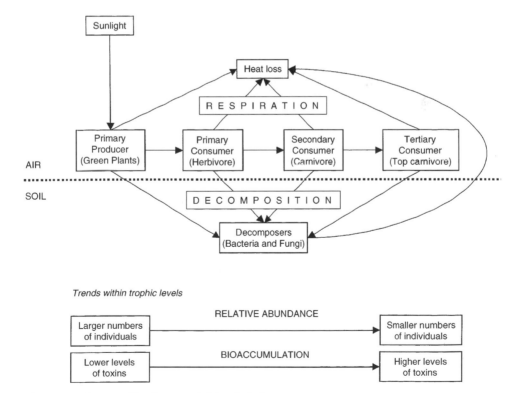

Figure 2.3 A generalised terrestrial food chain.

2.2.2   Energy flow

Most ecosystems use sunlight as a source of energy to produce plant material. Green plants use a process called photosynthesis to produce sugars from carbon dioxide in the air and water taken from the soil. These plants are then eaten by herbivores that eventually provide the food for carnivores that are in turn eaten by top carnivores. This sequence of feeding relationships is called a food chain and each level in the chain is called a trophic level.

Green plant ⇨ Herbivore ⇨ Carnivore ⇨ Top carnivore
Herbs   Herbivorous insects   Spiders   Blue tits

In nature food chains interconnect because most organisms do not simply eat one type of food. The result is called a food web. In systems where the primary producers are small, the relative numbers forming the various trophic levels form a pyramid of numbers, in which primary producers are the most common and tertiary comsumers (top carnivores) are very rare (Figure 2.4).

When pollutants enter food chains they may be concentrated as they are passed from one trophic level to the next. This process is called biomagnification and it results in high concentrations of some pollutants in top carnivores when only very low concentrations occur in the environment. If pollutants like mercury enter the human food chain the consequences for man may be fatal.

Disease organisms may also be transmitted from species to species along food chains. This phenomenon has been important in the transmission of BSE (bovine spongiform encephalopathy). The disease was passed to cows from the infected nervous systems of sheep that were used in cattle feed. In humans the disease manifests itself as a condition known as Creutzfeldt–Jacob disease (CJD).

The maintenance of the integrity of food chains is essential to the proper functioning of ecosystems and the security of the human food supply.

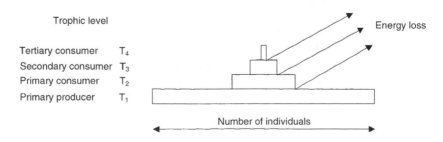

Figure 2.4 A pyramid of numbers.

2.2.3   Nutrient cycling

In addition to plants, herbivores and carnivores, ecosystems also need organisms that can break down dead materials in order to recycle the nutrients they contain. These organisms are bacteria and fungi and are collectively called decomposers. The soil also contains nitrogen-fixing bacteria that are capable of removing nitrogen from the air and using it to fertilise the soil.

Nutrients such as calcium and magnesium are taken up by the roots of green plants in the water they draw from the soil. These nutrients are then incorporated into the structure of the plant's tissues and passed on to herbivores when the plant is eaten. The nutrients are passed along the food chain but each time an organism dies (or sheds it leaves, antlers, or other body parts) nutrients are returned to the soil (Figure 2.5). Some nutrients are recycled back into bodies of water or into the air. These nutrient cycles are more or less closed, with chemicals alternating between the biological and the physical components of the ecosystem (Figure 2.6).

Figure 2.5 During their lives trees and other large organisms contain significant quantities of nutrients. When they die and decompose these nutrients are recycled back into the soil and other compartments of the ecosystem.

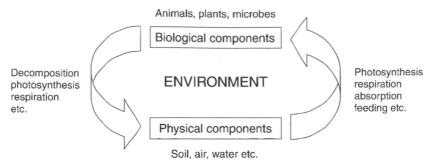

Figure 2.6 A generalised nutrient cycle.

Some pollutants may damage the decomposers in the soil. If these organisms are killed nutrient cycles will cease to function. In some soil types high concentrations of heavy metals may damage populations of nitrogen-fixing bacteria, thereby reducing soil fertility.

2.2.4   Species diversity

Ecosystems contain species that are not equally abundant: some are rare while others are common. Most ecosystems actually contain more rare species than common ones. There are usually a relatively small number of dominant species but many more species that survive in relatively small numbers. The protection of biological diversity is a major objective of environmental legislation at the international, European and domestic level.

In the UK as a whole, some 90 000 species have been recorded from terrestrial and freshwater habitats (Table 2.1). Even the simplest ecosystems contain large numbers of species. Studies of sites which have been cleared in urban areas in Sheffield have found a substantial fauna of invertebrate animals living on and in the soil surface of brick rubble within a few years. This fauna consisted of more than 50 species which included ground beetles, rove beetles, harvestmen, spiders, woodlice, snails, slugs, centipedes, millipedes, Lepidoptera larvae and pupae, ants and earthworms (Gilbert, 1991).

Established ecosystems are likely to contain many more species than this. Raveden Wood is part of the Smithills Estate, located on the northern fringe of Bolton (Greater Manchester) and covers an area of about 12.0 ha (29.6 acres). The area was probably planted with trees shortly after the establishment of Smithills Hall in the early fourteenth century and has remained as woodland ever since. The site is locally important because it is classified as Ancient Semi-natural Woodland and it is now part of Smithills Country Park. The Bolton Museum Natural History Unit has compiled a species list for the nature trail in the woodland:

Figure 2.7 The morphological diversity of moth species found in the UK.

(a) Privet hawkmoth
(b) Lime hawkmoth
(c) Lappet moth
(d) Hummingbird hawkmoth
(e) Poplar hawkmoth
(f) Broad-bordered bee hawkmoth

(g) Cinnabar moth
(h) Canary-shouldered thorn
(i) Swallowtailed moth
(j) Ground lackey moth
(k) Five-spot burnet
(l) Pebble hooktip

Sphinx ligustri
Mimas tiliae
Gasteropacha quercifolia
Macroglossum stellatarum
Laothoe populie
Hemaris fuciformis

Tyria jacobaeae
Ennomos alniaria
Ourapteryx sambucaria
Malacosoma castrensis
Zygaena trifolii
Drepana facataria

Table 2.1 Numbers of terrestrial and freshwater species recorded in the UK (adapted from UK Biodiversity Steering Group 1995 (Anon., 2000))

Group	Species
Bacteria	Unknown
Viruses	Unknown
Protozoa	>20 000
Algae	>20 000
Fungi	>15 000
Ferns	80
Bryophytes	1 000
Lichens	1 500
Flowering plants	1 400
Non-arthropod invertebrates	>3 000
Insects	22 500
Arthropods other than insects	>3 000
Freshwater fish	38
Amphibians	6
Reptiles	6
Breeding birds	210
Wintering birds	180
Mammals*	48
Total	>88 000

* Excludes certain feral and introduced species.

a total of 350 species in all. This list clearly excludes microscopic forms, e.g. nematodes (roundworms) bacteria, protozoa etc. but includes some species which may not reside in the woodland itself, e.g. foxes, sparrowhawks and barn owls. Since the unidentified forms must outnumber the non-residents it is likely that 350 is a significant underestimate of the total number of species present (see Table 2.2).

Temperate woodlands have the potential to contain many more plant species than found in Raveden Wood. Lichen surveys conducted in the area around Scales Wood and Buttermere village in the Lake District found 72 species while Longthwaite and Johnny's Woods were found to contain a total of 106 lichen species (Rose, Hawksworth and Coppins, 1970). Bryophytes are also common in the Lake District. Woods around Seatoller in Borrowdale contained 11 species of mosses and 29 species of liverworts (Pearsall and Pennington, 1973).

2.2.5   Population dynamics

Ecosystems do not have an infinite capacity to support animals and plants. For each species there is a 'carrying capacity' which is the

Table 2.2 Species identified in Raveden Wood, Bolton

Group	Number of species
Birds	42
Mammals	9
Fish	2
Amphibia	2
Insects	
Butterflies and moths	25
Alderflies	1
Bees	1
Beetles	7
Dragonflies	1
Stoneflies	4
True bugs	1
Lacewings	3
True flies	61
Leaf hoppers	1
Centipedes	2
Millipedes	1
Woodlice	2
Molluscs	8
Flowering plants	95
Ferns	7
Horsetails	2
Mosses	15
Liverworts	7
Fungi	51
Total	350

Data from Rodger, D. *Raveden Wood Management Plan.* No date.

optimum number that the system can support. This may be influenced by the amount of space, the availability of a particular nutrient, the number of predators and so on. Carrying capacities will vary from one year to the next. The number of individual organisms within any particular population depends upon the balance between births and immigrants (which cause an increase in number) and deaths and emigrants (which cause a decrease in number) (Figure 2.8).

When a species is introduced into a new habitat it may increase in number exponentially at first, eventually levelling off at the carrying capacity, exhibiting what is known as 'logistic growth' (Figure 2.9a). Few species actually grow like this, except where there is an 'empty' niche waiting to be filled. Starlings (*Sturnus vulgaris*) were introduced into North America at a time when urban environments were expanding. As a consequence the species was able to grow rapidly, spreading across the continent from east to west.

Figure 2.8 Factors affecting the size of a population of organisms.

Species with very similar ecological requirements may compete for resources such as food and shelter. In natural ecosystems evolution has ensured that most similar species have ecological requirements that are sufficiently different to allow them to coexist. However, when species are introduced into novel environments they may compete with the indigenous species and displace them. Such a displacement occurred in Britain as a result of the introduction of the grey squirrel (*Sciurus carolinensis*) (from North America) into habitats originally occupied by the native red squirrel (*Sciurus vulgaris*).

The introduction of new predatory species into an ecosystem may also result in the destruction of other species. For example, the introduction of domestic cats has damaged the ground nesting seabird colonies on some islands. Similarly introducing exotic herbivores, such as goats, may have a damaging effect on the vegetation and the other species that depend upon it. Introduced grazing species may initially show exponential growth only to exceed the carrying capacity of the environment, destroy the food supply, and then exhibit a population 'crash'. This type of growth is called 'boom and bust' growth (Figure 2.9b). Many species show stable oscillations of numbers from one year to the next, reflecting the numerous factors which affect their population size (Figure 2.9c). In a good year, when food is plentiful and the weather is mild, numbers may increase. In a bad year, when food is scarce and the weather is severe, numbers may decline.

In Britain, the introduced rabbit has had a dramatic effect upon the native vegetation. Selective grazing by herbivores removes some species from the ecosystem while encouraging the growth of others. Grazing by sheep and cattle will also dramatically affect plant diversity.

2.2.6   Ecological succession

Ecosystems do not remain unchanged for ever. When a new area of bare rock is colonised the first organisms to establish themselves are mosses and lichens (the pioneer community). These simple plants are then replaced by grasses and small shrubs, and then by larger shrubs and trees as the soil deepen. The animal community also changes as succession proceeds, and new niches are created.

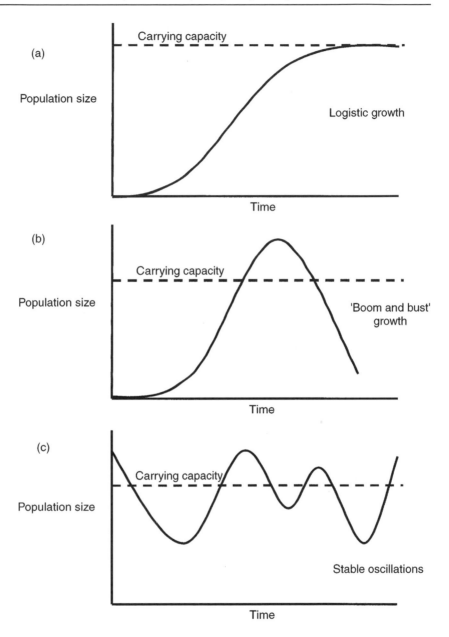

Figure 2.9 Population dynamics: (a) logistic growth; (b) 'boom and bust' growth; and (c) a population showing stable oscillations in numbers.

Each of these developmental stages is called a seral stage. The whole succession is called a sere. After perhaps a hundred years a climax community will become established. The nature of this community may depend upon the climate, the soil type, human activity and other factors. The climax community in most of Britain should be broad-

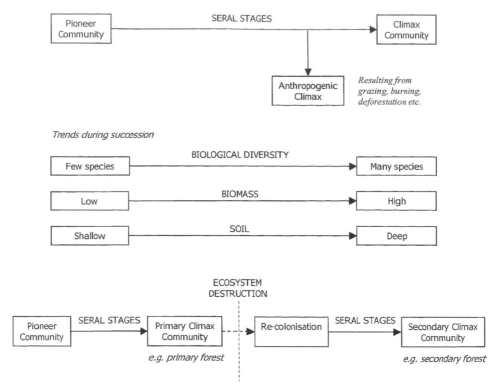

Figure 2.10 Ecological succession. As ecosystems develop the number and diversity of species present increase.

leaved woodland but in many areas human activity has removed this and prevents it from becoming re-established (Figure 2.10).

Often, the climax community is destroyed, perhaps by deforestation, and succession begins again. This time the soil is already present so succession proceeds at a fast rate. This process is called secondary succession and most of the forest in the UK is the result of this process.

2.2.7   Soil

The existence of terrestrial ecosystems depends upon soil. Soil contains organic and inorganic components, and also air and water. The soil acts as a substratum for plants, and their roots help to bind soil particles together so that they do not blow away. Soil left exposed on agricultural land or construction sites will dry out during the summer and simply blow away.

Soil provides plants with water and nutrients and contains an enormous reservoir of plant seeds. This seed bank is essential to the long-term survival of ecosystems because, when environmental conditions are favourable dormant seeds will germinate and replace plants that have died.

The soil is occupied by large numbers of animals, many of them microscopic in size, like nematodes and tiny arthropods (Panel 2.4). Soil takes many hundreds of years to develop. The surface is often covered with decaying leaf litter and below this is a layer of semi-decomposed organic matter called humus. This material helps to retain moisture in the soil and supplies nutrients to plants. The air between soil particles is important in supplying oxygen to the roots. Heavy machinery may cause soil compaction and smearing of soil particles. When soil becomes compacted or waterlogged it may be impossible for roots to respire.

PANEL 2.4   SOIL STRUCTURE AND ANIMALS

Major groups of soil animals

- Protozoa
  - Amoebae
  - Flagellates
  - Ciliates
- Rotifera (rotifers)
- Nematoda (roundworms)
- Platyhelminthes (flatworms)
- Annelida (segmented worms)
  - Oligochaeta
    - Lumbricidae (earthworms)
    - Enchytraeidae (potworms)
- Arthropoda
  - Arachnida
    - Oribateid mites
    - Mesostigmatid mites
    - Pseudo-scorpions
    - Araneida (spiders)
  - Crustacea
    - Isopoda (woodlice)
  - Insecta
    - Collembola (springtails)
    - Orthoptera (grasshoppers and crickets)
    - Coleoptera (beetles)
    - Diptera (flies)
    - Hymenoptera (ants)
    - Lepidoptera (butterflies and moths)
  - Myriapoda
    - Diplopoda (millipedes)
    - Chilopoda (centipedes)
- Mollusca
  - Gastropoda
    - Pulmonata (slugs and snails)

Soil profile – podsol

Vegetation

Upper limit of mineral soil

L Fresh leaf litter
F Fermenting, but identifiable litter
H Unidentifiable organic remains

A HORIZON
Topsoil

A1 Organically enriched mineral soil
A2 Zone of eluviation of humus, iron sesquioxides, clay

B HORIZON
Sub-soil

B1 Zone of humus enrichment

B2 Zone of susquioxide enrichment

C HORIZON
Parent rock

2.2.8 Ecological surveys

The key to a good ecological survey is appropriate sampling. It is clearly impossible to produce an inventory of every single specimen of every single species in a particular habitat. Instead, ecologists rely upon sampling methods to obtain data on small areas within the habitat that are representative of the habitat as a whole.

The sampling method used will depend upon the species or habitat that is being studied. Plants generally have fixed locations, so they may be studied using methods which examine small areas of the ground which have been selected at random. Since most animal species move around, individuals must generally be captured and there are various methods and pieces of equipment available for this purpose.

Sampling vegetation

Quadrats

Quadrats are used to study plant communities and some sedentary or slow-moving animal species, such as snails. A quadrat is essentially a square area marked out on the ground. Small areas can be sampled using a frame quadrat which is usually a metre square. Results may be expressed as percentage of the ground covered by each species. Larger areas may be marked out on the ground using surveying equipment and poles. The usefulness and accuracy of a study using quadrats depends upon various factors including:

- the size of the quadrat;
- the number of samples taken;
- the time of year; and
- the location of the quadrats.

Flowering plants, especially grasses, are extremely difficult to identify in the winter. Many species can only be identified by experts, and annuals may only be present as seed if the survey is conducted in winter. Rare species, by definition, are not likely to fall within the relatively small number of quadrats that is likely to be studied in a short ecological survey.

Transects

Line transects are used to sample areas of ground along a straight line across a habitat or series of habitats. Such a transect might be used in conjunction with a quadrat to study the changes in vegetation from the shore to the low water mark on a salt marsh. A quadrat would be used to sample vegetation, assessing percentage cover, at regular intervals

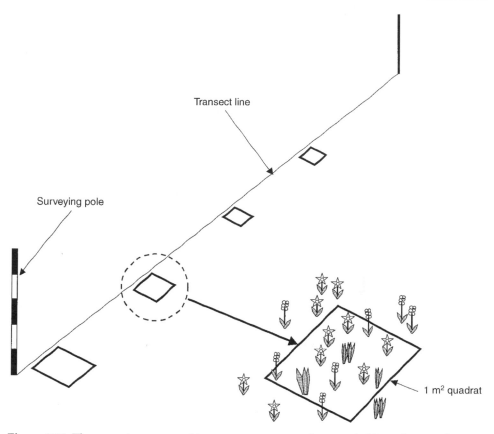

Figure 2.11 The organisms present in an ecosystem may be sampled by using quadrats arranged along a transect. This method is generally used for sampling plant communities but may also be used for other immobile or relatively immobile organisms such as snails in a woodland or limpets on a rocky shore.

along the transect (Figure 2.11). Vegetation classification systems are described in Appendix 8.

Sampling and counting animals

Mark–recapture techniques
Mark–recapture techniques may be used to estimate the size of a population of mobile organisms. These techniques involve the capture, marking, release and recapture of animals and allow estimates to be made of population size from the proportion of animals recaptured. There are several methods available. The simplest is called the Lincoln Index (Panel 2.5).

PANEL 2.5   THE LINCOLN INDEX – ESTIMATING THE POPULATION SIZE OF MOBILE ORGANISMS

Organisms which move, most of which are animals, are difficult to find and difficult to count. An estimate of population size can be calculated by the following method:

1. Capture a random sample of the population and record the number (first trapping occasion, T_1)
2. Mark the captured animals (using paint, coloured rings or some other suitable method)
3. Release the marked animals where they were captured
4. Allow the released animals to mix randomly with the remainder of the population (usually for several days)
5. Capture a second random sample of the population (second trapping occasion)
6. Record the total number of animals captured on this second occasion (T_2) and the number of recaptures (i.e. animals marked on the first trapping occasion and caught on the second trapping occasion, R)

The population size (N) is calculated, using a formula known as the Lincoln Index, as follows:

$$N = \frac{T_1 \times T_2}{R}$$

This method is very simple and makes a number of assumptions:

■ Each individual is equally likely to be captured
■ No marks are lost
■ Marked animals randomly mix in the population
■ Marks do not affect the chances of being recaptured
■ The population is closed (i.e. no births, deaths, immigration or emigration occur between the two trapping occasions)

Small mammals may be sampled, and counted, using Longworth traps. These should be camouflaged to avoid human interference and inspected regularly as small mammals soon perish if insufficient food is available.

Other methods of sampling animal populations
The presence of bats may be detected with electronic bat detectors.

Insects may be collected from vegetation by placing bins or trays under trees, or manually using a pooter (Figure 2.12a). Ground-living insects and other invertebrates may be captured with pitfall traps (Figure 2.12b). However, these must be emptied regularly to reduce the

(a)   Pooter

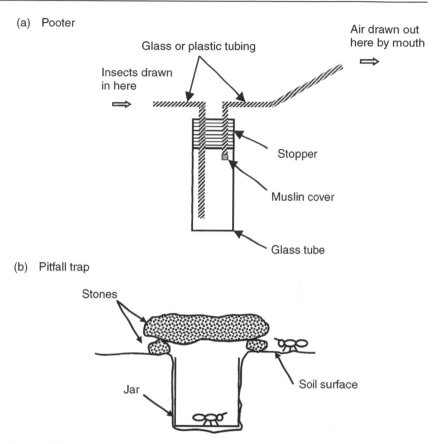

(b)   Pitfall trap

Figure 2.12 Some simple equipment used for sampling insects and other small invertebrates: (a) a pooter, used to collect small insects from vegetation by sucking them into the bottle; (b) a pitfall trap, used to catch ground-living insects etc., which will fall into the trap.

possibility of carnivorous species eating the herbivores. Moths may be captured with specially-designed light traps.

All sampling methods have their own particular advantages and disadvantages, and may produce misleading results in certain circumstances. However, it is important for non-ecologists to appreciate that these, and other, methods exist. An ecological survey that does not use appropriate sampling techniques and simply involves someone recording the most obvious species present in a habitat, merely by observation, is simply a token attempt. Unfortunately, such superficial studies sometimes appear in environmental statements.

Ecologists need to be aware of the legal constraints on catching and marking animals and taking vegetation samples. Licences may be required if certain species are to be captured or their sheltering places disturbed. Similar restrictions are placed on the taking or destruction of certain rare plants (see Chapter 4).

2.2.9   The distribution of species

The distributions of animal and plant species in the UK are mapped in 10 km squares. The records used to compile these maps are collected by County Recorders (often based in museums) and sent to the Biological Records Centre. Specialist societies collect distribution data for particular groups of species, for example the Mammal Society and the Botanical Society of the British Isles (BSBI). The BSBI has a network of recorders in each county and vice-county in Britain, and works with flora groups, botanical societies and local naturalists' societies.

Biological Records Centre and the National Biodiversity Network

The Biological Records Centre (BRC) was established in 1964 and is based at Monks Wood in Cambridgeshire. It is the national focus for the recording of species (other than birds) in the UK. The BRC works with voluntary recorders throughout Britain and Ireland and its database contains almost 12 million records of over 12 000 species.

The BRC is working with the a wide range of other organisations to develop a National Biodiversity Network (NBN). The NBN will co-ordinate the work of more than 60 000 volunteer and professional naturalists and will involve around 80 national biological societies and species recording schemes and over 70 local records centres.

The Lancashire Biodiversity Action Partnership is encouraging the reporting of wildlife sightings by producing record cards which observers are asked to complete and return when they see rare species such as the brown hare, song thrush, orange tip butterfly and bluebell. These cards are designed as postcards and are distributed through information centres (Figure 2.13).

2.3   NATURE CONSERVATION

2.3.1   Why bother to conserve nature?

Why should we bother to conserve nature? The following is a list of the reasons that conservationists traditionally put forward:

■ *Balance of nature*
The loss of biological diversity may upset the 'balance of nature'.

Figure 2.13 Recording leaflets distributed to the public by Lancashire Biodiversity Action Partnership in an attempt to monitor species abundance and distribution.

- *Biological indicators*
  The presence or absence of some animals and plants acts as an indicator of environmental quality. For example, songbirds, lichens, freshwater invertebrates, and fish.

- *Aesthetic value*
  Animals and plants are used in human art and as such have an aesthetic value without which our lives would be poorer.

- *Cultural value*
  Animals and plants are an important part of a very wide range of cultures.

- *Medical uses*
  Many animal and plant products are important in medicine, either because they provide us with the raw materials for drugs, or because they provide us with models with which to study disease.

- *Genetic value*
  Each time another species becomes extinct we lose its genes for ever. These genes may have been of some use to us in the future, for example, for breeding new crop plants or animals for food.

■ *Psychological value*
Many people fulfil a need to escape from the stress of living in urban areas by spending time in wilderness areas. Such areas are becoming increasingly important as leisure-time destinations.

■ *Ethical*
Plants and, especially, animals have an intrinsic 'right' to exist, and since we have control over their fate, we have an obligation to protect them.

■ *Economic value*
Certain materials found in nature have a clear economic value such as wood, meat and so on. Entire ecological systems may also have an economic value, for example, a wetland may remove organic wastes from a river system which would otherwise require the construction of a waste water treatment plant.

■ *Ecotourism*
Tourism in wilderness areas is important in generating income. In some areas, such as the Lake District National Park, this may make a very important contribution to the local economy.

2.3.2   Pollution and biological indicators of environmental quality

Chemical analysis of the environment can provide us with data on the quantities of particular elements or compounds present. For some purposes, such as drinking water quality monitoring, this approach is sensible and relatively cost effective. However, if we want to monitor the state of the environment continuously, biological indicators are often more useful than chemical measurements taken at particular points in time.

In the 1960s it became apparent that we were poisoning the natural environment to such an extent that many of our birds of prey were facing imminent extinction. Organochlorine insecticides (principally DDT) had entered the food chains and were concentrated as they moved successively from one trophic level to the next and eventually to the top carnivores. After the Second World War, organochlorine insecticides were extensively used to protect agricultural crops from pest damage. They were extremely successful but do not break down easily in the environment. Studies of egg shell thickness in birds of prey showed a marked reduction after 1947. The result was a high frequency of egg breakage and embryo death. When DDT was banned, bird of prey populations recovered.

Birds in general are good indicators of environmental quality. Scientists are concerned that there has been a marked decline in the populations of songbirds in Britain recently, probably indicating a deterioration in the environment. This is not due to a single factor as was the case with organochlorine insecticides. It is more likely to be the result of a series of factors such as habitat loss and the development of a more complex chemical environment.

River quality can be monitored by sampling the invertebrate fauna. Organic pollution, such as sewage, farm effluent or milk spillage, will reduce the oxygen levels in the water because it encourages the growth of bacteria and other microorganisms which use oxygen in respiration. The result is that many species are unable to survive while others thrive in the nutrient-rich water. *Tubifex* worms are minute relatives of the earthworm. They use haemoglobin to collect oxygen from water and are able to survive in large numbers near organic pollution sources. Fish, on the other hand, cannot survive in very low oxygen levels and will be killed by large inputs of organic materials. The presence of otters and water voles in rivers is a good indicator of the presence of clean water.

Lichens are a peculiar combination of a fungus and an alga living in a symbiotic relationship that benefits both (Figure 2.14). Some lichens are tolerant of high levels of air pollution while others are not. Since they grow very slowly lichens are good indicators of general pollution levels in the recent past. It is possible to produce maps of urban areas showing different pollution levels based on the abundance and diversity of the lichen species present. Some fungi are also sensitive to air pollution. *Rhytisma acerinum* causes a condition known as 'tar spot' in

Figure 2.14 Lichens growing on stone walls and trees may be used as indicators of air pollution as many cannot tolerate high levels of pollutants such as sulphur dioxide.

Figure 2.15 *Rhytisma acerinum* is a fungus which causes a condition known as 'tar spot' in sycamore (*Acer pseudoplatanus*) leaves. It cannot tolerate high levels of sulphur dioxide, so its presence generally indicates relatively 'clean air'. The distribution of tar spot has been used to map sulphur dioxide levels in Merseyside and Yorkshire.

sycamore leaves (*Acer pseudoplatanus*), but is absent in areas with high air pollution levels (Figure. 2.15).

A survey of selected ancient woodand and parkland sites in England and Wales found that the number of lichen species declined near major areas of air pollution. Lichens are particulary sensitive to sulphur dioxide in the air because it disrupts photosynthesis. At one rural site in southern England 256 species were recorded but in the London area this fell to only 2 or 3 species. The changes that have occurred within the lichen populations in Epping Forest (Essex) have been well documented. Up to the middle of the nineteenth century at least 120 corticolous (growing on trees) and lignicolous (growing on barkless wood) species were present, but by 1970–74 this had fallen to a mere 38 species (Hawksworth and Rose, 1976).

Man's effects on plant distribution are not always bad. In Shetland, and other generally treeless areas, fence posts have permitted some corticolous lichen species to extend their range. Other lichen species have been able to penetrate urban habitats by colonising man-made substrates like concrete, bricks, asbestos–cement, walls and tombstones.

Acid rain kills trees and tree death was any early indicator of the presence of this particular environmental problem. As acid rain lowers the pH of lakes it alters the composition of the populations of diatoms (microscopic plants). Since they are largely made of silica, diatoms persist in mud sediments for long periods after they have died. This has allowed scientists to infer changes in climate from changes in diatom composition in different layers of mud, representing different periods in the past. Studies of Scottish lakes have shown that their pH has fallen since about 1850, coinciding with an increase in industrialisation.

Apart from the importance of wildlife and wild places for their own sake and for human enjoyment, it is quite clear that animals and plants can act as early warning systems, unwitting indicators of the health of the environment, when we threaten to damage the ecosystems upon which we ultimately all depend.

2.3.3   Conservation management

The conservation of some ecosystems requires merely that they be left alone. Others, however, require active management if their conservation interest is to be maintained. Sometimes management is required in order to maintain a particular community of species: heathland is burned, grassland is grazed. Many ecosystems in Britain are unnatural in the sense that they are either artificially maintained by human interference or they contain non-indigenous species. Other systems have been 'created' by man and even where no recent interference is apparent the system is likely to have been disturbed by man within historic time.

Ecosystem management at the Gibraltar Point National Nature Reserve in Lincolnshire takes many forms: grazing by sheep and cattle, mowing, footpath repair (due to rabbit damage), removal of excessive algal growth from ponds, control of flooding, excavation of pools, cleaning of beaches, dune restoration, coppicing of sallows and removal of 'undesirable' flora (including cutting, hand removal and spraying with herbicide). Perhaps the most unusual management practice used on this reserve is the use of Hebridean sheep to create a favourable sward for natterjack toads (*Bufo calamita*) that have been introduced into a number of the ponds.

The natural environment of the British Isles is a mosaic of habitats. Some are ancient and almost untouched by man. Most contain vegetation that has been greatly influenced by human activity and a variety of introduced species. A landscape that should be dominated by trees has been reduced to grassland and moorland bisected by stone walls and hedgerows. Now even these artificial boundaries have an ecological value. Two hundred years of industrial activity have left land contaminated with heavy metals and scars in the landscape where sand and gravel have been extracted. Ecologists have restored these landscapes and many are now home to diverse ecosystems.

PANEL 2.6   GIBRALTAR POINT NATIONAL NATURE RESERVE

Some of the best areas for high species diversity are those where a mosaic of different habitats exists. Gibraltar Point National Nature Reserve (Lincolnshire) is located at the head of The Wash and is an important bird migration watch point. It covers an area of only 430 hectares but contains sand dunes, salt marsh, freshwater marsh and foreshore. A total of 718 species have been recorded on the reserve:

Group	Number of species
Plants	385
Birds	203
Mammals	24
Amphibians	4
Reptiles	1
Butterflies	25
Macrolepidoptera	27
Pyralidae	34
Dragonflies and damselflies	15
Total	718

This list does not include marine plants, lichens, fungi, most insect groups, non-insect invertebrate groups, or microorganisms but does include marine mammals seen from the shore.

Sources: *Annual Wildlife Report 1995. Gibraltar Point National Nature Reserve*, The Lincolnshire Trust for Nature Conservation.
Current Plant List. Gibraltar Point National Nature Reserve, Lincolnshire and South Humberside Trust for Nature Conservation Ltd.

The Hope Carr Nature Reserve was created by North West Water Ltd (now United Utilities) on the edge of Leigh, in Lancashire. The 15 ha (37 acre) site was originally part of Hope Carr Farm and was bought by the local sewage board in 1898. Fields which were once used to grow silage and others which were used for the disposal of sewage sludge have now been reclaimed. A variety of habitats, dominated by a network of wetland, and including lakes and ponds now support over 100 species of birds. Messingham Sand Quarry Nature Reserve, near Scunthorpe in Lincolnshire, contains a series of flooded lagoons left after sand excavations. It also contains reedbeds, woodland, marsh and a small remnant of heath. Over 160 bird species (of which 63 breed) have been recorded at the site, around 18 butterfly species and more than 200 species of moth.

2.3.4   Design of nature reserves

Nature reserves are created primarily for two reasons:

- as refuges for threatened species;
- to preserve ecosystems.

If they are badly designed, the rare species may die out or the ecosystem may not be sustainable. A reserve should be large enough to hold a viable population of any species it is trying to conserve. Genetic variability must be maintained so the protected area needs to contain a genetically viable population. If the population is too small, inbreeding will occur and may cause a decline in fertility in future generations. It may be necessary to consider the land use around the reserve because it may change in the future.

There are some important theoretical points to consider when designing nature reserves. The shape and size of a nature reserve affects its potential to retain individuals. If a reserve is very small an individual animal needs only to move a short distance before it finds itself outside of the protected area, so in general terms large reserves are probably better at protecting species than small reserves. Since species leave reserves across their boundaries, in theory a circular reserve is better than any other shape since this shape has the shortest perimeter per unit area. Long thin shapes are not generally good because animals are never far from the boundary. Chains of small reserves located close to each other are generally better than isolated small areas. If these small reserves are linked by protected corridors this will facilitate movement from one reserve to another. In urban areas such corridors may take the form of existing linear features of the environment such as disused railway lines, canals, river valleys or even hedges and rows or trees. A series of small reserves can also protect against catastrophes, e.g. damage by fire, flood, disease etc. However, if individual reserves are too small they may not be adequate for some species (see Figure 2.16).

Sometimes the shape and size of a reserve is dictated by a natural feature that is to be included, e.g. a lake or mountain. Where this is not the case a circular shape is to be preferred as it has the minimum possible 'edge'. Where two ecosystems meet there is a species-rich zone that is not representative of either ecosystem, creating an edge effect. This decreases the effective area of pure ecosystem in the reserve and is greater in small reserves than in large reserves. It is also greater in long, thin reserves than in reserves that are approximately circular. However, for other reasons, it may be preferable to have a number of small reserves rather than one large reserve.

Migratory species may be conserved by setting up reserves at feeding and breeding sites. Obviously, large grazing mammals need large reserves but a rare orchid or insect may be protected by a reserve measuring just a few square metres.

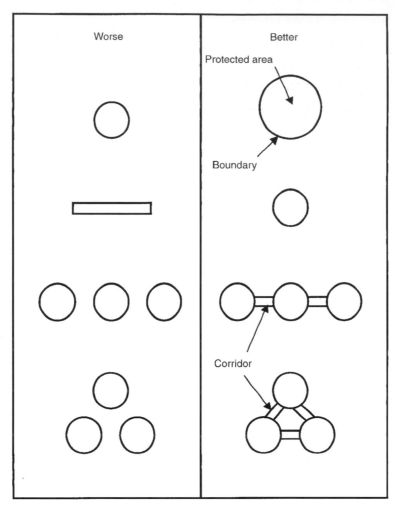

Figure 2.16 The theoretical design of nature reserves. Shapes which have a long perimeter enclosing a small area are considered poor as organisms may easily leave the protected area. Large, more or less circular reserves will have the lowest ratio of perimeter to area and are therefore considered to be a good shape. Corridors linking protected areas allow organisms to move safely from one to another.

It is often very difficult to establish how large a reserve should be to be effective.

2.3.5 Urban ecology

Urban ecology has developed in recent years as a discrete area of study (Gilbert, 1991). Canals, parks, gardens, walls, derelict land and even landfill sites have been studied. Suburban gardens are thought to

Table 2.3 Number of breeding bird species recorded in selected London parks (adapted from Hounsome, 1979)

Park	No. of breeding bird species
St James's/Green Park	22
Regent's Park/Primrose Hill	34
Richmond Park	56
Kew Gardens	39
Hampton Court	42

support the highest density of breeding birds of any habitat in Britain with nearly 600 pairs per km^2 (excluding house sparrows) reported in an area of interwar housing in London (Batten, 1972). City parks are also important habitats for birds. Results of surveys undertaken in parks in the London area (Hounsome, 1979) found high numbers of breeding species (Table 2.3)

Urban areas are also important as habitats for wild plants. A survey of 59 allotments in urban areas spread over six sites in Hertfordshire recorded 58 weed species (Willis, 1954). The City of Sheffield contains some 15 000 street trees, two-thirds of which are forest size (Gilbert, 1991). They include lime, plane, sycamore, ash, horse chestnut, elm, beech, and a few common alder, poplar and native oak. Towns and cities are important refuges for many species of plants and animals and of very great value to people who live and work in urban and suburban areas.

A small oasis of wildlife exists in the centre of the Trafford industrial estate in Manchester, in the form of a small ecology park. It contains a number of habitats including a large pond, woodland and grassland along with an education centre, and yet it is completely surrounded by warehouses and other industrial sites.

Urban ecology parks and country parks in urban areas make an important contribution to nature conservation and environmental education, and developers should be encouraged to incorporate conservation areas into their urban developments.

REFERENCES

Anon (2000) *The State of the Environment of England and Wales: The Land.* Environment Agency. The Stationery Office Ltd, London.

Batten, L.A. (1972) Breeding bird species diversity in relation to increasing urbanization. *Bird Study,* 19, 157–66.

Gilbert, O.L. (1991) *The Ecology of Urban Habitats.* Chapman & Hall, London.

Hawksworth, D.L. and Rose, F. (1976) *Lichens as Pollution Monitors.* Edward Arnold (Publishers) Ltd, London.

Hounsome, M. (1979) Bird life in the City. In *Nature in Cities* (ed. I. C. Laurie). John Wiley, Chicester.

Pearsall, W.H. and Pennington, W. (1973) *The Lake District*. William Collins Sons & Co. Ltd, Glasgow.

Rose, F., Hawksworth, D.L. and Coppins, B.J. (1970) A lichenological excursion through the north of England. *The Naturalist*. Reprinted in Pearsall, W.H. and Pennington, W. (1973) *The Lake District*. William Collins Sons & Co. Ltd, Glasgow, pp. 144–146.

Willis, S.J. (1954) Observations on the weed problem of allotments based on a survey. In *Proceedings of the British Weed Control Conference 1954*, pp. 71–74.

Identification guides

Arnold, E.N., Burton, J.A. and Ovenden, D.W. (1993) *Collins Field Guide. Reptiles and Amphibians of Britain and Europe*. Harper Collins Publishers, London.

Burrows, E.M. (1991) *Seaweeds of the British Isles*. Natural History Museum Publications, London.

Carter, D.J. and Hargreaves, B. (1994) *Collins Field Guide. Caterpillars of Britain and Europe*. Harper Collins Publishers, London.

Chinery, M. (1993) *Collins Field Guide. Insects of Britain and Northern Europe*. Harper Collins Publishers, London.

Courtecuisse, R. and Duhem, B. (1995) *Collins Field Guide. Mushrooms and Toadstools of Britain and Europe*. Harper Collins Publishers, London.

Delforge, P. (1995) *Collins Photo Guide. Orchids of Britain and Europe*. Harper Collins Publishers, London.

Duncan, U.K. (1970) *Introduction to British Lichens*. T. Buncle & Co. Ltd, Arbroath.

Harrison, C. and Castell, P. (1998) *Collins Filed Guide. Birds Nests, Eggs and Nestlings of Britain and Europe*. Harper Collins Publishers, London.

Hayward, P., Nelson-Smith, T. and Shields, C. (1996) *Collins Pocket Guide. Sea Shore of Britain and Northern Europe*. Harper Collins Publishers, London.

Hubbard, C.E. (1984) *Grasses. A Guide to their Structure, Identification, Uses and Distribution in the British Isles,* 3rd edn. Penguin Books Ltd, Harmondsworth, Middlesex.

Jermy, C. and Camus, J. (1991) *The Illustrated Field Guide to Ferns and Allied Plants of the British Isles*. Natural History Museum Publications, London.

Macdonald, D. and Barrett, P. (1995) *Collins Field Guide. Mammals of Britain and Europe*. Harper Collins Publishers, London.

Maitland, P.S. and Campbell, R.N. (1992) *The New Naturalist. Freshwater Fishes of the British Isles*. Harper Collins Publishers, London.

Peterson, R.T., Mountford, G. and Hollom, P.A.D. (1992) *Collins Field Guide. Birds of Britain and Europe,* 5th edn. Harper Collins Publishers, London.

Roberts, M.J. (1995) *Collins Field Guide. Spiders of Britain and Northern Europe*. Harper Collins Publishers, London.

Stace, C. (1997) *New Flora of the British Isles,* 2nd edn. Cambridge University Press, Cambridge.

Tolman, T. and Lewington, R. (1997) *Collins Field Guide. Butterflies of Britain and Europe*. Harper Collins Publishers, London.

Watson, E.V. (1981) *British Mosses and Liverworts,* 3rd edn. Cambridge University Press, Cambridge.

See Appendix 8 for guides to vegetation classification systems.

3

The organisation and administration of nature conservation in the UK

The administration of nature conservation in the UK is performed by a complex combination of statutory and non-statutory bodies. It has a long and complex history. The aim of this chapter is to explain the role of government departments and agencies in conservation, and the history of the development of the law at a national, European and international level. It also considers the role of non-statutory bodies in lobbying on behalf of wildlife and providing specialist information to inform government decision making.

3.1   CENTRAL GOVERNMENT AND ITS AGENCIES

3.1.1   Department for Environment, Food and Rural Affairs

The Department for Environment, Food and Rural Affairs (DEFRA) was formed in 2001. It has taken on the responsibilities (in England) of the disbanded Ministry of Agriculture, Fisheries and Food together with those of the Environmental Protection Division of the former Department of Environment, Transport and the Regions (DETR). The remaining responsibilities of the DETR passed to the Department for Transport, Local Government and the Regions. The DETR was originally formed in June 1997 by the merger of the Department of the Environment and the Department of Transport. In May 2002, the Prime Minister created a new Department for Transport.

While there is clearly some merit in forming a closer association between those sectors of government concerned with the countryside and those concerned with environmental protection, this reorganisation has been criticised for separating the responsibility for the environment into one department while transport and urban issues will be dealt with by another.

The aims of DEFRA are to enhance the quality of life through promoting:

- a better environment;
- thriving rural communities;
- diversity and abundance of wildlife resources;
- a countryside for all to enjoy; and
- sustainable and diverse farming and food industries.

The department's key tasks which are likely to have an impact on wildlife conservation are:

- to establish DEFRA as the leading voice of government for sustainable development, environmental protection and the renewal of rural areas, including the farming community;
- to protect and improve the environment and to integrate the environment with other policies across government and internationally;
- to conserve and enhance the diversity and abundance of English wildlife and to protect globally threatened animals and plants;
- to set the future direction of the rural economy;
- to define the vision of a sustainable, diverse and dynamic farming industry and promote a strategy for achieving this, through further reform of the Common Agricultural Policy;
- to secure reform of the Common Fisheries Policy in order to preserve a healthy marine environment and sustainable fishing for the future.

DEFRA sponsors a number of agencies and non-departmental public bodies (NDPBs) whose work may have an impact on wildlife and nature conservation, including:

- Centre for Environment, Fisheries and Aquaculture Science
- Pesticides Safety Directorate
- Environment Agency
- Royal Botanic Gardens, Kew
- Sea Fish Industry Authority
- Advisory Committee on Pesticides
- Advisory Committee on Releases to the Environment of Genetically-modified Organisms
- Royal Commission on Environmental Pollution

The Wildlife and Countryside Directorate (WACD) is the DEFRA Directorate responsible for wildlife and countryside policy. It sponsors the following NDPBs:

- Countryside Agency
- English Nature

- Joint Nature Conservation Committee
- National Forest Company

The objective of the Directorate is to enhance opportunity in rural areas, improve enjoyment of the countryside and conserve and manage wildlife resources. It aims to achieve this by:

- conserving English landscapes and increasing opportunities for recreation;

PANEL 3.1 WILDLIFE AND COUNTRYSIDE DIRECTORATE (WACD)

The objective of WACD is to enhance opportunity in rural areas, improve enjoyment of the countryside and conserve and manage wildlife. Its responsibilities range from improving access to the countryside and tackling wildlife crime, to assuring that the UK fulfils its European and international conservation commitments.

The Directorate is divided into four divisions, each of which is made up of a number of branches:

Countryside Division
- National Park sponsorship and landscape protection branch
- Access to open countryside and coastal policy coordination
- Common land branch
- Countryside Agency sponsorship and recreation branch
- Rights of way branch

European Wildlife Division
- Natura 2000 branch
- SSSI review team
- Habitats conservation team
- Species conservation team
- Sponsorship and procurement branch
- Scientific Advisor's unit
- Biodiversity Action Plan Secretariat

Global Wildlife Division
- CITES and zoos policy branch
- Wildlife crime and inspectorate unit
- Wildlife licensing, enforcement and information systems branch
- International conservation policy branch

Rural Development Division
- Rural White Paper team
- Agriculture and the environment
- Rural regeneration and forestry

- coordinating the implementation of the Countryside and Rights of Way Act 2000 (which improves access to the countryside and increases protection for Sites of Special Scientific Interest);
- conserving and enhancing the diversity and abundance of English wildlife in accordance with EU and international commitments;
- protecting globally threatened species by regulating trade, developing international wildlife agreements, tackling wildlife crime and promoting high standards in zoos; and
- promoting rural development, as well as ensuring that agricultural, forestry and regeneration policies are environment-friendly.

3.1.2   The statutory nature conservation agencies

The Environmental Protection Act 1990 created the Countryside Council for England (English Nature) and the Countryside Council for Wales as the statutory bodies responsible for nature conservation in England and Wales respectively. The equivalent organisation for Scotland is Scottish Natural Heritage, created by the Natural Heritage (Scotland) Act 1991. The Countryside and Rights of Way Act 2000 formally changed the name of the Countryside Council for England to English Nature. These three organisations are collectively referred to as the Nature Conservancy Councils (following the definition established by the Wildlife and Countryside Act 1981, s. 27A). However, the terms 'Nature Conservancy Councils (NCC)', 'nature conservation body', 'appropriate nature conservation body' and 'statutory conservation agencies' have been used here interchangeably as they appear in various legislative instruments and other government documents.

The UK statutory conservation agencies are now:

- English Nature
- Countryside Council for Wales
- Scottish Natural Heritage
- Environment and Heritage Service (Northern Ireland)

In addition, there is a Countryside Agency in England. The mechanisms available to the statutory conservation agencies (and others) to protect wildlife and the countryside include:

- establishment of protected areas;
- creation and enforcement of specific offences against wildlife and habitats;
- control of certain activities by licensing;
- encouragement of environment-friendly management practices by the provision of advice, support and education;
- financial compensation and incentives;
- contributing to planning decisions.

English Nature

English Nature (formerly the Nature Conservancy Council for England) is the statutory body responsible for nature conservation in England. It was created under the Environmental Protection Act 1990, s. 128, but was renamed 'English Nature' under the Countryside and Rights of Way Act 2000, s. 73. English Nature advises the Secretary of State on nature conservation in England. Its responsibilities include:

■ identification and notification of Sites of Special Scientific Interest (SSSIs);
■ designation and management of National Nature Reserves (NNRs);
■ identification and management of Marine Nature Reserves (MNRs);
■ promotion of nature conservation;
■ provision of advice and information about nature conservation;
■ support and conduct of research.

Countryside Agency

The Countryside Agency was established in 1999 to supersede the Countryside Commission in England, and includes part of the former Rural Development Commission. The agency promotes the conservation and enhancement of the countryside in England. It also undertakes activities aimed at stimulating the creation of jobs and the provision of essential services in the countryside.

The Agency is responsible for the designation of:

■ National Parks
■ Areas of Outstanding Natural Beauty (AONBs)
■ National Trails
■ Heritage Coasts
■ National Forest and Community Forests (in partnership with the Forestry Commission)

Countryside Council for Wales (Cyngor Cefn Gwlad Cymru)

The Countryside Council for Wales was established by the Environmental Protection Act 1990, s. 128, and is the government's statutory advisor on nature conservation in Wales. It is funded by the National Assembly for Wales and accountable to the First Secretary.

The Council's responsibilities include:

■ designation of National Parks;
■ identification and notification of Sites of Special Scientific Interest (SSSIs);

■ designation and management of National Nature Reserves (NNRs);
■ identification and management of Marine Nature Reserves (MNRs);
■ designation of Areas of Outstanding Natural Beauty (AONBs).

Scottish Natural Heritage

Scottish Natural Heritage (SNH) was established in 1992 under the Natural Heritage (Scotland) Act 1991. It is funded by the Scottish Executive and provides advice on nature conservation in Scotland to all whose activities affect:

■ wildlife;
■ landforms;
■ features of geological interest.

SNH also seeks to develop and improve facilities for the understanding and enjoyment of the countryside. Its conservation responsibilities include:

■ designation of National Scenic Areas (NSAs);
■ identification and notification of Sites of Special Scientific Interest (SSSIs);
■ designation and management of National Nature Reserves (NNRs);
■ identification and management of Marine Nature Reserves (MNRs).

It should be noted that National Parks and AONBs are not statutory designations in Scotland.

Environment and Heritage Service (Northern Ireland)

This organisation is an executive agency of the Department of the Environment for Northern Ireland and is responsible, within Northern Ireland, for:

■ designation of Areas of Special Scientific Interest (ASSIs);
■ designation of National Nature Reserves (NNRs);
■ designation of Marine Nature Reserves (MNRs).

It also performs a similar function to English Heritage in relation to historic buildings in Northern Ireland (see Section 3.1.5).

Joint Nature Conservation Committee

The Joint Nature Conservation Committee (JNCC) was created under the Environment Act 1995 to deal with those aspects of nature conservation that affect the UK as a whole and its international obligations.

The Committee advises the government and others on international conservation issues and disseminates knowledge on these subjects. The Committee provides guidance to English Nature, Scottish Natural Heritage, the Countryside Council for Wales and the Department of the Environment for Northern Ireland. It is also responsible for establishing common standards for the monitoring of nature conservation and research.

3.1.3   National Parks Authorities

National Parks Authorities (NPAs) were created by the Environment Act 1995 to replace the National Parks boards and committees. They are the sole local planning authorities for their areas and are therefore able to influence land use and development within the National Parks. The duties of NPAs in relation to National Parks include:

- conservation and enhancement of natural beauty, wildlife and cultural heritage;
- promotion of opportunities for public understanding and enjoyment;
- fostering the economic and social well-being of communities within National Parks.

The NPAs are made up of representatives from the relevant local authorities and others nominated by the Secretary of State or the National Assembly for Wales, as appropriate to the geographical location.

3.1.4   Wildlife Inspectorate

The Inspectorate is part of the Global Wildlife Division within DEFRA. It consists of a small team based at its headquarters in Bristol and a panel of about 80 part-time consultants located throughout the UK and recruited mainly for their expertise in the identification of animal and plant species. The Inspectorate liaises with the police and HM Customs and Excise, as the statutory enforcement authorities for the majority of wildlife law.

The work of the Inspectorate involves:

- inspections of premises of keepers of birds listed on Schedule 4 of the Wildlife and Countryside Act 1981;
- inspections of premises of persons who apply for licences to release barn owls (*Tyto alba*) and certain other species into the wild;
- identifying and determining the origin of certain species imported into the UK;
- monitoring compliance with conditions imposed on licences; and

- assisting the police and other enforcement agencies with species identification and other evidence.

Wildlife Inspectors have powers of entry to premises of:

- keepers of birds listed on Schedule 4 of the Wildlife and Countryside Act 1981;
- importers and/or exporters of endangered species; and
- persons applying for licences to release barn owls or certain other species into the wild.

3.1.5   Heritage

Heritage in England

The Historic Buildings and Monuments Commission for England (English Heritage) was established under the National Heritage Act 1983. In 1999 it merged with the Royal Commission on the Historical Monuments of England (RCHME) and is now responsible for the National Monuments Record (which includes over 12 million photographs, maps and drawings gathered since the formation of the RCHME in 1908).

The duties of English Heritage include:

- providing expert advice and skills;
- awarding grants to secure the preservation of listed buildings, archaeological sites and ancient monuments;
- encouraging the re-use of historic buildings to aid regeneration of urban and rural areas;
- managing buildings and monuments in its care;
- promoting access to, and enjoyment of, ancient monuments and historic buildings.

Heritage in Scotland

Historic Scotland performs a similar function to English Heritage. The Ancient Monuments Board for Scotland advises Scottish ministers on the exercise of their functions in providing protection for monuments of national importance under the Ancient Monuments and Archaeological Areas Act 1979.

The Historic Buildings Council for Scotland advises Scottish ministers on matters related to buildings of special architectural or historic interest, and in particular to proposals for awards and grants for repair.

The Royal Commission on the Ancient and Historical Buildings of Scotland is responsible for the surveying and recording of ancient and historical monuments connected with the culture, civilisation and living conditions of people in Scotland. It compiles and maintains the

National Monuments Record of Scotland and the national record of the archaeological and historical environment.

Heritage in Wales

Cadw (Welsh Historic Monuments) performs a similar function to English Heritage by supporting the preservation, conservation, appreciation and enjoyment of the built heritage in Wales. The Ancient Monuments Board for Wales advises the National Assembly for Wales on its statutory functions in relation to ancient monuments. The Historic Buildings Council for Wales advises the National Assembly on the built heritage through Cadw.

The Royal Commission on the Ancient and Historical Monuments of Wales is empowered by royal warrant to survey, record, publish and maintain a database of ancient, historical and maritime sites and structures and landscapes in Wales. It is also responsible for the National Monuments Record of Wales which:

- supplies archaeological information to the Ordnance Survey;
- coordinates archaeological aerial photography in Wales;
- sponsors the regional Sites and Monuments Records.

Heritage in Northern Ireland

The built heritage in Northern Ireland is the responsibility of the Environment and Heritage Service. It is responsible for listing buildings and scheduling monuments on the advice of the Historic Buildings Council for Northern Ireland and the Historic Monuments Council for Northern Ireland. The criteria for evaluating buildings are similar to those used in England

3.1.6 Other government agencies

Environment Agency

The Environment Agency was established by the Environment Act 1995. It is a non-departmental public body sponsored by the Department for Environment, Food and Rural Affairs and the National Assembly for Wales. In Welsh the agency is 'Asiantaeth yr Amgylchedd'. The equivalent agency for Scotland is the Scottish Environment Protection Agency which receives funding from the Scottish Executive.

The Environment Agency has assumed the functions of:

- the National Rivers Authority;
- the waste regulation authorities;

- the waste disposal authorities;
- Her Majesty's Inspectorate of Pollution.

Its main responsibilities relate to water pollution, air pollution and waste disposal. However, the agency is also responsible for the management and use of water resources, including freshwater fisheries, flood control and the control of coastal erosion. It has some specific responsibilities in relation to protected areas.

PANEL 3.2   THE NATURE CONSERVATION ROLES OF THE ENVIRONMENT AGENCY

The Environment Agency's principal aim is to protect and enhance the environment and in doing so, to contribute towards the objective of sustainable development in England and Wales. The Agency's remit covers:

- 15 million hectares of land
- 36 000 kilometres of river
- 5000 kilometres of coastline
- 2 million hectares of coastal waters

Although the Environment Agency's functions are generally considered to be related to the control of environmental pollution, it also has a significant role to play in nature conservation, including:

- Monitoring of river water quality
- Monitoring of bathing water quality
- Freshwater fisheries protection
- Sand dune protection and restoration
- Recreation

The Agency's long-term objectives in relation to wildlife include working with partners to achieve:

- Restoration of degraded habitats, especially rivers, estuaries and wetlands
- Establishment of wildlife corridors with no artificial barriers to wildlife movement
- Delivery of the UK's Biodiversity Action Plan, thereby removing threats to priority species
- Appropriate fish communities in rivers, estuaries, lakes and canals
- The protection and restoration of habitats, species and natural processes by the encouragement of appropriate urban and rural land use
- Acceptance and support of the management of land for wildlife and landscape benefits as a normal activity of rural life
- A broad consensus on how biodiversity should be managed against a background of climate change
- A reduction in threats to the genetic integrity of native wildlife

Scottish Fisheries Protection Agency

This body is an executive agency of the Scottish Executive Rural Affairs Department. It enforces fisheries law and regulations in Scottish waters and ports.

Maritime and Coastguard Agency

The Maritime and Coastguard Agency is an executive agency of the newly created Department for Transport. One of the agency's responsibilities is to minimise the risk of pollution of the marine environment from ships, and, where pollution occurs, to minimise the impact on UK interests. Regional Principal Counter Pollution and Salvage Officers are responsible for monitoring and dealing with pollution incidents in their regions. The agency can provide advice on shoreline clean-up.

Planning Inspectorate

The role of the Planning Inspectorate is discussed in Chapter 8.

Highways Agency

The Highways Agency was established in 1994, and is an executive agency of the Department for Transport. The Agency maintains, operates and improves the network of trunk roads and motorways in England on behalf of the Secretary of State. This network currently includes over 10 000 km of roads. One of the key objectives of the Agency is to 'minimise the impact of the trunk road network on both the natural and built environment'.

Natural Environment Research Council (NERC)

The Natural Environment Research Council funds and manages scientific research and training in Earth systems science, including terrestrial and aquatic sciences. NERC's research centres include:

- British Geological Survey
- Centre for Ecology and Hydrology
- Plymouth Marine Laboratory

The Council also has a number of collaborative centres including:

- Dunstaffnage Marine Laboratory
- Centre for Population Biology
- Environmental Systems Science Centre
- Sea Mammal Research Unit

Centre for Ecology and Hydrology (CEH)

The Centre for Ecology and Hydrology is a centre of NERC and is the leading body for research, survey and monitoring in terrestrial and freshwater environments in the UK. It has laboratories and field facilities at nine sites around the UK. One of the main aims of the CEH is to describe and understand the dynamics of terrestrial and freshwater ecosystems through monitoring, experimentation and modelling.

Forestry Commission

The Forestry Commission of Great Britain is responsible for the protection and expansion of Britain's forests and woodlands. Its mission is to 'Protect and expand Britain's forests and woodlands and increase their value to society and the environment'. One of its principal objectives is to conserve and improve the biodiversity, landscape and cultural heritage of forests and woodlands. The Forestry Commission is the largest landowner in the UK and has been carrying out national woodland surveys since 1924. The latest National Inventory of Woodlands and Trees was completed in July 2000.

The Forestry Commission has two executive agencies:

■ Forest Enterprise manages 800 000 hectares of forests and woodlands owned by the nation.

Figure 3.1 The Forestry Commission owns large areas of land in Britain. Although its primary task is to provide wood, the land under the Commission's control includes many important wildlife areas.

- Forest Research conducts scientific research and surveys to inform the development of forestry policies and practices and to promote sustainable forest management.

The Forest Authority is the regulatory, grant aiding and advisory arm of the Commission and is responsible for issuing felling licences.

Forest Service (Northern Ireland)

The Forest Service is an executive agency of the Department of Agriculture and Rural Development for Northern Ireland. It designates and administers Forest Nature Reserves and has a similar role to the Forestry Commission in Great Britain.

Deer Commission for Scotland

The Commission is responsible for furthering the conservation and control of deer in Scotland. It has the statutory duty (and the necessary powers) to prevent damage to agriculture, forestry and the habitat by deer.

British Geological Survey (BGS)

The British Geological Survey was established in 1835. Its mission is to:

- collect geoscience data relevant to the UK landmass and continental shelf;
- deliver products and services to meet the needs of government, industry and the public;
- undertake high-quality, impartial research in the geosciences.

The BGS is a component survey of the Natural Environment Research Council. It provides the geoscience knowledge and advice necessary to support informed decision making in the public and private sectors. The BGS provides services to a wide range of businesses and other organisations, including central government and its agencies, local authorities, the Environment Agency, the water industry, the minerals industry, and the waste management industry. It also provides services to the oil, gas, energy and power industries, civil engineers, and the insurance, finance and legal sectors.

The National Geosciences Information Service (NGIS) provides access to a wide range of collections of bibliographic and cartographic material, records, rock, mineral and fossil samples and digital databases.

The Ordnance Survey

The Ordnance Survey of Great Britain and the Ordnance Survey of Northern Ireland produce maps (and other products) of England, Wales, Scotland and Northern Ireland. They include maps focusing on the National Parks, and a CD-ROM on National Parks which contains information on geology and biodiversity. Specialist maps showing the locations of designated conservation sites in England are available from English Nature.

The Natural History Museum

The Natural History Museum, London, (Figure 3.2) provides a wide range of library and information services. Its collections include zoological, botanical and geological specimens. The Department of Library and Information Services is a national reference library for the life and earth sciences, specialising in taxonomy and systematics. It

Figure 3.2 The Natural History Museum, London, houses important collections of specimens of UK animal and plant species. The museum contains a large collection of books and other documents, and its staff are able to provide a range of identification services.

holds over one million books and 25 000 periodical titles, together with photographs, news cuttings, drawings and other documents in its libraries and archives, which are:

- General and Zoology Libraries
- Ornithology and Rothschild Libraries
- Botany Library
- Entomology Library
- Earth Sciences Library
- Museum Archives

The library catalogue may be found at *www.library.nhm.ac.uk*. In addition to library services, the museum also provides identification services.

Water companies

Although the water companies are obviously not conservation agencies as such, they have been included here because they are required by law to take account of conservation in the carrying out of their functions, and are regulated by the government.

PANEL 3.3   NATURE CONSERVATION AND UNITED UTILITIES

United Utilities is the largest manager of water and waste water assets in the UK. It owns approximately 57 500 hectares of land in the North West of England, much of which is in National Parks and Areas of Outstanding Natural Beauty.
  The company:

- Manages 17 nature reserves
- Planted 184 500 trees in 2000/2001, and over 1 million since 1996
- Controls around 14 600 hectares of catchments designated as SSSIs
- Will undertake biodiversity audits of four of its estates
- Financed the planting of reedbeds
- Supports 70 priority species and habitats on its land
- Works with regional Biodiversity Action Plan (BAP) groups
- Is a member of the North West Biodiversity Forum
- Supports the 'Red Alert' project to assist the recovery of red squirrels
- Participates in the 'North West Water and the Wildlife Trusts Otters and Rivers Project'
- Increased the numbers of fish and birds living in the Mersey estuary by treating all major waste water discharges
- Supports a reforestation project in the Philippines

Source: *Social & Environmental Impact Report 2001*, United Utilities.

The activities of the water companies in relation to the environment are regulated by the Environment Agency. In addition, these companies are required to exercise their powers so as to further the conservation and enhancement of natural beauty and the conservation of flora, fauna and geological or physiographical features of special interest (Water Industry Act 1991, s. 3(2)(a)).

North West Water (now part of United Utilities) has established 17 nature reserves on its land. It has undertaken restoration projects in moorland areas that have involved airlifting moorland plants into remote locations by helicopter to prevent trampling and has restored former sludge lagoons to woodland.

3.1.7   Local government and police

Local authorities

Local authorities (LAs) have a number of responsibilities in relation to nature conservation including:

- establishment of Country Parks;
- establishment of Local Nature Reserves (LNRs) in England, Scotland and Wales;
- establishment of Local Authority Nature Reserves (LANRs) in Northern Ireland;
- protection of limestone pavements by limestone pavement orders;
- protection of hedgerows by hedgerow retention notices;
- protection of trees and woodlands under tree preservation orders.

Although they have no statutory responsibilities in relation to wildlife conservation, local authority environmental health departments may be called upon to deal with pest species such as rats, mice, birds or insects.

Local planning authorities play an important role in determining the nature of the local landscape and in the protection of conservation sites such as SSSIs through the application of the planning system (see Chapters 5, 8 and 9). LAs have officers responsible for trees, hedgerows and footpaths. Some LAs have also appointed Biodiversity Officers.

The Local Land Charges Register maintained by the local authority holds details of property and land. It contains, among other things, details of existing and proposed tree preservation orders, conservation areas, details of SSSIs, and conditional planning applications from 1 August 1977.

Police

The police have numerous powers in relation to wildlife crime, which vary with the legislation which is being relied upon. Every British police force has appointed a specialist Wildlife Liaison Officer (WLO) who specifically deals with wildlife crime. Most WLOs undertake their wildlife duties on a part-time basis, alongside their other operational duties. However, in some parts of the country full-time WLOs have been appointed in response to the rising levels of wildlife crime.

The work of the WLO involves liaising with local badger groups, the RSPB and other conservation organisations, and may also include giving lectures to school children and other groups. The types of crime dealt with by WLOs include:

- badger crime;
- bird crime;
- illegal international trade in endangered species;
- cruelty;
- illegal poisoning;
- disturbance to wildlife.

The Metropolitan Police Service has a Wildlife Crime Unit which is part of its Special Operations Command. It maintains contact with many conservation organisations, government departments, other police forces and individuals involved in the protection of wildlife. It also acts as a focus for information and for inquiries from the public.

Partnership for Action Against Wildlife Crime (PAW)
This is a multi-agency body made up of all of the statutory and non-governmental organisations involved in wildlife law enforcement in the UK. Its main objective is to promote the enforcement of wildlife legislation, particularly through the networks of Police Wildlife Liaison Officers and Customs Wildlife and Endangered Species Officers.

National Wildlife Crime Intelligence Unit (NWCI)
The NWCI was established in April 2002 as part of the Specialist Intelligence Branch of the National Criminal Intelligence Service (NCIS) in response to an increase in wildlife crime. The NCIS was formed in 1992 to develop intelligence to combat serious and organised crime at the national and international level. The NWCI is funded by DEFRA, the Association of Chief Police Officers, the Home Office and the Scottish Executive. It will help the fight against wildlife crime by:

- providing law enforcement with actionable intelligence to target crimes and criminals;
- acting as a national focal point for gathering and analysing intelligence on serious wildlife crime at regional, national and international level;
- identifying trends and patterns in wildlife crime, and links to other types of serious crime;

- developing sources to gather intelligence;
- providing a nucleus of expertise and knowledge;
- establishing links with domestic and international agencies dealing with wildlife crime.

3.2   NON-STATUTORY CONSERVATION ORGANISATIONS

The organisations described below have been grouped, for convenience, into those which are primarily concerned with particular species or groups of species, those concerned with protecting habitats and those that have a more general interest in wildlife and environmental protection.

3.2.1   Species-based organisations

Mammal Society

The Mammal Society is an organisation of professional and amateur mammalogists. The society works to protect British mammals and halt the decline of threatened species. Its members conduct research and advise on all issues affecting British mammals. The Society collects data on the distribution of British species and publishes distribution maps. It also published a series of information leaflets on British mammals and a number of books on identification, conservation and other aspects of mammal biology.

British Trust for Ornithology (BTO)

The British Trust for Ornithology was established in 1933 as an independent scientific research trust. It investigates the population biology, movements and ecology of wild birds in the British Isles and conducts very large-scale bird surveys using volunteers employing specially designed census methods. Some BTO volunteer members are nest recorders while others are licensed bird ringers. The activities of the BTO are important in monitoring long-term changes in British bird populations. The BTO is not a campaigning organisation or pressure group so its data is considered objective and unbiased.

Bat Conservation Trust

The Bat Conservation Trust is devoted to the conservation of bats and their habitats and has both professional and amateur members. It supports a network of over 90 voluntary bat groups and has established a National Bat Monitoring Programme which is conducting field surveys and monitoring roosting and hibernation sites. The Trust also operates a telephone helpline for anyone who finds bats on their property and publishes a series of information leaflets.

National Federation of Badger Groups

The objectives of the National Federation of Badger Groups (NFBG) are to promote the conservation, welfare and protection of badgers, their setts and habitats. It supports a network of 85 voluntary badger groups throughout Britain with a combined membership of about 20 000. The NFBG works closely with the police, the RSPCA and other organisations and can provide expert advice on all badger issues.

Plantlife

Plantlife is a charity which acts directly to stop common plants becoming rare in the wild, rescues wild plants that are close to extinction, and protects sites of exceptional botanical importance. It carries out practical conservation work and also influences government policy and legislation.

There are 22 Plantlife Reserves in Britain, including flower meadows, limestone pavement, chalk grassland, blanket bog and ancient woodland. Plantlife conducts plant surveys and publishes guidance on management techniques for various plant species.

Plantlife founded and provides the secretariat for Plant Europa, a European network of organisations working in plant conservation. It also convenes and administers the Fungus Conservation Forum, which promotes the conservation of fungi in the UK, and Plantlife Link, a forum of botanical societies and conservation organisations. Plantlife works on conservation projects with a wide range of partners, including the statutory conservation agencies.

Froglife

Froglife is a charity concerned with the conservation of native amphibians and reptiles in Britian and Ireland. It supports voluntary conservation groups and projects and provides the secretariat for the Herpetofauna Groups of Britain and Ireland. Froglife operates a telephone helpline to provide advice on amphibians and reptiles and can give advice on pond maintenance. The organisation publishes a range of materials on conservation and mangement.

Butterfly Conservation Society

The British Butterfly Conservation Society Ltd is the largest insect conservation organisation in Europe. It has created and manages a network of nature reserves of special importance to butterflies and moths, and it encourages butterfly-friendly farming and gardening. The organisation advises landowners and managers how to conserve and restore habitats for butterflies and moths, and it lobbies government to influence planning and policy decisions. It also organises local and national surveys, and carries out research.

Zoological gardens

Zoos have an important role to play in the conservation of native species. Chester Zoo has been involved in captive breeding programmes for the barn owl (*Tyto alba*), the harvest mouse (*Micromys minutus*) and the sand lizard (*Lacerta agilis*). It has also established areas of wild flowers within its grounds.

Zoos may provide information on the identification of animals and animal products to customs officers investigating cases of the importation of endangered species whose import is banned under CITES and may provide holding facilities for such species when they are seized.

3.2.2   Habitat-based organisations

Many important wildlife areas and scenic places are in private ownership. A wide range of charitable organisations own and manage some of the best wildlife sites in the UK. These sites benefit from the protection of property law but in many cases are also covered by statutory designations, such as SSSI. In some cases they may also be protected sites under European or international law.

Royal Society for the Protection of Birds (RSPB)

The Royal Society for the Protection of Birds was founded in 1889 to combat the plumage trade. It is now Europe's largest conservation charity and has over one million members. The society works for the conservation of biodiversity, especially birds and their habitats, and advises the government on conservation issues. It produces a wide range of educational materials including bird identification guides.

Officers of the society work with other organisations and the police to collect evidence against and prosecute individuals who commit crimes against birds. The RPSB may give expert evidence to public inquiries into development projects which may affect important habitats for birds. However, it is takes a neutral stance on issues relating to animal welfare, falconry and field sports.

The RSPB manages over 150 nature reserves throughout the UK. Not all of the Society's reserves are large wilderness areas. In January 2000 the RSPB used a £90 000 National Lottery grant to buy a 120 hectare (296 acre) farm on Islay in the Inner Hebrides to help protect the chough population that feeds on the grubs found in cow pats.

Wildfowl and Wetlands Trust (formerly the Wildfowl Trust)

The Wildfowl and Wetlands Trust (WWT), formerly the Wildfowl Trust, was founded by Peter Scott in 1946 at Slimbridge in Gloucestershire. Some of the best wetland areas in the UK are owned by the Wildfowl and Wetlands Trust, including Martin Mere (Lancashire), Washington (Tyne and Wear), Welney (Norfolk) and Caerlaverock (Dumfries). The WWT has nine centres in all, covering around 2000 hectares.

The Trust undertakes captive breeding programmes for rare species and conducts scientific research. It also produces educational materials.

Marine Conservation Society (MCS)

The Marine Conservation Society is a charity dedicated to the protection of the marine environment and its wildlife. The MCS provides advice to the government, the EU and industry on marine conservation issues. It is working with the JNCC to establish a map of marine habitats around the UK using volunteer divers.

National Trust

The National Trust is a charity that was founded in 1895 to preserve places of historic interest or natural beauty permanently for the nation. The Trust protects over 200 historic houses and gardens and around 50 industrial monuments and mills. In total it owns more than 244 000 hectares (603 000 acres) of countryside and 909 km (565 miles) of coast.

The National Trust has the unique statutory power to declare land inalienable. Such land cannot be sold, mortgaged or compulsorily purchased against the Trust's wishes, without intervention by Parliament. This means in effect that the Trust is custodian of woodland and forests, farmland, moorland, downs, islands, nature reserves, archaeological remains and villages and all of its other properties for ever.

The Trust owns a number of important SSSIs. They include Blake's Wood and Hatfield Forest in East Anglia, and an upland heath called Long Mynd in Shropshire. Sandscale Haws, near Barrow, is an expanse of dunes and marshes of international importance, harbouring the rare natterjack toad (*Bufo calamita*) and over 500 species of wild flowers.

The National Trust is responsible for the conservation and management of approximately one-quarter of the Lake District National Park and also owns Bridestones Moor within the North York Moors

National Park. Some Trust properties are home to large herds of deer such as Lyme Park in Cheshire which incorporates a medieval deer park containing approximately 450 red deer and 90 fallow deer.

The Trust owns 17 km (10.5 miles) of Yorkshire's coastline including Cayton Bay and Hayburn Wyke near Scarborough, sand dunes and pine woodland at Formby and 8.9 km (5.5 miles) of the White Cliffs of Dover. Islands owned by the Trust include the Farne Islands in Northumberland, Lundy in the Bristol Channel, Brownsea Island in Poole Harbour and Northey Island in the Blackwater estuary on the Essex coast.

Woodland Trust

The aim of the Woodland Trust is to keep ancient woodlands safe from development. All of the woods owned by the Trust are managed with trees, wildlife and people in mind, and where new woods are planted only native seedlings are used. Where desirable, exotic conifers are removed allowing oak, beech and ash to regenerate. The Trust has created or saved over 1000 woodlands and planted over 2.5 million trees.

Wildlife Trusts

The Wildlife Trusts partnership is a network of 47 local Wildlife Trusts which manages around 2500 nature reserves in the UK coving some 75 600 hectares. The Wildlife Trusts lobby government for better protection for wildlife and also undertake practical conservation work.

3.2.3   Environmental pressure groups

Council for the Protection of Rural England (CPRE)

The Council for the Protection of Rural England campaigns to protect the English countryside. The CPRE is concerned with issues such as agriculture, forestry, quarrying, transport issues, wind turbines, noise, major retail developments and other planning issues. It produces a number of publications on these and other issues.

Royal Society for the Prevention of Cruelty to Animals (RSPCA)

The Royal Society for the Prevention of Cruelty to Animals is a large and complex charitable organisation based in Sussex. The primary concern of the RSPCA is cruelty to animals, regardless of whether they are pets, farm animals or wild species. In some areas the Society maintains its own veterinary facilities and accommodation for sick, injured and unwanted animals. It employs over 300 inspectors, has 109 branches, 105 clinics and animal centres, 4 veterinary hospitals and 3 specialist wildlife hospitals.

RSPCA inspectors' duties range from dealing with abandoned animals to investigating cases of animal cruelty in cooperation with the police. Inspectors and other staff may also be called upon to rescue distressed wildlife. The Society may assist the police in prosecuting offenders or it may bring prosecutions itself. The RSPCA also lobbies government and campaigns on behalf of wildlife especially in relation to cruelty issues.

Other organisations

Some large environmental organisations, which were not established specifically to protect wildlife and habitats in the UK, have been party to a number of important legal cases concerning wildlife and the environment. These groups include Greenpeace, Friends of the Earth and the World Wide Fund for Nature (formerly the World Wildlife Fund).

3.3   AN OUTLINE OF NATURE CONSERVATION LAW

3.3.1   A brief history of the law in the UK

Early attempts to protect animals in the national laws of modern western society were based on either a welfare approach or a species approach. Both of these approaches have developed alongside the growth in voluntary organisations concerned with animal welfare and the conservation of wildlife and the natural environment.

In England, in the eleventh century, William the Conqueror imposed the death penalty for the killing of a royal deer, although this was clearly to protect his own hunting rights rather than the 'rights' of the deer. This penalty was rescinded by Henry III.

A statute of 1472, passed during the reign of Edward IV, required the removal of all keddels (weirs) throughout England except those by the sea coast to allow the free passage of ships and boats,

> … and also in safeguard of all the fry of fish spawned within the same.

This requirement to remove weirs was also contained within the Magna Carta (1215) but fish were not mentioned in the text and the protection of fish stocks may have been purely incidental to the protection of navigation. Close seasons for salmon fishing were first introduced into English law in 1285 and the statute appears to identify persons who would perform the duties that are today performed by water bailiffs:

> ... there shall be assigned overseers of this Statute, which being sworn shall oftentimes see and inquire of the offenders.

The modern law restricting the use of certain types of nets for fishing has its origins in a law that was established over 600 years ago. In 1389 the use of nets called 'stalkers' was forbidden and in 1423 the fastening and hanging of nets continually across the Thames by day and night was outlawed. In 1558 protection was given to all immature fish and size limits were imposed for pike, salmon, trout and barbel. In more recent times, the Salmon Fishery Act of 1861 listed prohibited methods of taking fish, including the use of poisons, spears and nets with below minimum mesh dimensions.

In Britain in the Middle Ages birds such as the heron were protected as quarry for falconry, a royal sport, and there is a long history of deer being protected in Royal Parks.

In 1772 the Game (Scotland) Act was passed and this was followed by a series of Acts dealing with game and quarry species. The government then turned its attention towards the protection of birds and by 1939 there were 16 complete Acts for the Protection of Birds in force in Britain, with references to birds in several others.

The Royal Society for the Prevention of Cruelty to Animals began life as the Society for the Prevention of Cruelty to Animals and was founded in England in 1824 to promote the humane treatment of work animals, such as horses and cattle, and of household pets. Two years earlier Richard Martin MP had obtained an Act to prevent the cruel treatment of cattle. In 1875 the Society for the Protection of Animals Liable to Vivisection was founded in England and in the next year the Cruelty to Animals Act 1876 was passed, becoming the first national anti-vivisection law. However, the Act covered procedures on vertebrates only. The Society changed its name to the Anti-Vivisection Society in 1897, and subsequently became the National Anti-Vivisection Society.

The first national organisation founded in the UK specifically concerned with protecting wildlife was the Selbourne Society for the Protection of Birds, Plants and Pleasant Places, in 1885. This was a logical extension of the RSPCA in trying to extend humanity to all wildlife. The Society for the Protection of Birds was founded in Manchester in 1889 specifically to stop the slaughter of thousands of egrets, herons and birds of paradise each year for their plumes. Branches were set up overseas and the branch in India secured the first measure against the plumage trade: an order from the Indian government in 1902 that banned the export of bird skins and feathers. In 1904 the Society became the Royal Society for the Protection of Birds.

Current wildlife and nature conservation law is fragmented. It has evolved over a long period of time and has often been reactive in nature rather than part of an overall government plan. The wildlife legislation enacted in the United Kingdom in the first three quarters of the twentieth century was largely concerned with individual species, for example seals, deer and badgers.

The first town and country planning law appeared in the Housing, Town Planning etc. Act 1909 and was some of earliest urban planning legislation in the world. In contrast, Britain was slow to establish a system of National Parks. Yellowstone National Park in Wyoming was the world's first national park, designated in 1872. It was not until 1949 that the National Park system was established in English law (National Parks and Access to the Countryside Act). The Nature Conservancy was established by Royal Charter in the same year.

Since the 1970s, UK conservation law has been influenced by European law, in particular the Wild Birds Directive and the Habitats Directive. The Wildlife and Countryside Act 1981, which protects, among other things, endangered species and habitats, was passed in order to implement the Wild Birds Directive, but did not enter into force until September 1982.

A number of important pieces of legislation have been proposed by private members, such as the Animal Protection Act 1822, sponsored by Richard Martin MP, and more recently two separate attempts to ban fox hunting, by Michael Foster MP and Ken Livingstone MP, and the Marine Wildlife Conservation Bill 2001, by John Randall MP.

In 1994 the Conservation (Natural Habitats etc.) Regulations were passed to implement the Habitats Directive. Cruelty to wildlife has been an area that has been particularly neglected and it has taken almost three-quarters of a century for the protection from cruelty afforded to domestic animals to be extended to wild mammals (Wild Mammals (Protection) Act 1996).

In 2000 Parliament passed the Countryside and Rights of Way Act which, as well as increasing access to the countryside, improved the protection to Sites of Special Scientific Interest and Areas of Outstanding Natural Beauty.

A chronological list of the major legislation affecting wildlife and nature conservation in the UK is presented in Appendix 1.

3.3.2 European law

The European Community began to take action to protect the environment in 1972 with four successive action programmes. During this period some 200 pieces of Community legislation were adopted, principally concerned with pollution control and waste management, but including measures to conserve nature. A clear environmental policy was established for the first time by the Treaty on European Union (Maastricht Treaty) which came into force in 1993. Article 174(2) of the EC treaty requires that EC action provides a high level of environmental protection.

The principle of sustainable development is now enshrined as one of the European Community's aims in the Treaty of Amsterdam 1997. The Fifth Environment Action Programme (*Towards Sustainability*) established the principles of a strategy of voluntary action for the period 1992–2000. This marked the beginning of a 'horizontal' Community approach, taking into account all of the causes of pollution. This approach to environmental policy was confirmed by the Commission following its Communication on integrating the environment in European Union policies (1998) and by the European Council in Vienna in December of the same year. Community institutions are now obliged to consider the environment in all their other policies.

As Community environmental policy has developed, the range of legislative instruments available has been expanded. It includes a financial instrument (the *Life* programme) and technical instruments, such as a system for the assessment of the effects of public and private projects on the environment (environmental impact assessment). In addition the European Environment Agency was created in 1993, with its headquarters in Copenhagen. The Agency gathers and disseminates environmental data. Although its role is purely supervisory, its work is crucially important in the adoption of new measures and in the assessment of the impact of decisions already adopted.

The introduction of environmental accounting and environmental taxes, following the 'polluter pays' principle, are areas into which Community legislation will diversify. However, the effective implementation of such environmental legislation is likely only when incentives for economic operators have been introduced.

The member states of the Community have adopted a number of measures intended to have a beneficial effect on the natural environment. These include Directives introducing quality standards to bathing waters and to control air pollution, and should indirectly bring long-term benefits to many species. In Europe, more than 150 bird species and approximately 1000 species of plants are severely threatened or close to extinction. A number of Directives are specifically concerned with nature conservation.

The most important Community nature conservation legislation includes:

■ Directive 78/659/EEC of 18 July 1978 on the quality of fresh waters needing protection or improvement in order to support fish life
■ Directive 79/409/EEC of 2 April 1979 on the conservation of wild birds

- Regulation 348/81/EEC of 20 January 1981 on common rules for imports of whales or other cetacean products
- Directive 83/129/EEC of 28 March 1983 concerning the importation into member states of skins of certain seal pups and products derived therefrom
- Directive 92/43/EEC of 21 May 1992 on the conservation of natural habitats and of wild fauna and flora
- Regulation 338/97/EC of 9 December 1996 on the protection of species of wild fauna and flora by regulating trade therein
- Regulation 2087/2001/EC of 24 October 2001 suspending the introduction into the Community of specimens of certain species of wild fauna and flora

In addition, the Community has signed all of the most recent multilateral agreements aimed at environmental protection, including for example, the Convention on Biological Diversity.

The Sixth Environment Action Programme of the European Community 2001–2010

Environment 2010: Our Future, Our Choice is the Sixth Environment Action Programme of the European Community and it gives a strategic direction to the Commission's environmental policy for the next decade. As the Community prepares to expand its boundaries this new programme identifies four priority areas which will provide the environmental component of its strategy for sustainable development:

- climate change;
- nature and biodiversity;
- environment and health;
- natural resources and waste.

The Action Programme introduces five new approaches. They are to:

- ensure the implementation of existing environmental legislation;
- integrate environmental concerns into all relevant policy areas;
- work closely with business and consumers to identify solutions;
- ensure better and more accessible information on the environment for citizens;
- develop a more environmentally conscious attitude towards land use.

The Programme recognises the crucial role of land-use planning and management in the development of environmental problems. Some of these are direct, such as the destruction of habitats and landscapes; others are indirect, for example, the generation of traffic leading to increased air pollution. Conflict between conservation and land use are of particular concern in urban and coastal areas.

The Community Directive on Environmental Impact Assessment (EIA) and the proposal on Strategic Environmental Assessment (SEA) will help to ensure that environmental concerns are better integrated into planning decisions.

The Commission intends to extend other initiatives such as the Sustainable Cities Network and the pilot programme on Integrated Coastal Zone Management. It also proposes to launch a specific programme aimed at architects, planners, government officials, developers, environmental groups and citizens to encourage best practice in urban planning and the development of sustainable cities.

The Common Agricultural Policy (CAP) has been widely recognised as incompatible with many of the aims of nature conservation. The Commission now recognises that there is increasing scope for the encouragement of environmentally positive land management within the framework of the CAP, by the use of agri-environment programmes. This is particularly important in the implementation of the Community's network of protected nature conservation sites, known as Natura 2000 (see Section 7.1.2).

The Sixth Action Programme declares that healthy and balanced natural systems are essential for supporting life on Earth, but it also recognises that nature provides us with resources such as air, water, food, fibres, medicines and building materials. The Programme appreciates the scientific and aesthetic value of nature and recognises that we have a responsibility to preserve the intrinsic value of nature for ourselves and for future generations.

Future Community policy towards the protection of biodiversity will build on existing policies and instruments using a multitrack approach. The main elements will be:

- the establishment of the Natura 2000 network of protected sites;
- the contribution of the *Life* programme's nature projects;
- the Community Biodiversity Strategy;
- Community legislation protecting water, reducing air pollution, acidification and eutrophication and mandating environmental assessment of projects and (in future) land-use plans and programmes;
- the development of agri-environment measures and rural development plans within the Common Agricultural Policy, and the possibility of enforcing compliance with environmental protection requirements by withdrawing or cutting direct payments;
- the revision of the Common Fisheries Policy after 2002, leading to the greater integration of environmental concerns;
- the Commission has proposed recommendations for the implementation of Integrated Coastal Zone Management.

The EC Directives which have the greatest impact on nature conservation are the Wild Birds Directive, the Habitats Directive and the EIA Directive. These are discussed in detail in Chapters 7 and 9.

3.3.3 International law

The UK is a party to a number of international wildlife and nature conservation treaties in its own right or by virtue of its membership of the EU. The provisions of these treaties are embodied in EC and UK domestic law, and are to be found in a variety of Acts of Parliament and Statutory Instruments. The treaties include:

- 1946 International Convention for the Regulation of Whaling (Whaling Convention)
- 1971 Convention on Wetlands of International Importance Especially as Waterfowl Habitat (Ramsar Convention)
- 1972 Convention Concerning the Protection of the World Cultural and Natural Heritage (World Heritage Convention)
- 1973 Convention on International Trade in Endangered Species of Wild Fauna and Flora (CITES – Washington Convention)
- 1979 Convention on the Conservation of Migratory Species of Wild Animals (Bonn Convention)
  - The Agreement on the Conservation of Cetaceans in the Black Sea, Mediterranean Sea and Contiguous Atlantic Area (ACCOBAMS) 1996*
  - The Agreement on the Conservation of African–Eurasian Migratory Waterbirds (AEWA) 1995
  - The Agreement on the Conservation of Bats in Europe (EUROBATS) 1991
  - The Agreement on the Conservation of Small Cetaceans of the Baltic and North Seas (ASCOBANS) 1991
- 1979 Convention on the Conservation of European Wildlife and Natural Habitats (Berne Convention)
- 1992 Convention on Biological Diversity (Biodiversity Convention)

*Not yet signed.

The effects of these treaties on UK law are discussed in Part II.

3.4 UK GOVERNMENT POLICY

In its last manifesto the Labour Party indicated its intention to examine the protection of SSSIs. In September 1998 the government issued a consultation document entitled *Sites of Special Scientific Interest (SSSI) – Better Protection and Management*. After receiving 560 responses from individuals and a wide range of organisations the government's response was published in *A Framework for Action* in August 1999. In 2000 the government passed the Countryside and Rights of Way Act which makes significant improvements to a number of aspects of nature conservation law, especially the protection of SSSIs and fulfils some of the UK's obligations under the Biodiversity Convention. In the same year, two White Papers were published concerned with the urban and rural environments.

3.4.1 The Urban and Rural White Papers

In November 2000 the government published two parallel White Papers describing how it intended to help rural and urban communities achieve their full potential. Both documents refer to measures intended to improve the protection of the environment in general, including parks, open spaces and the countryside.

In 1998 the government asked the Urban Task Force, chaired by Lord Rogers of Riverside, to examine the causes of urban decline. In response to their report the government produced a White Paper entitled *Our Towns and Cities: The Future – Delivering an Urban Renaissance* (Anon., 2000a). This document envisions a transformation of urban areas which will encourage people to move back to the towns and cities.

The White Paper recognises that the local environment needs stronger protection and that a return to urban living would help to protect the countryside from development pressure. It also recognises that good urban design and planning make it practical to live in a more sustainable way, with less noise, pollution and traffic congestion.

Local councils will be required to draw up community strategies to guide their social, economic and environmental development. Local strategic partnerships will involve the council, the community, service providers, voluntary groups, business leaders and other stakeholders. The planning system will be used to bring previously-developed 'brownfield sites' and empty property back into constructive use. Further action will be taken to improve parks and open spaces, and new sources of funding will be made available to help communities improve their local environment.

The White Paper announced the establishment of an urban policy unit and a cabinet committee that will follow up its work. An Urban Summit will be held in 2002, hosted by ministers, to examine the government's progress.

A second White Paper was published in November 2000 entitled *Our Countryside: The Future – A Fair Deal for Rural England* (Anon., 2000b). This recognised that there has been pressure for unwelcome poor quality development in the countryside which has encroached on some valued landscapes, and that wildlife habitats and wildlife diversity have declined. In the White Paper the government declared its vision of a countryside in which we all take the opportunities which change brings to build sustainable rural communities in an improved countryside environment; of rural areas evolving in ways which enhance landscape and biodiversity; and a protected countryside, in which the environment is sustained and enhanced.

The White Paper announced that the newly created England Rural Development Programme (ERDP) would invest £1.6 billion in the countryside by 2006. Money for agri-environment schemes would double and the ERDP would promote environmentally sensitive practices. It also pledged £152 million for a new Rural Enterprise Scheme to help farmers diversify, and additional funds for woodland grants.

The government declared its intention to open up more rights of access to mountain, moor, heath and registered common land and to protect and improve the network of historic rights of way. It announced stronger protection for our most valued landscapes in National Parks and Areas of Outstanding Natural Beauty, along with new guidance for local wildlife sites. This is to be complemented by plans to restore disappearing wildlife species and habitats through the development of new biodiversity strategies.

The White Paper drew attention to plans to favour the recycling of brownfield sites which, as well as helping the revitalisation of urban areas, will also reduce development pressure in the countryside, and suggested that planning authorities should try to maintain the distinctive local features of the countryside. At the same time, it announced a proposal to build 3000 new homes every year in small settlements.

Many of the commitments made in *Our Countryside: The Future* were brought into law by the Countryside and Rights of Way Act 2000.

3.4.2 Countryside and Rights of Way Act 2000

At the end of November 2000 Parliament passed the Countryside and Rights of Way Act. This is the most important piece of conservation legislation to be passed in the UK since the Wildlife and Countryside Act 1981. Its main provisions were to:

■ create greater rights of public access to the countryside;
■ allow the regulation of traffic for the purpose of conserving an area's natural beauty;
■ control the use of mechanically propelled vehicles elsewhere than on roads;
■ amend the law relating to various aspects of nature conservation, including Sites of Special Scientific Interest, Ramsar sites, limestone pavements and Areas of Outstanding Natural Beauty;
■ give a statutory basis to UK Biodiversity Action Plans.

The Act is a substantial piece of legislation (165 pages) consisting of four parts and including 15 schedules. It makes a large number of amendments to the Wildlife and Countryside Act 1981 and a wide range of other laws. The main effects of these changes on wildlife and nature conservation law are discussed under the appropriate sections elsewhere, but particularly in Chapters 4 and 5.

Current UK policy in relation to the protection of wildlife and habitats is based largely upon the implementation of Biodiversity Action Plans and Agenda 21. As these concepts arose as a result of the Earth Summit in 1992 they are dealt with in Chapter 7.

A list of selected primary legislation, consultation and policy documents affecting nature conservation issued by the former Department of the Environment, Transport and the Regions between April 1997 and July 2001 is presented in Appendix 2.

REFERENCES

Anon. (2000a) *Our Towns and Cities: The Future – Delivering an Urban Renaissance*. Cm 4911, The Stationery Office Ltd, London.
Anon. (2000b) *Our Countryside: The Future – A Fair Deal for Rural England*. Cm 4909, The Stationery Office Ltd, London.

PART II
Species and Habitat Protection

Part II describes the national, European and international law which protects wildlife and wild places in the United Kingdom. It also considers some aspects of the protection of archaeological and cultural heritage, as this cannot be separated from the environment in which it exists and the living things with which it coexists. Part II concludes with an account of the special protection given to certain trees and hedgerows.

4   Species protection under UK law   93

5   Habitat and landscape protection under
     UK law   183

6   The protection of trees and hedgerows   234

7   European and international wildlife law   255

4

Species protection under UK law

Wildlife law in the UK has developed as a series of individual statutes and regulations covering particular species or groups of species. In 1981 the Wildlife and Countryside Act (WCA 1981) consolidated much of the earlier law (Figure 4.1) but some animals (notably badgers, deer and seals) are still covered by individual statutes. Game species and fish are also covered by separate legislation. Trade in endangered species and the protection of individual species under European law are discussed in Chapter 7.

The species approach to wildlife protection makes it difficult for the lay person to access and keep up to date with the law. Typically, species are grouped into lists that enjoy similar degrees of protection. These lists are occasionally amended as the threats to particular species change with time, so it is important that the most up-to-date list is consulted. The WCA 1981 uses this approach, listing species in a series of schedules. The identification of species is a further problem and may only be possible with specialist advice.

Certain activities that may affect wildlife may be licensed under, for example, the WCA 1981, the Protection of Badgers Act 1992 and the Conservation (Natural Habitats etc.) Regulations 1994 (SI 1994/2716). These activities may include photography, scientific research and conservation, and during 1999/2000 Scottish Natural Heritage issued 679 such licences (Anon., 1999a).

4.1 THE PROTECTION OF BIRDS

4.1.1 Introduction

UK and European laws single out birds for special consideration. Over 400 bird species have been regularly recorded in the UK. Some breed here while others are seasonal migrants. Occasional rarities are also sometimes seen. In November 2001 bird-watchers descended on Seil

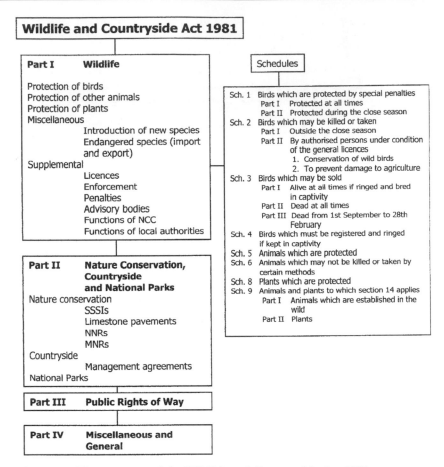

Figure 4.1 The structure of the Wildlife and Countryside Act 1981.

Island near Oban off the west coast of Scotland, in the hope of seeing a snowy egret (*Egretta thula*) that had been swept 4000 miles off its migration course south from Mexico by Hurricane Michelle (Anon., 2001a). This species is a native of Central and South America, but many other 'accidentals' have been recorded from Europe, Asia, North America and Oceania.

In most British ecosystems the most conspicuous animals are the birds. For this reason they are useful indicators of the health of the environment. Birds of prey are particularly useful indicators of environmental quality because they are at the top of the food chain and therefore susceptible to those pollutants that are prone to biomagnification within the ecosystem (see Section 2.3.2).

Birds have been persecuted for food, for their plumage and because some species are hunted for sport or killed as pests. The loss of habitat and the poisoning of food chains have also led to the demise of many species. In the past particular species have been singled out for persecution while others have been protected by folklore. Within recent memory wren hunts were held in many parts of the British Isles on St Stephen's Day (26 December) during which groups of youths in motley dress would beat the hedgerows in an attempt to kill as many wrens (*Troglodytes troglodytes*) as possible. In Irish folklore it was unlucky to kill a swan because it was believed that they embodied the souls of people. Similar beliefs existed in the Western Isles of Scotland (recorded from the early eighteenth century) (Armstrong, 1970)

Writing in 1898 the naturalist W.H. Hudson described the loss of songbirds from the capital in his book *The Birds of London* (Hudson, 1898). A hundred years ago bird populations were depleted by egg and skin collectors, and as gulls migrated up the Thames from the coast they were greeted by a volley of shots from gunmen waiting on the bridges. Now, one hundred years on, populations of many bird species in Britain are in decline again (Figure 4.2).

The population of barn owls (*Tyto alba*) in England and Wales fell from 12 000 breeding pairs to just 3800 pairs by 1985: a decrease of over 70 per cent in 50 years. There are now thought to be around 5000 pairs in Britain. This decline was caused by changes in farming practices including the use of combine harvesters and removal of hedges (reducing the numbers of suitable prey organisms) along with the draining of ponds and use of pesticides. Fifty per cent of all barn owl deaths are road kills (see Figure 8.4).

In the UK the largest organisation concerned with wild bird conservation is the Royal Society for the Protection of Birds (RSPB). The RSPB owns and manages a large number of nature reserves including one at Silverdale in Lancashire, one of the few remaining homes of the bittern (*Botaurus stellaris*).

The Wildfowl and Wetlands Trust (formerly the Wildfowl Trust) was set up by Sir Peter Scott in 1946 to conserve our native species of ducks and geese. It now owns and manages several nature reserves including one at Slimbridge in Gloucestershire and another at Martin Mere in Lancashire. These reserves are important wintering grounds for migrant species and provide protection to our resident birds.

At the beginning of the year 2000 the RSPB and the British Trust for Ornithology published *The State of the UK's Birds*, a report on population changes of regular breeding bird species between 1970 and 1999 (Anon., 2000a). The survey found disturbing declines in 44 species from a range of habitats including, farmland and woodland species and some urban species. Other species, like the nuthatch (*Sitta europaea*), blackcap (*Sylvia atricapilla*) and great spotted woodpecker (*Dendrocopos major*) have shown population increases, some possibly benefiting from milder winters, which may be the result of climate change (Figure 4.2).

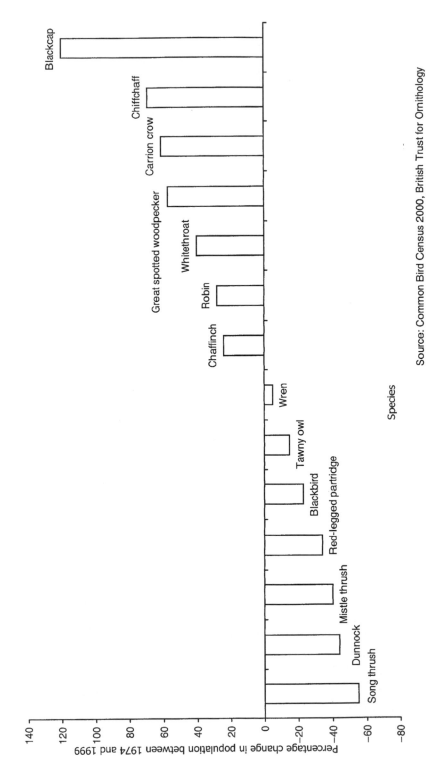

Figure 4.2 Population changes in selected bird species between 1974 and 1999.

Source: Common Bird Census 2000, British Trust for Ornithology

Urban bird problems

A survey of urban starling (*Sturnus vulgaris*) populations published in the 1960s (Potts, 1967) stated that the number of built-up areas with roosts on masonry increased from none in 1896 to fifteen in 1965. From 1845 the number of urban roosts steadily increased, with the birds moving progressively from using trees to resting on structures like bridges and industrial plant, and finally to roosting on buildings. Each morning starlings leave their urban roosts and move to their suburban and rural feeding grounds, returning at dusk (Cramp *et al.*, 1964; Spencer, 1966).

The first licences to catch house sparrows (*Passer domesticus*) were issued in 1963 (Thearle, 1968). As a result of 1002 operations against house sparrows by servicing companies at urban and industrial sites between 1963 and 1967 over 50 000 sparrows were killed using stupefying baits. These operations also killed 244 birds from species that were (at that time) protected and a further 1233 birds from non-target species, including 950 starlings.

Now, some 35 years later, starlings and sparrows have experienced significant declines in their numbers. Surveys conducted in London parks (Moss, 2001) found over 2500 sparrows in Kensington Gardens in 1925; in 2001 there were none (Figures 4.3 and 4.4).

Changes to the urban landscape have significantly affected some species. The starling roost that once existed in the centre of Plymouth did not re-establish itself after the wartime air raids because the new

Figure 4.3 House sparrows (*Passer domesticus*) have disappeared from London parks, including Kensington Gardens.

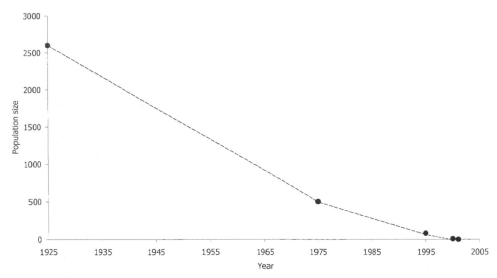

Figure 4.4 The number of house sparrows (*Passer domesticus*) recorded in Kensington Gardens, London, between 1925 and 2001.

buildings constructed in the city centre were unsuitable for roosting (Potts, 1967). Post-war reconstruction, combined with changes in building design, has undoubtedly had a significant effect on urban starling populations. Birds are excluded from the exterior of many city centre buildings by netting (Figure 8.5) and other deterrent devices, but some structures, such as piers at seaside resorts may still provide suitable roosting sites (Figure 4.5)

4.1.2   Classification and natural history

Classification

Birds species recorded in the United Kingdom belong to 58 families within 20 different orders as follows.

Class:	Aves	
Order:	Gaviiformes	Divers
Family:	Gaviidae	Divers
Order:	Podicipediformes	Grebes
Family:	Podicipedidae	Grebes
Order:	Procellariiformes	Tubenoses
Family:	Hydrobatidae	Storm petrels
	Procellariidae	Shearwaters
Order:	Pelecaniformes	Cormorants and allies
Family:	Phalacrocoracidae	Cormorants
	Sulidae	Gannets

Figure 4.5 Starlings (*Sturnus vulgaris*) have been driven from many of their traditional urban roosting places by changes in building design and the use of protective measures such as bird netting. The South Pier at Blackpool still provides a safe roosting place for some of the local population.

Order:	Ciconiiformes	Herons and allies
Family:	Ardeidae	Herons
	Ciconiidae	Storks
Order:	Anseriformes	Ducks, geese, swans
Family:	Anatidae	Ducks
Order:	Faloniformes	Day-hunting birds of prey
Family:	Accipitridae	Eagles, buzzards and allies
	Falconidae	Falcons
	Pandionidae	Ospreys
Order:	Galliformes	Pheasants and allies
Family:	Tetraonidae	Grouse
	Phasianidae	Pheasant, partridges and quails
Order:	Gruiformes	Rails and allies
Family:	Rallidae	Crakes, rails and coots
	Otididae	Bustards
Order:	Charadriiformes	Waders
Family:	Haematopodidae	Oystercatchers
	Charadriidae	Plovers and lapwings
	Scolopacidae	Snipes and sandpipers
	Phalaropidae	Phalaropes
	Recurvirostridae	Avocets and curlews
	Burhinidae	Stone curlews
	Glareolidae	Pratincoles

Order:	Lariformes	Gulls and terns
Family:	Laridae	Gulls
	Sternidae	Terns
	Stercoraridae	Skuas
Order:	Alciformes	Auks
Family:	Alcidae	Auks
Order:	Columbiformes	Pigeons and doves
Family:	Columbidae	Pigeons and doves
Order:	Cuculiformes	Cuckoos
Family:	Cuculidae	Cuckoos
Order:	Strigiformes	Owls
Family:	Strigidae	Owls
	Tytonidae	Barn owls
Order:	Caprimulgiformes	Nightjars
Family:	Caprimulgidae	Nightjars
Order:	Apodiformes	Swifts
Family:	Apodidae	Swifts
Order:	Coraciiformes	Kingfishers and allies
Family:	Alcedinidae	Kingfishers
Order:	Piciformes	Woodpeckers
Family:	Picidae	Woodpeckers
Order:	Passeriformes	Songbirds
Family:	Hirundinidae	Swallows
	Bombycillidae	Waxwings
	Alaudidae	Larks
	Motacillidae	Wagtails
	Cinclidae	Dippers
	Troglodytidae	Wrens
	Laniidae	Shrikes
	Prunellidae	Accentors
	Sylviidae	Warblers
	Regulidae	Goldcrests
	Muscicapidae	Flycatchers
	Turdidae	Thrushes
	Paradoxornithidae	Bearded tits
	Aegithalidae	Long-tailed tits
	Paridae	Titmice
	Sittidae	Nuthatches
	Certhiidae	Tree-creepers
	Emberizidae	Buntings
	Fringillidae	Finches
	Ploceidae	Weavers
	Sturnidae	Starlings
	Oriolidae	Orioles
	Corvidae	Crows

Natural history

Birds are homeothermic (endothermic) and maintain their body temperature independently of the environment. This allows them to remain active in cold weather and has enabled them to spread through-out a diverse range of habitat types.

The most obvious signs of the presence of birds are their nests, but it is important to distinguish between used and unused nests. The presence of some species may be apparent from their song even though the birds themselves may be rarely observed. Some bird census techniques involve counting the number of individuals heard singing while walking along a transect.

Most birds species produce a single clutch of eggs each season, but for other species two or three broods are normal. Blackbirds (*Turdus merula*), sparrows and finches have a long breeding season and regu-larly produce more than one brood. Blackbirds may breed from March until late summer and may raise three broods, each taking about two weeks to incubate and two weeks to fledge. Many sea birds produce a single egg each season, but the stock dove (*Columba oenas*) frequently attempts five broods, its dependence upon arable weed seeds allowing it a very long breeding season (from mid-February to mid-November). The blue tit (*Parus caeruleus*) has been known to lay 22 eggs in a single clutch (Carwardine, 1995).

Many birds that are normally single-brooded may replace a clutch of eggs that is lost in the early stages of incubation. However, some groups, such as divers and petrels, are unable to do this. Time, the weather and food availability may affect whether an extra brood is reared. A clutch may be lost through predation, disturbance or theft. The longer the bird has sat on the eggs the more reluctant it is to leave them, probably because of the lack of time to produce another brood.

The timing of nest building depends upon a number of factors including the weather and the response of the female to courtship. The golden eagle (*Aquila chrysaëtos*) is Britain's earliest nester, and begins building in the autumn before it lays its eggs. The latest nester is the hobby (*Falco subbuteo*), which usually uses the nest of another species and lays its eggs in June or July. Most birds in the UK lay their eggs between February and September (see Appendix 3). The young of some species leave the nest immediately after hatching and the nest may then be abandoned. However, other species are tied to the nest for some time after producing their young (see Panel 4.1). It is important that any operations that are damaging (or potentially damaging) to bird habitats are confined to periods when nests, eggs or young are not vulnerable.

Nests may be elaborate structures built in trees, hedgerows, build-ings, on the ground or on cliffs, or they may be no more than a simple scrape in the soil (see Panel 4.1). Some birds nest in single pairs while some seabirds breed in colonies of tens of thousands. Some build a new nest each year while others return to the same nest or may even use the nests of other species. Old nests may be used as roosts long after breeding activity has ceased.

PANEL 4.1   TYPES OF BIRD NESTS AND NESTING BEHAVIOUR

Some bird species construct a single nest while others, like the wren, construct many, using one and keeping the others in reserve. In some species, the male selects the nest site while in others this task is performed by the female. In some species both sexes are involved. The structure of nests varies from a simple hollow in the ground to a complex structure made of twigs and mud.

No structured nest
- Tree holes are used by the tawny owl, green woodpecker, nuthatch, great tit, wryneck, starling and stock dove
- The guillemot lays its eggs on a cliff ledge
- The nightjar lays its eggs on a clear space on the ground, while the stone curlew uses a simple depression in the ground

Nests which require construction
- Nidifugous birds (nest-leaving species)
  These species produce chicks that can run and leave the nest on the day of hatching; they produce simple nests sometimes consisting of scrapes in the ground but of greater complexity when built on water:
  - Ducks
  - Geese
  - Swans
  - Waders
  - Game birds
  - Grebes
  - Divers

Many of these species rely upon camouflage for safety
- Nidicolous birds (nest-attached species)
  The chicks of these species are born naked and helpless, and may stay in the nest until fledged. They build intricate nests, reaching their greatest complexity in the passerines (songbirds), for example:
  - Blackbirds and thrushes build cup-shaped nests
  - House sparrows build a dome over a cup-shaped nest
  - Magpies build a canopy of twigs
  - Long-tailed tits constuct a bottle-shaped nest of lichen-covered moss
- Intermediates
  The nidifugous and nidicolous forms are two extemes, and there are various intermediate types. Their young are able to leave the nest within a few days, and they build more elaborate nests, for example:
  - Gulls
  - Terns
  - Skuas

4.1.3 Birds and the law

The law in the UK frequently deals with birds separately to other animals even though, zoologically, they are animals.

Wild birds are protected under the Wildlife and Countryside Act 1981 in England, Wales and Scotland (although amendments made to the Act by the Countryside and Rights of Way Act 2000 do not apply in Scotland). In Northern Ireland the Wildlife (Northern Ireland) Order 1985 (SR 1985/171 (NI2)) provides corresponding protection.

The degree of protection given to birds under WCA 1981 depends upon the species and, in some cases, the time of year. Protected birds are listed in two schedules. Schedule 1 lists the rarer species such as birds of prey, rare songbirds, seabirds, waders, swans and other species. Schedule 2 lists wildfowl species which may be taken outside the close season (with a very small number of such species appearing in Part II of Schedule 1) (see Appendix 4). Schedule 3 lists species which may be sold, and Schedule 4 lists species which must be registered and ringed if kept in captivity.

The WCA 1981 (s. 27(1)) defines a 'wild bird' as 'any bird of a kind which is ordinarily resident in or is a visitor to Great Britain in a wild state ...' This definition excludes poultry and game birds (except for the purposes of sections 5 and 16, which relate to illegal methods of killing and taking wild birds, and the power to grant licences for activities that would otherwise be unlawful, respectively).

Wildlife and Countryside Act 1981 s. 1

(1) ... if any person intentionally –
  (a) kills, injures or takes any wild bird;
  (b) takes, damages or destroys the nest of any wild bird while that nest is in use or being built; or
  (c) takes or destroys an egg of any wild bird,
  he shall be guilty of an offence.

(2) ... if any person has in his possession or control –
  (a) any live or dead wild bird or any part of, or anything derived from, such a bird; or
  (b) an egg of a wild bird or any part of such an egg,
  he shall be guilty of an offence.

No offence is committed under subsection (2)(a) if the bird or egg has not been killed or taken, or has been killed or taken without contravention of the relevant provisions. A person commits no offence under subsection (2)(b) if the bird, egg or other thing has been sold to him without contravention of the relevant provisions (s. 1(3)).

Disturbance of birds listed in Schedule 1 is prohibited under the Act, including disturbance caused through recklessness (a provision added by the Countryside and Rights of Way Act 2000).

Wildlife and Countryside Act 1981 s. 1

(5) ... if any person intentionally or recklessly –
  (a) disturbs any wild bird included in Schedule 1 while it is building a nest or is in, on or near a nest containing eggs or young; or
  (b) disturbs dependent young of such a bird,
  he shall be guilty of an offence.

The offences in s. 1 do not apply to certain wildfowl species. Section 2 of the Act allows for the killing or taking of birds listed in Part I of Schedule 2 (certain wildfowl species), and the taking or destruction of their eggs or nests outside the close season. The Secretary of State may by order vary the close season for any species (s. 2(5)).

Areas of special protection for birds may be designated by the Secretary of State by order (s. 3(1)). Such special areas were previously designated by Sanctuary Orders under the Protection of Birds Acts. Birds in areas so designated are still afforded extra protection since these remain in force under WCA 1981. Some forty areas of special protection have been designated, but the details of protection vary from site to site. For example, in the Humber Estuary, protection is given to all birds but not eggs, and entry is prohibited, except by permit, between 1 September and 20 February (Order No. 1532, 1955). However, entry to the Farne Islands, Northumberland, is prohibited at all times except by permit; protection is given to all birds and eggs; and it is also an offence to disturb any bird while it is on or near a nest containing eggs or young (Order No. 402, 1980).

Exceptions to sections 1 and 3 allow for the destruction of birds, their eggs and nests to protect animal health; to prevent the spread of disease; to preserve public health or public or air safety; or to prevent serious damage to livestock, foodstuffs for livestock crops, fruit, vegetables, fisheries, inland waters or growing timber. Exceptions also apply to the taking of an injured bird for the purposes of tending it and releasing it; the destruction of birds that are so seriously disabled that there is no serious chance of recovery; and the killing of birds as an incidental result of lawful operations (s. 4).

Wildlife and Countryside Act 1981 s. 4

(2) Notwithstanding anything in the provisions of section 1 or any order made under section 3, a person shall not be guilty of an offence by reason of –

    (c) any act made unlawful by those provisions* if he shows that the act was the incidental result of a lawful operation and could not easily have been avoided.

* e.g. the killing of any wild bird, destruction of nests or eggs.

Where it is lawful to kill or take wild birds, certain methods of killing or taking them are prohibited. The use of the following is generally prohibited, except under licence, by s. 5(1):

- spring, trap, gin, snare, hook and line;
- any electrical device for killing, stunning or frightening;
- any poisonous, poisoned or stupefying substance;
- any gas, smoke or chemical wetting agent;
- any net, baited board, bird-lime or similar substance;
- any bow or crossbow;
- any explosive (other than ammunition for a firearm);
- any automatic or semi-automatic weapon;
- any shotgun with an internal muzzle diameter of more than $1\frac{3}{4}$ inches;
- any artificial lights, mirrors, dazzling devices or night sights;
- any sound recording;
- any live bird or other animal used as a decoy, which is tethered, blind, maimed or injured;
- the use of any mechanically propelled vehicle in immediate pursuit of a wild bird for the purpose of killing or taking it.

It is also an offence (s. 5(1)(f)) to cause or knowingly permit to be done any of the acts listed in s. 5. Exceptions to these offences are listed in subsection (5).

Section 6 provides offences in relation to the sale, transport for sale, and advertising for sale of live or dead wild birds, eggs or anything derived from them. Birds which may be sold are listed in Schedule 3.

Bird species listed in Schedule 4 must be registered, ringed or marked (s. 7) in accordance with the Wildlife and Countryside (Registration and Ringing of Captive Birds) Regulations 1982 (SI 1982/1221). Regulation 3(1) requires the Secretary of State to maintain a register of birds to which these regulations apply. Keepers of birds covered by these regulations must ring every bird to which they apply with a ring obtained from the Secretary of State (reg. 5(1)). Section 8(1) of the WCA 1981 creates offences in relation to the keeping of birds in cages of inadequate size.

It is an offence to liberate captive birds for the purpose of being shot immediately after their release (s. 8(3)(a)); or to promote, arrange, assist in, receive money for or take part in such an event; or for the owner or occupier of any land to permit land to be used for this purpose (s. 8(3)(b)).

Licences may be granted under s. 16 of the Act for various activities which would otherwise give rise to an offence under the Act. These activities are described below under the law relating to wild animals (Section 4.2.3). The General Licences issued under the Act with respect to birds are for:

- killing of birds to prevent serious damage to agriculture;*
- killing of birds to preserve public health/air safety and to conserve wild birds;*
- keeping birds in Larsen traps (to protect crops, livestock, growing timber, fisheries, collections of wild birds, and to conserve wild birds etc.);
- sale of gulls' eggs (of the three species listed below**);
- eggs in nest boxes (permits removal between 1 August and 31 January);
- taking of mallard eggs (before 31 March in England and Wales, and 10 April in Scotland);
- sale of captive-bred native birds (except those on Part I Schedule 3 WCA 1981, birds or prey, owls and certain wildfowl species);
- sale of wildfowl (certain species or their eggs; must have been bred in captivity);
- sale of dead birds and derivatives (captive-bred, or not illegally taken);
- exhibition of captive birds (bred in captivity);
- veterinary surgeons (WCA 1981 Schedule 4 species kept for treatment without registration);
- keeping disabled birds (WCA 1981 Schedule 4 species kept when disabled by RSPCA and SSPCA staff without registration);
- semi-automatic weapons (permits use by authorised persons to preserve public health, air safety and to prevent serious damage to agriculture);
- artificial light (permits use for killing feral pigeon, house sparrow or starling by authorised persons to preserve public health, air safety and to prevent serious damage to agriculture);
- killing of birds on airfields (applies to nests and eggs of common gull, black-headed gull, lapwing and oystercatcher on certain airfields);

■ keeping captive-bred birds in show cages (allows confinement in small cages for training purposes, for no more than one hour in every 24 hours).

*These licences apply to the following species:

Carrion crow	*Corvus corone*
Collared dove	*Streptopelia decaocto*
Great black-backed gull**	*Larus marinus*
Lesser black-backed gull**	*Larus fuscus*
Herring gull**	*Larus argentatus*
Jackdaw	*Corvus monedula*
Jay	*Garrulus glandarius*
Magpie	*Pica pica*
Feral pigeon	*Columba livia*
Rook	*Corvus frugilegus*
House sparrow	*Passer domesticus*
Starling	*Sturnus vulgaris*
Woodpigeon	*Columba palumbus*

Details of the application of these licences may be obtained from the issuing department (DEFRA, National Assembly for Wales, or the Scottish Executive Rural Affairs Department as appropriate) (see Information sources at the end of this book).

Section 18 makes it unlawful to attempt to commit an offence under Part I of the Act and s. 19 describes police powers in relation to Part I offences. Offences relating to the introduction of non-native species of birds are discussed in Section 4.8.2.

Protected bird species that appear in the schedules to the Wildlife and Countryside Act are listed in Appendix 4.

Case law

In 1998 the Royal Society for the Protection of Birds (RSPB) received 738 reports of crimes against birds in the UK. These were categorised as follows:

■ Shooting or destroying birds	36%
■ Illegal trapping, possession or sale of birds	17%
■ Egg collecting	18%
■ Poisoning	12%

Other offences included the illegal import and export of birds. Birds of prey accounted for some 68 per cent of all birds shot or destroyed and 56 per cent of the birds illegally trapped, possessed or sold. In 1998 there were 46 successful prosecutions for crimes against birds and fines of £34 125 were issued (Anon., 2000a).

First conviction for killing a hen harrier
On 25 May 2001 a gamekeeper was fined £2000 by Elgin Sheriff Court for killing a rare bird of prey (Harris, 2001). He was filmed by the RSPB

PANEL 4.2   WILDLIFE AND COUNTRYSIDE ACT 1981 – SCHEDULES CONTAINING BIRDS

■ Schedule 1   Birds which are specially protected

Part I   Specially protected birds
Rare birds which are afforded special protection and cannot intentionally (or recklessly in England and Wales only) be disturbed when nesting

Part II   Birds and their eggs specially protected during the close season

■ Schedule 2   Birds which may be killed or taken

Part I   Outside the close season
Sporting or quasi-sporting birds which may be shot during the winter for a limited period only

Part II   By authorised persons at all times
This part has been deleted from the Act. The thirteen species originally listed are now subject to control under General Licences

■ Schedule 3   Birds which may be sold

Part I   Alive at all times if ringed and bred in captivity

Part II   Dead at all times

Part III Dead from 1 September to 28 February

■ Schedule 4   Birds which must be registered and ringed if kept in captivity

■ Schedule 9   Animals and plants to which section 14 applies
This includes bird species which are established in the wild in Britain and which may not be released from captivity

All bird species are fully protected throughtout the year, except those listed on Schedule 2, Part I, game birds, and birds covered by a General Licence.

---

as he shot a young hen harrier (*Circus cyaneus*) with a double-barrelled shotgun on a grouse moor in Morayshire in July 2000. The nest was being watched because the RSPB felt that it was particularly vulnerable, as a clutch of eggs had previously been destroyed. The gamekeeper claimed that he acted out of frustration because he had discovered 23 dead grouse in the area which he was attempting to restore to its status as a grouse moor. The Sheriff did not accept the killing had been done in the heat of the moment. This was the first conviction in the UK for killing a hen harrier. Only about 570 pairs were thought to be present in the UK at the time.

Raymond Holden v. Lancaster Justices (1998)
Raymond Holden ran a hatchery and protected his birds using 11 decoy jackdaws (*Corvus monedula*) in traps. In order to prevent these birds from escaping or injuring themselves he had their primary wing feathers clipped, but they would grow back after moulting. Chorley Magistrates considered that, as the birds had their wings clipped annually, they were permanently flightless and therefore maimed. Mr Holden was convicted under section 5(1)(d) of the Wildlife and Countryside Act 1981 after a prosecution brought by the RSPCA.

On appeal, Lord Justice Brooke and Mr Justice Rougier (Queen's Bench Divisional Court) held that the justices were not entitled to look into the future to determine whether the birds were maimed, but could only consider the state of the birds at the time the information was laid. Maiming had historically denoted permanent deprivation of a member or mutilation or crippling (*R v. Jeans* (1884)). Since there was no permanency in clipping feathers, the birds were not maimed within the meaning of WCA s. 5(1)(d).

Royal Society for the Prevention of Cruelty to Animals v. Craig Cundey (2001)
Although licences had been granted to kill wild birds, this did not provide a defence to a charge of attempting to kill wild birds contrary to s. 18 WCA 1981 where attempts to kill the birds were not for one of the purposes in s. 16 of the Act.

The Royal Society for the Protection of Birds and the Wildfowl & Wetlands Trust Ltd (Petitioners) v. The Secretary of State for Scotland (Respondent) (2001)
The Secretary of State for Scotland had permitted the shooting of protected geese in a special protection area because they were damaging crops. The court found that the Secretary of State had acted unlawfully since he had not taken into account the effects of the resulting disturbance on the population of geese in that area.

4.1.4   Mitigation measures

Birds of prey

Artificial nest sites are important to the recovery of species of birds of prey. The Hawk and Owl Trust has installed 3000 nestboxes and nesting

PANEL 4.3

Black redstart   *Phoenicurus ochruros*

Identification/size
Size of a sparrow. Male is coal black and ash grey, with white area on the wings, and russet tail. Female dark grey with no white area on wings.

Habitat
Nests on derelict sites, building sites, quarries and rocky areas, power stations, marshalling yards, etc.

Distribution
Mainly a passage migrant. Recorded from London and along the south coast, Midlands and East Anglia.

Ecology and behaviour
Commonly nests in urban areas in rock crevices, holes in walls, on roof beams. Feeds on berries and insects. Rare, perhaps around 100 pairs, but established as a breeding bird in Britain.

baskets and each year over 1000 of these artificial nest sites are occupied. Barn owls (*Tyto alba*) and peregrine falcons (*Falco perigrinus*) will nest in boxes, merlins (*F. columbarius*) and long-eared owls (*Asio otus*) will use baskets, and ospreys (*Pandion haliaetus*) and red kites (*Milvus milvus*) will nest on elevated platforms. The Trust specially selects sites which provide good feeding habitat, such as on farms or golf courses, on river banks or in woods. Where there are no suitable trees or farm buildings nest boxes are erected on top of tall wooden poles.

Designs for nest boxes suitable for a wide variety of birds species are described in Bolund (1987) (see Panel 4.7).

4.1.5   Case studies

Black redstarts, peregrine falcons and the Millennium Dome

In January 1999 three pairs of black redstarts (*Phoenicurus ochruros*) arrived on the site of the unfinished Millennium Dome. The high profile nature of the construction site, and concerns that the Dome would not be finished in time for the millennium celebrations, resulted in the event being reported on the front page of *The Times* (Owen, 1999). The species is protected by the Wildlife and Countryside Act 1981 and is listed in Schedule 1 (Birds which are specially protected). At the time there were thought to be only 70 breeding pairs in the country, up to a third of which lived in London, some being resident in nearby Deptford Creek. The species is rare now but flourished on derelict sites after the Second World War. Black redstarts had previously nested in the area and The New Millennium Experience Company reassured wildlife groups that it would take their rarity into account if they decided to nest on the site. Some work had been stopped in one area of the construction site in the previous summer when a pair of black redstarts had been found.

In June 2001 a pair of peregrine falcons (*Falco peregrinus*) made their nest at the top of one of the Dome's supporting masts. Their presence could have made it difficult to demolish the building when this was considered as a possible fate for the structure (Anon., 2001b). At the time just three pairs of peregrines were known to nest in London.

Under the WCA 1981 it is an offence to disturb a Schedule 1 bird while building a nest or while it is in or on a nest containing eggs or young. It is also an offence to disturb dependent young. Although the Act affords these species the highest level of protection, it does nevertheless, allow disturbance as the incidental result of a lawful operation that could not reasonably have been avoided.

The red-throated divers and the Orkney wind turbine

In April 2000 a £2 million wind turbine project was almost brought to a standstill by the presence of red-throated divers (*Gavia stellata*) on an RSPB reserve bordering the construction site. Orkney Islands Council had given planning permission to Orkney Sustainable Energy for the 210 ft high turbine to be erected on Burgar Hill on Orkney on condition that work was finished before the start of the birds' mating season. The machine has a 236 ft rotor blade and stands beside a smaller turbine with a 210 ft blade. The combined output of the two turbines provides around a quarter of the island's electricity needs. The work was finished just one day before the deadline and the RSPB reported that the divers were preparing to nest (English, 2000).

In April 2001 the government announced plans to construct a total of 540 wind turbines at 18 off-shore sites around the coastline of England and Wales. Concern has been expressed at one site around the north Norfolk coast which is close to an Area of Outstanding Natural Beauty. Other proposed sites include Blackpool, Southport, Rhyl, Liverpool, Whitehaven, Skegness and a site near the Gower Peninsula

Figure 4.6 A wind farm in an industrial setting in Liverpool Docks.

in South Wales. Most sites will be some three miles out to sea, but some will be around half this distance from land. The Council for the Protection of Rural England has expressed concern about the impact of the proposed wind farms on the landscape.

In December 2001 the government announced its intention to construct the largest wind farm in the world on the island of Lewis in the Outer Hebrides. It will consist of 300 wind generators capable of producing an equivalent output to a nuclear power station. The project is to be undertaken by AMEC and British Energy and will undoubtedly attract considerable opposition from conservationists. The wind generators are to be constructed on the island, creating much-needed employment.

Osprey death

In spring 2001 the Scottish Society for the Prevention of Cruelty to Animals investigated the death of an osprey after it was found caught in power lines near the River Braan in Perthshire, with an illegal trap hanging from its leg (Anon., 2001c).

Ruddy duck cull

In February 1999 the Environment Minister, Michael Meacher, authorised a trial cull of ruddy duck in the West Midlands, Anglesey and Fife in an attempt to control their numbers. The species was originally introduced from the United States and has since escaped and spread across Europe. The ruddy duck (*Oxyura jamaicensis*) interbreeds with the rare white-headed duck (*O. leucocephala*) and its future is now threatened. The White-Headed Duck Task Force recommended to the Minister that the ruddy duck should be exterminated within ten years (Hawkes, 1999).

Operation Easter: targeting egg collectors

Operation Easter is a nationwide operation set up in 1997 and aimed at the most determined egg collectors known to the police and the RSPB. It is coordinated by Tayside Police and has collated information on over 100 egg collectors, who are invariably male. On 29 March 2002 police seized osprey eggs believed to have been taken from a nest in Scotland. In May 2001 a man was convicted of taking common sandpiper and chaffinch eggs and fined £3100. A second man was convicted in May 2001 of illegal possession of over 1200 eggs and received a fine of £1500. The same individual was convicted in June 2001 for possession of goshawk and goosander eggs and sentenced to four months imprisonment: the first person to be imprisoned under amendments made to the WCA 1981 by the CRoW Act 2000 (Anon., 2002).

Unlawful killing of cormorants to protect fisheries

A vermin controller, Terence Day, was convicted of killing two cormorants and attempting to kill a third in July 2001. Day was fined £250, his firearms licence was revoked, and his shotgun seized. Where it can be shown that cormorants are causing severe damage to fisheries licences are available from DEFRA authorising their destruction, where no other satisfactory solution is available (Anon., 2002)

4.2   THE PROTECTION OF ANIMALS

4.2.1   Introduction

In UK law the term 'animal' often excludes birds and provisions relating to birds appear separately. This section is concerned with the protection given to animals (other than birds) under the Wildlife and Countryside Act 1981.

4.2.2   Classification and natural history

Mammals

There are at least 69 species of wild mammals living in and around the British Isles. In addition, another 29 migratory species of bats and

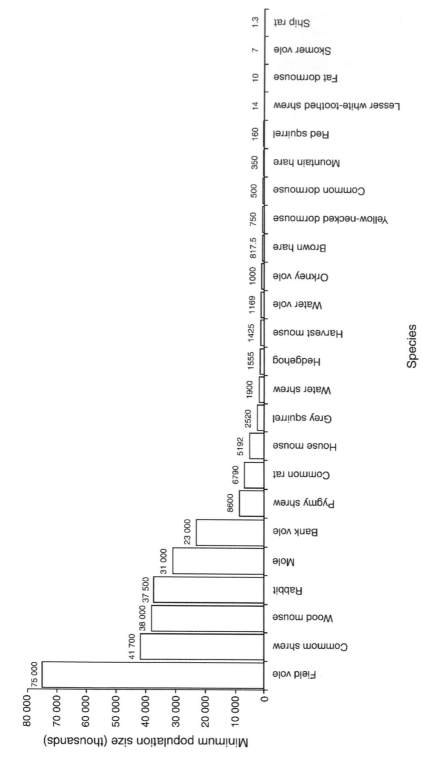

Figure 4.7 Population sizes of rodent, lagomorph and insectivore species. (Source: based on data in Harris *et al.*, 1995.)

marine mammals are occasionally recorded. Our mammalian fauna is similar to that of the rest of western Europe, but we have lost some species through extinction and we have gained others as a result of introductions and escapes.

The limestone caves of Derbyshire contain layers of clay and rubble that hold evidence of a once rich mammalian fauna. Bison, reindeer, wolf, lynx and the brown bear once roamed the Peak District (Pigott, 1975). Reindeer (*Rangifer tarandus*) were still present in Scotland in the tenth century and were then reintroduced in the seventeenth century, again in 1820 and then once again in 1951. The brown bear (*Ursus arctos*) survived in England to c. 1000 and in Wales and Scotland to c. 1057. There were still beavers (*Castor fiber*) in Wales and Scotland five hundred years ago and wolves (*Canis lupus*) survived in Scotland until 1743 (Fisher, 1970).

Mammalian species have been introduced into Britain from as far away as North and South America, the Far East and Australasia. Some have become pests, like the rabbit (*Oryctolagus cuniculus*), grey squirrel (*Sciurus carolinensis*), American mink (*Mustela vison*) and coypu (*Myocaster coypus*). Others have added an interesting diversity to the fauna, such as the muntjac deer (*Muntiacus reevesi*) and the edible dormouse (*Glis glis*).

Britain is host to a number of mammal populations of European importance. We have two-thirds of all of the grey seals (*Halichoerus grypus*) in the world and about half of Europe's otter (*Lutra lutra*) population. Apart from being important for their own sake and because the larger species (such as deer) are some of the more impressive elements of the fauna, mammals are also significant components of the food chain for many other species. In particular, many small mammals are an important food source for rare predatory birds like the barn owl and the red kite.

Classification

Class:	Mammalia	
Order:	Insectivora	
Family:	Erinaceidae	
Species:	*Erinaceus europaeus*	Hedgehog
Family:	Talpidae	
Species:	*Talpa europaea*	Mole
Family:	Soricidae	
Species:	*Sorex araneus*	Common shrew
	S. minutus	Pygmy shrew
	Neomys fodiens	Water shrew
Order:	Rodentia	
Family:	Muridae	
Species:	*Clethrionomys glareolus*	Bank vole

		Microtus agrestis	Field vole
		M. arvalis orcadensis	Orkney vole
		Arvicola terrestris	Water vole
		Micromys minutus	Harvest mouse
		Apodemus sylvaticus	Wood mouse
		A. flavicollis	Yellow-necked mouse
		Mus musculus	House mouse
		Rattus rattus	Ship rat
		R. norvegicus	Black rat
Family:	Gliridae		
Species:	*Muscardinus avellanarius*	Dormouse
	Glis glis	Edible dormouse
Family:	Sciuridae		
Species:	*Sciurus vulgaris*	Red squirrel
	S. carolinensis	Grey squirrel
Order:	Lagomorpha		
Family:	Leporidae		
Species:	*Oryctolagus cuniculus*	Rabbit
	Lepus europaeus	Brown hare
	L. timidus	Mountain hare
Order:	Carnivora		
Family:	Canidae		
Species:	*Vulpes vulpes*	Red fox
Family:	Mustelidae		
Species:	*Meles meles*	Badger
	Lutra lutra	Otter
	Martes Martes	Pine marten
	Mustela putorius	Polecat
	M. erminea	Stoat
	M. nivalis	Weasel
	M. vison	American mink
Family:	Felidae		
Species:	*Felis silvestris*	Wildcat
Order:	Pinnepedia	Two seal species (see Section 4.6.2)
Family:	Phocidae		
Order:	Chiroptera	Sixteen bat species (see Section 4.3.2)
Family:	Rhinolophidae		
Family:	Vespertilionidae		
Order:	Artiodactyla		
Family:	Cervidae	Six deer species (see Section 4.5.2)
Order :	Cetacea	Various whale, dolphin and porpoise species are observed in coastal waters and occasionally in estuaries

Natural history

Like birds, mammals are homeothermic (endothermic) and maintain their body temperature independently of the environment. Mammals are widely distributed throughout the terrestrial, freshwater and marine habitats of the British Isles. Species such as the fox, badger, rabbit and hedgehog are widespread and common in the British Isles. Other species are confined to particular habitats or regions and some are relatively rare, such as the wildcat and pine marten.

Marine mammals are largely confined to coastal areas although some species may swim up rivers. Large concentrations of wild deer are confined to Scotland. The mountain hare is largely confined to Ireland, Scotland and the Pennines, while the brown hare is widespread throughout England, Scotland and Wales. The polecat is largely confined to Wales but has begun colonising adjacent areas of England. The common dormouse is found mainly in southern England, Wales and Cumbria and the wildcat only occurs in remote areas in Scotland. Some species, such as the fox and grey squirrel thrive in urban areas.

Some mammals are active during the daylight hours but many species have become nocturnal and are rarely seen. Small mammals are difficult to find, as are woodland species, and those that inhabit subterranean shelters. Most species are solitary or live in small groups. The only species likely to be seen in large concentrations are rabbits, deer, seals and roosting bats.

British mammals include insectivores, seed eaters, grazers, carnivores and omnivores. The presence of mammals is more likely to be

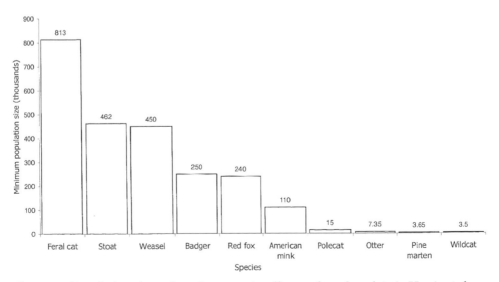

Figure 4.8 Population sizes of carnivore species. (Source: based on data in Harris *et al.*, 1995.)

Figure 4.9 Mole (*Talpa europaea*) hills. Most British wild mammals are very difficult to see, their presence often being detected only by the signs they leave.

apparent from their tracks, trails and signs than from observations of the animals themselves (Figure 4.9). Squirrels, deer and various other species cause distinctive types of damage to trees. Some species produce easily recognisable faecal pellets, while others construct characteristic shelters. Some of these signs may be useful when conducting censuses. For example, otter populations are monitored by counting otter spraints along fixed lengths of river bank.

Mammals produce their young in spring or summer when temperatures are favourable and food is abundant. Some small mammals have more than one litter each year. For example, the water vole (*Arvicola terrestris*) has three or four litters each year with up to ten young in each litter. Gestation lasts 21 days and the young are independent after four weeks. Larger mammals tend to have fewer and more dependent young. The red deer usually has a single young (rarely twins) which suckles for up to nine months and stays with its mother for about a year. Some species reproduce out of sight within shelters, while others give birth in the open.

Reptiles

Classification
The class Reptilia contains lizards and snakes. The native British fauna includes just three species of each.

Class:	Reptilia		
Order:	Squamata		
Family:	Viperidae		
Species:	*Vipera berus*	Adder (viper)*	Native and widespread in Britain
Family:	Colubridae		
Species:	*Natrix natrix*	Grass (ringed) snake	Native in England and Wales
	Coronella austriaca	Smooth snake	Rare. Surrey, Hampshire and Dorset only
Family:	Anguidae		
Species:	*Anguis fragilis*	Slow worm**	Native and widespread in Britain
Family:	Lacertidae		
Species:	*Lacerta vivipara*	Common lizard	Native and widespread in Britain
	L. agilis	Sand lizard	Native (rare) Lancashire, New Forest, Dorset
	L. muralis	Wall lizard	Surrey (introduced from Europe)

* The adder is highly poisonous and its venom may be fatal to man.
** The slow worm is a legless lizard and therefore appears snake-like.

In addition to these terrestrial and freshwater species, the UK coastline is regularly visited by marine turtles. Leathery turtles (called leatherbacks in the USA) are washed up on the west coast of Britain. They are attracted to the large quantity of plastic bags in our coastal waters, which they mistake for jellyfish, and may prove fatal if ingested.

The turtle species which have been regularly recorded are:

- Common loggerhead   *Caretta caretta*
- Kemp's loggerhead   *C. kempi*
- Hawksbill   *Eretmochelys imbricata*
- Leathery turtle   *Dermochelys coriacea*

These turtles are not native species but are carried to our waters by ocean currents from the Caribbean.

Natural history
Reptiles are poikilothermic (ectothermic), their body temperature being determined by the environmental temperature. In order to raise their temperature they must gain heat from the environment by lying on a warm rock or basking in the sun. They hibernate in winter, sometimes in holes, crevices or under piles of leaves. Diet depends upon the

species, but reptiles will eat a wide range of animals including small insects, spiders, centipedes, worms, molluscs, small fish, amphibians, birds and rodents, and even other reptiles.

Reproduction occurs between May and October, depending upon the species. Some reptiles produce eggs which hatch outside the female's body (ovoviviparous), while others produce eggs which hatch inside the female so that she gives birth to fully-formed young (viviparous). The number of young depends upon the species but may range from 2 or 3 to 30.

There are no snakes in Ireland. The only native reptile in Ireland is the common lizard which occurs locally. Some reptile species have a wide distribution in Britain but others, such as the sand lizard are rare, this species being confined to sand dunes on the Lancashire coast and a small number of other locations.

Amphibians

Classification

The class Amphibia contains frogs, toads and newts. There are six (perhaps seven) native British species but others have been introduced at various times.

Class:	Amphibia		
Family:	Ranidae		
Species:	*Rana temporaria*	Common frog	Native (introduced to Ireland)
	R. esculenta	Edible frog	Introduced from Europe
	R. ridibunda	Marsh frog	Introduced from Europe
	R. lessonae	Pool frog	Native
	R. catesbeiana	North American bullfrog	Introduced
Family:	Bufonidae		
Species:	*Bufo bufo*	Common toad	Native
	B. calamita	Natterjack toad	Native
Family:	Discoglossidae		
Species:	*Alytes obstetricans*	European midwife toad	Introduced from Europe
Family:	Pipidae		
Species:	*Xenopus laevis*	Clawed toad	Introduced
Family:	Salamandridae		
Species:	*Triturus cristatus*	Great crested newt	Native
	T. helveticus	Palmate newt	Native
	T. vulgaris	Common newt	Native
	T. alpestris	Alpine newt	Introduced
	T. carnifex	Italian crested newt	Introduced

Natural history

Like reptiles, amphibians are poikilothermic (ectothermic), their body temperature being determined by the environmental temperature. The adult forms possess lungs and can live on land and in water, but the tadpole stage has gills and is confined to water.

The most obvious sign of the presence of amphibians is spawn or eggs found in bodies of stagnant water or on aquatic vegetation. All amphibians reproduce in water during the spring and are heavily dependent upon the existence of ponds. Their larval development occurs in water and involves a gill-breathing tadpole stage which is relatively short-lived. Newts lay individual eggs while frogs and toads produce spawn. The common frog may produce spawn containing up to 6000 eggs.

Amphibians feed on insects, molluscs, worms and other invertebrates. During the winter they hibernate. Some frog and toad species are nocturnal.

All the introduced species have become established and given rise to breeding colonies except the North American bullfrog. This species can mature and survive in Britain but summer temperatures are too low to allow successful reproduction.

A study of the five widespread native amphibian species (the common frog, common toads and the newts) concluded that all except the common frog have declined since the 1980s. The main cause of these declines appears to have been habitat loss (Hilton-Brown and Oldham, 1991).

Fish

Classification of British freshwater fish

Class:	Agnatha	Jawless fish
Order:	Cyclostomata	Hagfish and lampreys
Family:	Petromyzonidae	Lampreys
Class:	Osteichthyes	Bony fish
Order:	Acipenseriformes	Sturgeons and paddlefishes
Family:	Acipenseridae	Sturgeons
Order:	Clupeiformes	Herrings, sardines, salmon, trouts
Family:	Clupeidae	Shads
Family:	Salmonidae	Salmon, trout, charr
Family:	Coregonidae	Whitefish
Family:	Thymallidae	Grayling
Family:	Osmeridae	Smelt
Order:	Esociformes	Pikes and allies
Family:	Esocidae	Pike
Order:	Cypriniformes	Carp, characins and gymnotids
Family:	Cyprinidae	Carps

PANEL 4.4

Natterjack toad   *Bufo calamita*

Identification/size
Length 6–8 cm. Yellow stripe down back.

Habitat
Occurs locally on sand dunes and heaths.

Distribution
Mainly found in south and east England. Also in the north-west, including the Sefton Coast.

Ecology and behaviour
Nocturnal. Load voice. Crawls, does not jump. Feeds on insects, worms etc. A good swimmer, climber and burrower. Female lays 3000–4000 eggs on aquatic plants. Metamorphoses after 42–49 days. Young climb into trees when 1 cm long.

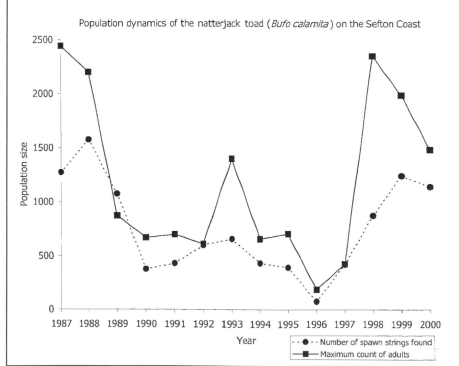

Population dynamics of the natterjack toad (*Bufo calamita*) on the Sefton Coast

Family:	Cobitidae	Loaches
Family:	Siluridae	Catfish
Order:	Anguilliformes	Eels and allies
Family:	Anguillidae	Eels
Order:	Gasterosteiformes	Sticklebacks, seahorses and pipefish
Family:	Gasterosteidae	Sticklebacks
Order:	Gadiformes	Codfish and allies
Family:	Gadidae	Cods
Order:	Perciformes	Perch and allies
Family:	Percidae	Perch
Family:	Serranidae	Bass
Family:	Cottidae	Sculpins
Family:	Gobiidae	Gobies
Family:	Mugilidae	Mullets
Order:	Pleuronectiformes	Flatfish
Family:	Pleuronectidae	Flatfish

Natural history

All fish are poikilothermic (ectothermic), that is their body temperature is determined by the environmental temperature. They breathe by passing oxygenated water through their gills.

Fish are widely distributed from high mountain streams to river estuaries, with some 370 species found in the British Isles and surrounding seas. Anglers divide freshwater fish into 'game fish' (salmon, trout and other members of the Salmonidae) and 'coarse fish' (all other species).

The eggs of freshwater fish are usually attached to aquatic vegetation and fertilised externally. Some species, such as sticklebacks (*Gasterosteus aculeatus*), guard the eggs aggressively. Some species occupy relatively small ranges while others, for example, salmon (*Salmo salar*) migrate over long distances. Newly hatched salmon are called 'alevins'. Once the yolk-sac is absorbed they become 'parr'. After two years most parr turn silvery in colour and become 'smolts' and migrate to the sea. A year or so later they return as 'grilse' and ascend the rivers to their spawning grounds. These are gravelly beds in the headwaters of rivers and are called 'redds'. Here, the salmon buries its eggs in the gravel. After spawning, the salmon return downstream as 'kelts'. Some die while others return to spawn again.

The movements of migratory species are obviously susceptible to any obstruction of river flow. Where obstacles to movement are created it is essential that fish ladders (a series of linked, stepped pools) are built into the design.

Fish are highly susceptible to organic pollution. Pollutants such as sewage or manure will remove oxygen from water. High temperatures may also cause physiological problems for fish as warm water holds less oxygen than cold water.

Invertebrates

The invertebrate species found in the UK are too numerous to list here. They occur in the soil, freshwater, marine water and even at high altitude in the atmosphere. Many are important in decomposition, especially in the soil, while others are important in the pollination of flowering plants. Some species are solitary, while others, such as wasps and ants, live in large social groups. Some invertebrate species live for many years while some insects, such as mayflies, may exist as adults for just an hour.

The most visible invertebrates include molluscs (slugs, snails, octopuses), annelids (segmented worms), arthropods (spiders, insects, lice and crabs). Those that attract greatest concern are the insects, especially bees, butterflies and moths, some of which are extremely rare.

The are only 51 remaining colonies of the high brown fritillary (*Argynnis adippe*), and just 40 colonies of the heath fritillary (*Mellicta athalia*). The large blue (*Maculinea arion*) became extinct in 1979 but has been successfully reintroduced to seven sites in England. All three species are listed in Schedule 5 of the WCA 1981 along with a number of other species of butterflies and other insects (see Appendix 4).

Classification of insects native to the British Isles

Phylum:	Arthropoda	Animals with hard exoskeletons and paired, jointed appendages
Class:	Insecta	Insects
Sub-class:	Apterygota	Wingless insects (which have never had wings during the course of evolution)
Order:	Thysanura	Silverfish and other bristetails
Order:	Diplura	Two-pronged bristletails (very small soil-living species)
Order:	Protura	Minute soil-living species
Order:	Collembola	Springtails
Sub-class:	Pterygota	Winged insects (but some have lost their wings during the course of evolution)
Division:	Exopterygota	Wings develop externally. Young stages are called nymphs and resemble miniature adults
Order:	Ephemerotpera	Mayflies
Order:	Odonata	Dragonflies
Order:	Plecoptera	Stoneflies
Order:	Orthoptera	Grasshoppers and crickets
Order:	Dermaptera	Earwigs
Order:	Dictyoptera	Cockroaches and mantids
Order:	Psocoptera	Booklice
Order:	Mallophaga	Biting lice
Order:	Anoplura	Sucking lice

Order:	Hemiptera	True bugs
Order:	Thysanoptera	Thrips
Division:	Endopterygota	Wings develop internally. Young stages are very different from adults and are called larvae. They metamorphose into adults after passing through a pupal stage
Order:	Neuroptera	Alder flies, snake flies, lacewings
Order:	Coleoptera	Beetles
Order:	Strepsiptera	Stylopids
Order:	Mecoptera	Scorpion flies
Order:	Siphonaptera	Fleas
Order:	Diptera	True flies (2-winged)
Order:	Lepidoptera	Butterflies and moths
Order:	Trichoptera	Caddis flies
Order:	Hymenoptera	Bees, wasps, ants, ichneumons etc.

The vast majority of animal species in the UK are invertebrates, but very few are protected by the law.

4.2.3   Animals and the law

All bats, reptiles (including marine turtles), and amphibians are completely protected under the Wildlife and Countryside Act 1981, along with the rarest mammals, fish, butterflies and other species (Schedule 5). Mammals are also protected from cruel acts by the Wild Mammals (Protection) Act 1996.

Deer, seals and badgers are dealt with separately as they are covered by their own legislation, but the provisions of the WCA 1981 which apply to 'wild animals' or 'animals' may equally apply to these species. Bats are treated separately below, not because they are protected by any special legislation, but because they are of particular conservation interest and their roosts are frequently threatened by building operations. Furthermore, one section of the WCA 1981 specifically refers to bats (s. 10(5)).

A wild animal is defined by the s. 27(1) WCA 1981 as 'any animal (other than a bird) which is or (before it was killed or taken) was living wild.'

Wildlife and Countryside Act 1981 s. 9

(1) ... if any person intentionally kills, injures or takes any wild animal included in Schedule 5, he shall be guilty of an offence.

(2) ... if any person has in his possession or control any live or dead wild animal included in Schedule 5 or any part of, or anything derived from, such an animal, he shall be guilty of an offence.

Section 9(1) does not apply to wild animals within a dwelling-house (s. 10(2)) (but see Section 4.3.3 below for the law relating to bats). Disabled wild animals may be tended and released when they have recovered (s. 10(3)(a)) or killed humanely if seriously disabled with no chance of recovery (s. 10(3)(b)).

An authorised person is not guilty of an offence if he kills or injures a wild animal if this is necessary to prevent serious damage to live-stock, foodstuffs for livestock, crops, vegetables, fruit, growing timber or any other form of property or to fisheries (s. 10(4)). However, if it was known that the action would need to be taken and no application was made for a licence as soon as reasonably practicable after this became obvious, this defence does not apply (s. 10(6)).

Subsection 9(2) does not apply if the wild animal was not killed or taken, or had been killed or taken lawfully (s. 9(3)(a)). This subsection does not apply if the wild animal or other thing in an individual's possession or control had been sold lawfully (s. 9(3)(b)).

As well as protecting scheduled animals from injury, death or taking from the wild, the Act also protects them from disturbance in, and destruction of, their shelters (s. 9(4)).

Wildlife and Countryside Act 1981 s. 9

(4) ... if any person intentionally or recklessly –
   (a) damages or destroys, or obstructs access to, any structure or place which any wild animal included in Schedule 5 uses for shelter or protection; or
   (b) disturbs any such animal while it is occupying a structure or place which it uses for that purpose,
he shall be guilty of an offence.

The selling, transporting for the purpose of sale, or advertising for sale, of Schedule 5 species is also prohibited (s. 9(5)).

Under the WCA 1981 a 'wild animal' is defined as any animal (other than a bird) which is or (before it was taken or killed) was living wild (s. 27(1)). In any proceedings for an offence under subsections (1), (2) and (5)(a) of section 9, the wild animal in question shall be presumed to be wild unless the contrary is shown (s. 9(6)).

Section 10(1)(a) creates an exemption from s. 9 offences for actions done in pursuance of a requirement by the Minister of Agriculture, Fisheries and Food (now DEFRA) and the Secretary of State under s. 98 of the Agriculture Act 1947 or the Secretary of State under s. 39 of the Agriculture (Scotland) Act 1948. Anything done under or in pursuance

---

Wildlife and Countryside Act 1981 s. 9

(5) ... if any person –
  (a) sells, offers or exposes for sale, or has in his possession or transports for the purpose of sale, any live or dead wild animal included in Schedule 5, or any part of, or anything derived from, such an animal; or
  (b) publishes or causes to be published any advertisement likely to be understood as conveying that he buys or sells, or intends to buy or sell, any of those things,
  he shall be guilty of an offence.

---

of an order made under the Animal Health Act 1981 is also exempt from s. 9 offences.

No act is made unlawful under s. 9 if it is done in order to tend or humanely kill a disabled animal (s. 10(3)(a) and (b)) or if it was the incidental result of a lawful operation and could not easily have been avoided (s. 10(3)(c)).

The Countryside and Rights of Way Act 2000 added an offence of intentionally or recklessly disturbing certain marine animals, following concern about their harassment around our coastline (see Panel 4.5). This is particularly important to marine species as it is difficult for them to find refuges when disturbed.

Wildlife and Countryside Act 1981 s. 9

(4A) ... if any person intentionally or recklessly disturbs any wild animal included in Schedule 5 as –
  (a) a dolphin or whale (Cetacea),* or
  (b) a basking shark (*Cetorhinus maximus*),
  he shall be guilty of an offence.
* The order Cetacea includes all species of whales, dolphins and porpoises.

A procedure for the systematic reporting of stranded whales, dolphins and porpoises to the Natural History Museum, London, came into operation in 1913 and still operates under the auspices of the museum. The Cetacean Strandings Co-ordinator records all cetaceans stranded on the shores of the UK and reports any trends to DEFRA. Around 200 strandings are reported every year (see Panel 4.6).

PANEL 2.5

Basking shark   *Cetorhinus maximus*

Identification/size
Second largest fish in the world and the largest animal to occur regularly in Britain. May grow to over 10 metres in length.

The largest fish ever recorded in British waters was a basking shark measuring 11.12 metres and weighing approximately 8 tonnes. It was washed ashore at Brighton, East Sussex, in 1806. (Source: *Guinness Book of Records*, 1998.)

Habitat
Occurs in temperate waters.

Distribution
Worldwide. Found around the coast of the British Isles, especially the Isle of Man.

Ecology and behaviour
Plankton feeder. Moves into British coastal waters from June to early September. Location at other times of the year unknown. Thousands of basking sharks are harpooned each year. Their tails and fins are used to make shark-fin soup.

Section 11 prohibits certain methods of killing or taking wild animals. These generally include:

- self-locking snares designed to cause injury;
- bows, crossbows, explosives (other than ammunition for a firearm);
- decoys;
- trap, snare, electrical device, poison, stupefying substance;
- automatic or semi-automatic weapon;
- illuminating or sighting devices for night shooting;
- artificial light, mirror, dazzling device;
- gas or smoke;
- pursuit by mechanically propelled vehicle.

PANEL 4.6   CETACEAN STRANDINGS

Between 1991 and 1999 a total of 1887 cetaceans were stranded around the shores of the UK, ranging in size from small dolphins to large whales.

On 28 September 2001 a 14 metre long female Sei whale (*Balaenoptera borealis*) was washed ashore alive at Cockerham Sands, Lancashire. It died shortly afterwards and was then washed out to see again, drifting northwards across Morecambe Bay to Chapel Island, Ulverston. The decomposing carcase was eventually washed up at Greenodd Sands where it was photographed on 13 October 2001. This species is a deep sea whale and, consequently, recordings of strandings are rare.

The coast around Scotland has been described as one of the biggest 'sperm whale graveyards' in the world. Eight sperm whales (*Physeter catodon*) were reported to be trapped in Scapa Flow in the Orkney Islands on 17 April 1998 (Anon., 1998a). Sperm whales are an ocean species and would not normally frequent shallow coastal waters. The reasons for this are unclear but the following have been suggested:

- Changes in the distribution of food caused by climatic change.
- Increases in sperm whale numbers as a result of the moratorium on whaling in the 1980s, so that more whales accidentally swim into shallow water.
- Disturbance by the industrial oil operations off the Shetland Isles.
- Increased marine pollution causing deaths, which result in bodies being washed ashore.

Notice that all of these possible explanations are the result of human activity on the marine environment.

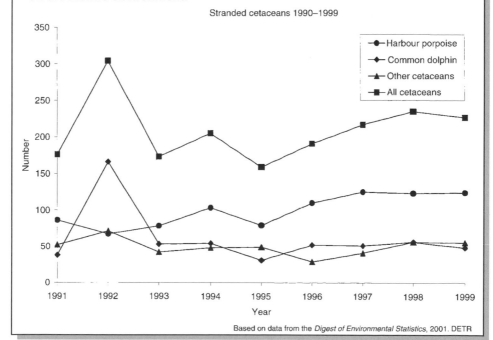

Stranded cetaceans 1990–1999

Based on data from the *Digest of Environmental Statistics*, 2001. DETR

Causing or knowingly permitting the use of any of the above methods is an offence (s. 11(1)(d) and s. 11(2)(f). Licences may be issued by the 'appropriate authority' under s. 16 for some activities prohibited under WCA 1981 for the following purposes:

- scientific, research or educational;
- ringing or marking, or examining rings or marks;
- conservation, including breeding and reintroduction of species;
- protecting any zoological or botanical collection;
- falconry or aviculture;
- any public exhibition or competition;
- taxidermy;
- photography;
- preserving public health, or public or air safety;
- preventing spread of disease;
- preventing serious damage to livestock, foodstuff for livestock, crops, vegetables, fruit, growing timber, or any other form of property, or to fisheries or inland waters.

These licensing arrangements apply to wild birds, wild animals and wild plants as defined by the WCA 1981, with different licensing arrangements applying to specific sections of the Act (see s. 16(1)–(11)). The 'appropriate authority' is defined by s. 16(9) and is either the Secretary of State, the NCC or the agriculture minister depending upon the nature of the licensed activity. Other licences may be issued in relation to, for example, deer, seals or badgers under the appropriate legislation (see Information sources at the end of this book).

Some animal species are protected by the Habitats Directive whose provisions have been implemented in domestic legislation by the Conservation (Natural Habitats etc.) Regulations 1994 (SI 1994/2716) (see Chapter 7 and Appendix 4).

4.2.4   Mitigation measures

Some very simple measures may be taken to encourage animal populations. Many insects and other invertebrates will live in piles of decaying branches and twigs. Hedgehogs and other small mammals may also shelter in wood piles. Dead trees should be left in place (provided they do not represent a hazard) as they provide shelter for birds and small mammals.

Small metal sheets are useful hiding places for reptiles as they heat up when exposed to sunlight and provide dark, warm shelters for snakes and lizards.

Butterflies and other insect species may be attracted by planting appropriate wild flower species, and artificial honeycomb will attract bees.

PANEL 4.7   COMMERCIALLY AVAILABLE WILDLIFE FEEDERS AND SHELTERS

Wildlife populations may be restored or enhanced by increasing the carrying capacity of the habitat by supplementary feeding and increasing the number of sheltering places. Some of the commercially available feeders, nest boxes and other wildlife products are listed below.

Insects
- Red mason bee nest box
- Insect nesting block
- Earwig tube
- Lacewing box
- Underground bumble bee box
- Overground bumble bee box

Birds
- Suspended nut and seed feeder
- Ground feeder (for ground-feeding species)
- Fruit feeder
- Bird table
- Nest box (for use by small songbirds)
- Starling box
- Robin box
- House martin nest
- Swallow nest
- Swift box
- Treecreeper box
- Nuthatch box
- Sparrow terrace (designed to accommodate several pairs of sparrows)
- Brick box (for installation in house walls etc.)
- Nest builder (e.g. suspended cotton balls for use as nest-lining material)
- Water dish
- Bird bath

Mammals
- Hedgehog dome (for use during hibernation)
- Bat box
- Bat brick (for installation in house walls, bridges etc.)

These products are available from Jacobi Jayne & Company, Canterbury.

Ponds can be stocked with fish and will provide habitats for amphibians, waterbirds and some small mammals.

A number of feeding devices and shelters for a variety of animal species are commercially available (see Panel 4.7 and Figure 4.10).

Figure 4.10 A bird feeder used to attract a variety of small songbirds to a visitor centre in the West Pennine Moors. Birds may be observed through a tinted window.

4.2.5   Case studies

Chapters 8 and 9 describe various projects which have involved the translocation of species and habitats, and the mitigation measures used in a number of major developments.

4.3   THE PROTECTION OF BATS

4.3.1   Introduction

Populations of bats in the UK have suffered significant declines in recent years, largely as a result of the loss of suitable roosting places and the decline in insect populations caused by increased insecticide use. Loss of old farm buildings and changes in modern building design have reduced the availability of roof spaces, while vandals have destroyed roosts, and tourists and cavers have disturbed cave-living colonies. Where bats have access to roof spaces they may be poisoned by contact with insecticides used to protect timber and they may be found hiding in broken window frames when double glazing is fitted.

4.3.2   Natural history

Classification

Class:	Mammalia	
Order:	Chiroptera	
Family:	Rhinolophidae (horseshoe bats)	
Species:	*Rhinolophus ferrumequinum*	Greater horseshoe bat
	Rhinolophus hipposideros	Lesser horseshoe bat
Family:	Vespertilionidae (typical insect-eating bats)	
Species:	*Myotis daubentonii*	Daubenton's bat
	Myotis mystacinus	Whiskered bat
	Myotis brandtii	Brandt's bat
	Myotis nattereri	Natterer's bat
	Myotis bechsteinii	Bechstein's bat
	Myotis myotis	Greater mouse-eared bat (extinct)
	Plecotus auritus	Long-eared bat
	Plecotus austriacus	Grey long-eared bat
	*Pipistrellus pipistrellus**	Pipistrelle
	*Pipistrellus pygmaeus**	Pipistrelle
	*Pipistrellus nathusii***	Nathusius's pipistrelle
	Barbastella barbastellus	Barbastelle
	Eptesicus serotinus	Serotine
	Nyctalus noctula	Noctule
	Nyctalus leisleri	Leisler's bat

*These species have only recently been separated on the basis of the sound frequency they utilise and were originally distinguished as the 45 and 55 kHz phonotypes.
**Rarely recorded in south-east England.

Almost a quarter of all the species of mammals in the world are bats. All bats are classified within the order Chiroptera. There are 16 species found in the UK, making up about a third of our land mammal species. All bats enjoy considerable protection under the law.

Distribution

Some species of bat occur in localised areas in Britain while others are widely distributed. The pipistrelle (*Pipistrellus*) occurs throughout Britain and Ireland, whereas the greater horseshoe bat (*Rhinolophus ferrumequinum*) is largely confined to south-west England and South Wales. Some species are relatively common while others are very rare. The pipistrelle is not considered to be threatened but the barbastelle

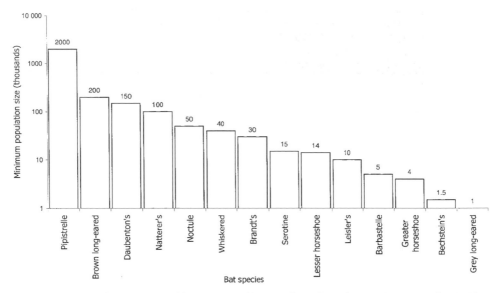

Figure 4.11 Population sizes of bat species. (Source: based on data in Harris *et al.*, 1995.)

(*Barbastella barbastellus*) is rare, and the greater mouse-eared bat (*Myotis myotis*) is now considered extinct in Britain.

Behaviour

All European bats hibernate for a shorter or longer time depending upon the climate. Bats are active at night and rest in roosts during the day. Bats are generally nocturnal but some species may be seen during the day. The period of activity during the day varies between species. The pipistrelle emerges soon after sunset, while the greater horseshoe bat appears late in the evening.

Problems caused by disturbing hibernating bats

Bats feed on insects and must therefore find a means of surviving the winter when their food is in very short supply. The solution is hibernation, during which they allow their body temperature and metabolic rate to fall, thereby conserving energy. Bats build up fat reserves during the summer as an energy source to see them through the winter. However, if they are disturbed during hibernation they may become aroused and as a consequence their vital energy sources may become depleted to the point where they die of starvation. In less severe cases they may survive but fail to breed in the summer.

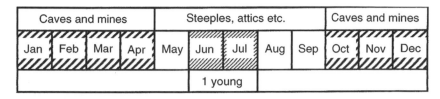

Caves and mines				Steeples, attics etc.					Caves and mines		
Jan	Feb	Mar	Apr	May	Jun	Jul	Aug	Sep	Oct	Nov	Dec
					1 young						

///// hibernation

///// breeding

Figure 4.12 Breeding and hibernation in the lesser horseshoe bat (*Rhinolophus hipposideros*).

Signs

Bats are small and rarely seen. The smallest British bat is the pipistrelle weighing between 3 and 8 grams. Our largest bat is the noctule, weighing up to 40 grams. The most likely signs of the presence of bats are piles of droppings at the entrance to roosts. The remains of insects taken on the wing may also be found on the ground. Tracks of bats are rarely found on the ground.

Breeding

Bats found in the British Isles generally only give birth to a single young each year, although twins have been recorded in some species. Young bats are born from late May to early July, depending upon the species.

Studies of the greater horseshoe bat have shown that as colony size declines there are fewer individuals available to generate the heat necessary for successful reproduction. These bats normally roost in clusters of many thousands to conserve body heat. When there are fewer individual bats in the cluster each must use more energy to generate heat, leaving less energy available in the female for the developing foetus. Clearly, as colony numbers decline, so too does the potential for numbers to recover.

4.3.3   Bats and the law

There is no specific legislation protecting bats in the UK, but all species are protected under the Wildlife and Countryside Act 1981. Schedule 5 (Animals which are protected) includes all horseshoe bats (Rhinolophidae) and all typical bats (Vespertilionidae). In effect this means all bat species that occur in the UK. Bats are also protected

PANEL 4.8

Greater horseshoe bat   *Rhinolophus ferrumequinum*

Identification/size
Head and body 5–6.5 cm long with tail 3–4 cm. Broad wings with wingspan of about 37 cm. Grey to reddish-brown fur, creamy on the belly. Ears large and pointed. Horseshoe-shaped, fleshy nose.

Habitat
Light woodland, scrub and open country.

Distribution
Occurs mainly in south-west England and South Wales.

Ecology and behaviour
Lives in social groups. Nocturnal. Flies at height of 0.5–3 metres from the ground. Roosts in roofs, church towers and ruins in the summer. Hibernates in mines, caves and cellars between September and April. Feeds on flying insects, especially moths. Mates in autumn or spring. Usually produces a single young.

by the Conservation (Natural Habitats etc.) Regulations 1994 (see Section 7.1.3).

Bats are covered by the same provisions as other animals under WCA 1981 ss. 9–11. Some of the main offences and defences are reiterated here but see Section 4.2.3 for a more detailed account. References to 'bats' below are used for convenience, however, the WCA 1981 refers to 'wild animals'.

It is unlawful to kill, injure or take any bat; or to intentionally or recklessly damage, destroy or obstruct access to any shelter used by any bat; or to disturb any bat which is occupying any sheltering place.

It is an offence to possess or control any live or dead bat, or any part or anything derived from a bat (s. 9(2)). Subsection (2) does not apply if the bat was not killed or taken, or had been killed or taken lawfully. This subsection does not apply if the bat or other thing in an individual's possession or control had been sold lawfully. It is an offence to possess or control any bat or to offer any bat for sale.

Bats may be captured and released if found in the dwelling area of a house. Disabled bats may be tended and released when they have recovered (s. 10(3)(a)) or killed humanely if seriously disabled with no chance of recovery (s. 10(3)(b)).

An authorised person is not guilty of an offence if he kills or injures a bat if this is necessary to prevent serious damage to livestock, food-stuffs for livestock, crops, vegetables, fruit, growing timber or any other form of property or to fisheries (s. 10(4)). However, if it was known that the action would need to be taken and no application was made for a licence as soon as reasonably practicable after this became obvious, this defence doe not apply (s. 10(6)).

In subsections (1), (2) and (5)(a) of section 9, references to wild animals include all species of bats (see Section 4.2.3). In any proceedings for an offence under these subsections the bat in question shall be presumed to be wild unless the contrary is shown. Under the WCA 1981 a 'wild animal' is defined as one which is or was (before it was taken or killed) living wild.

The only specific reference to bats within the WCA 1981 is found in s. 10(5). This section removes defences which apply to other animals with respect to action taken against them within a dwelling house (s. 10(2)) or acts which are the incidental result of a lawful operation which could not be avoided (s. 10(3)(c). Section 10(5) requires that the Nature Conservancy Council be notified of any action to be taken against bats within a dwelling house (except in the dwelling area) or any lawful operation that might affect bats, allowing them a reasonable time to give advice. This effectively means that any building operations that may disturb bats must be notified to the NCC even if planning permission has been obtained.

Wildlife and Countryside Act 1981 s. 10(5)

A person shall not be entitled to rely on the defence provided by subsection (2) or (3)(c) as respects anything done in relation to a bat otherwise than in the living area of a dwelling house unless he has notified the Nature Conservancy Council for the area in which the house is situated or, as the case may be, the act is to take place or the proposed action or operation and allowed them a reasonable time to advise him as to whether it should be carried out and, if so, the method to be used.

As with other animal species, s. 10(1)(a) creates an exemption from s. 9 offences for actions done in pursuance of a requirement by the Minister of Agriculture, Fisheries and Food (now DEFRA) and the Secretary of State under s. 98 of the Agriculture Act 1947 or the Secretary of State under s. 39 of the Agriculture (Scotland) Act 1948. Anything done under or in pursuance of an order made under the Animal Health Act 1981 is also exempt from s. 9 offences. Prohibited activities which may be licensed are listed in s. 16 (see Section 4.2.3)

4.3.4   Mitigation measures

The presence of bats can be encouraged by a number of measures.

Attracting insects

Bats feed on insects that fly after dark. They are attracted to night-scented flowers and floodlit areas where moths and other insects are attracted to the light. Woodpiles will also provide a suitable habitat for many insects, such as beetles, that may provide food for some bat species.

Bat roost bricks and bat boxes

Modern building structures can be made more accessible to bats by the incorporation of specially made access bricks. Marshalls Clay Products (Dewsbury, West Yorkshire) produces a Bat Access Brick that has a slot cut from its bottom surface (Figure 4.13). When incorporated into the outer wall of a building it allows bats to gain access to the wall cavity and roof spaces. The company also produces similar units made of stone.

In cooperation with the Bat Conservation Trust and British Waterways Technical Services Department, Marshalls Clay Products has developed a Bat Roost Unit. This unit provides a rough-surfaced cavity of $110 \times 150 \times 215$ mm to which entry is gained through an access brick. This module can be set within a wall structure in new building projects and also used in repairs to bridge arches and abutments, in both brick and stone structures. British Waterways and British Rail have used significant numbers of these bat access units in their continuous maintenance programmes for tunnels and bridges.

Changes to building methods and the demolition of old buildings must have contributed to the decline of our native bat species. Could this decline be reversed by adding a requirement within the Building Regulations to fit bat access bricks and roost units to new buildings?

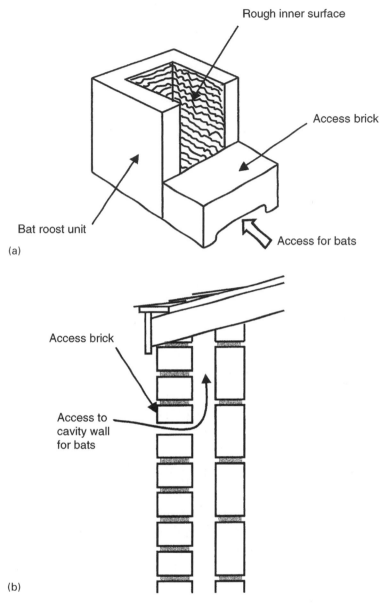

Figure 4.13 Bat bricks: (a) incorporating a roosting chamber which may be built into a wall or other structure; (b) a brick allowing access to the cavity wall of a building via the outer wall.

Bat boxes are available commercially but may also be easily constructed. Essentially, they are similar to bird nesting boxes, but instead of a small entry hole in the front of the box, a narrow slit is provided in the base.

Wood treatment

Non-toxic wood treatments must be used in roof spaces to which bats may gain access. Some products have 'bat-friendly' notices printed on the container. The statutory nature conservation agencies can provide information on suitable products.

4.3.5 Case studies

Reckless disturbance of bats

Wynbrook Ltd of Nottinghamshire and two of its managers were prosecuted on 25 September 2001, for intentionally or recklessly damaging a bat roost in Leicestershire. The company had demolished a building despite knowing that a bat survey had established that it was being used by brown long-eared and Natterer's bats. Each of the defendants was fined £200. This was the first occasion on which the amendment to s. 9(4) WCA 1981, which introduced the offence of recklessly disturbing sheltering places of Schedule 5 animals (introduced by CRoW 2000), had been tested in the courts (Anon., 2002).

First arrest for destruction of bat roost

In August 2001 a man from the Wirral became the first person to be arrested for destroying a bat roost site. However, he was only issued with a caution (Anon., 2002).

Lack of understanding of licence terms

Three defendants were tried for damaging barns containing a resting place of Natterer's bats in September 2001, contrary to s. 39(1)(d) of the Conservation (Natural Habitats etc.) Regulations 1994. The owner of the barns pleaded guilty and was fined £500. The architect and the builder were acquitted after pleading not guilty on the grounds that they did not understand the terms of a licence issued by English Nature.

Bats in mine shafts, at Manchester Airport and on the canal network

Scientists from Leeds University have discovered that six bat species are using disused mine shafts as roosts in the Yorkshire Dales. They include all five cave-roosting species and one species that does not normally live in caves (Anon., 2001d). Provision has been made for bats in a number of recent development projects, including the second runway at Manchester Airport (Section 9.3.1) and work done by British Waterways on the canal network (Panel 5.2).

4.4 THE PROTECTION OF BADGERS

4.4.1 Introduction

Badgers have long been persecuted because they are both suspected of transmitting disease to farm animals and taken for badger-baiting (the organised use of badgers to fight dogs). The National Federation of Badger Groups estimates that 50 000 badgers are killed each year on British roads and a further 10 000 are killed in badger baiting (Anon., 1999b). Others are killed by farmers and gamekeepers.

Approximately 9000 badgers were killed by diggers each year in Britain in the mid-1980s. However, in the 1990s badger digging had declined to less than half that recorded in the 1980s. Badger digging is still a significant problem in the north of England, and there is no evidence of a significant reduction in this part of the country, with over 25 per cent of main setts showing signs of digging. Twenty per cent of the main setts blocked in the 1990s involved the use of oil drums, wire or other objects which prevented the badgers from re-opening the entrance (Anon., 2000b).

In the south of England the badger has long been accused of spreading bovine tuberculosis to domestic cattle but as yet there is no conclusive evidence of this. In 1998 a badger culling trial began in which a detailed comparison will be made of tuberculosis in cattle living within areas occupied by badgers and cattle in control areas where badgers have been removed. The results of the study will not be known for several years, but the project has already run into difficulties. Protesters have interfered with traps and during the 2001 foot-and-mouth disease outbreak badger culling temporarily ceased.

4.4.2 Natural history

Classification

Class:	Mammalia
Order:	Carnivora
Family:	Mustelidae
Species:	*Meles meles*

Distribution

Badgers are common and widespread throughout the British Isles. They prefer hilly country with a sandy soil, but also favour deciduous

PANEL 4.9

Badger   *Meles meles*

Identification/size
Black and white stripes on the head, white border on ears, black legs and belly, silvery grey back. Long claws on front feet. Head and body up to 90 cm long with tail up to 24 cm long.

Habitat
Deciduous and mixed woodland, hedgerows, gardens and parks.

Distribution
Widespread in Britain but most common in the south-west of England, the border counties between England and Wales, and some of the counties in the north and north-west of England.

Ecology and behaviour
Mainly nocturnal. Live in social groups in underground setts but usually forage separately, following well-used tracks. Varied diet from insects to small mammals, carrion, roots and fruits. Still common but threatened, often killed on roads and taken for badger baiting. Persecuted in some areas because suspected of transmitting TB to livestock.

woodland interspersed with fertile grassland which supports high densities of earthworms. They are particularly common in the south-west of England, in the border counties between England and Wales and in the north and north-west of England. They are generally absent from very large conurbations, mountainous areas and lowlands which are wet or liable to flooding (Neal and Cheeseman, 1996).

Behaviour

Badgers are generally nocturnal but may occasionally be encountered above ground in daylight. They are usually found in wooded areas, but also occur in treeless localities, and are often seen in suburban gardens.

Badgers build a sett in the slope of a hill, sunken road, sandpits, quarries and other high banks into which they can burrow. Badgers are highly social animals and up to 15 may live in the same sett. They do not hibernate but in very cold weather they may remain underground for many days. Badgers eat insects, birds, small mammals, roots and other plant material.

Badger setts

Badger setts vary in size and complexity (Figure 4.14). A sett may have a single entrance leading into a large chamber. Some are just a few metres long and may consist of three or four tunnels with perhaps two entrances. Others are extremely large with a complex network of interconnecting tunnels and chambers, the entire structure sometimes occupying three storeys. One excavated sett was found to cover an area of 704 square metres, with 80 entrances, 354 metres of tunnels and 20 chambers (Roper *et al.*, 1991). Another, covering a slightly larger area (740 square metres) but with a similar tunnel length (360 metres), had only 38 entrances but a total of 78 chambers (Leeson and Mills, 1977).

Setts are not static structures but change considerably over time. Badgers continue to excavate new tunnels long after the size of the sett has exceeded their needs. They may end up like cities in which some

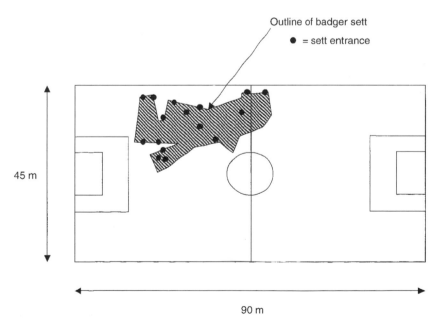

Figure 4.14 The outline of a badger sett illustrated in Neal and Cheeseman (1996) is shown here superimposed on an Association Football pitch for scale.

parts are allowed to become derelict while new tunnels and chambers are excavated elsewhere in the complex. Sometimes old entrances that have not been used for some time (perhaps a year or so) may be cleared out and that section of the sett may then be renovated. Some setts are believed to be hundreds of years old and will have been occupied by many generations of badgers.

Signs

Badgers regularly use the same paths and runs. Where they pass through long grass there may be obvious areas of flattened vegetation. It is believed that some of these paths may have been used by badgers for centuries. Such paths often lead to drinking places or latrines that may contain mounds of droppings.

The entrance to an occupied sett may show signs of activity such as freshly deposited soil and vegetation that have been removed from the living areas. Piles of ferns and grass (used as bedding) which have been removed from the sett by badgers may have a faint musty smell. Loose badger hairs (black, silver and white) may also be found in the vicinity of the sett on twigs and fences. Badgers may leave deep scour marks on wooden posts after sharpening and cleaning their claws. Near the sett badgers may remove the bark from the lower part of sycamore trees to obtain the sweet sap. They may also scrape shallow holes (snaffles) to get at roots. Droppings resemble those of dogs in shape and are always firm. Their colour, size and content vary but they often contain the undigested remains of insects such as beetles.

Breeding

Mating occurs from mid-July to about the end of August. Badgers exhibit delayed implantation so fertilised eggs do not implant until the end of the year. Young are born from the end of February to the end of April: usually 2–3 cubs, that first appear above ground at 6–8 weeks old. Exceptionally, delay in implantation may result in young being born as much as a year after mating. Such a delay may occur if a female is psychologically stressed, for example, by being captured.

4.4.3   Badgers and the law

Badgers are protected under the Protection of Badgers Act 1992. This Act extends to England, Wales and Scotland, but not Northern Ireland. In addition, the Wildlife and Countryside Act 1981 and the Wildlife (Northern Ireland) Order 1985 proscribe certain methods of taking wild animals and this includes badgers. The Wild Mammals (Protection) Act 1996 prohibits cruelty to all wild mammals.

Protection of Badgers Act 1992 s. 1

(1) A person is guilty of an offence if ... he wilfully kills, injures or takes, or attempts to kill, injure or take, a badger.
(2) ... he shall be presumed to have been attempting to kill, injure or take a badger unless the contrary is shown.
(3) A person is guilty of an offence if ... he has in his possession or under his control any dead badger or any part of or anything derived from, a dead badger.

Where there is evidence that an offence has been committed under s. 1(1) the onus shall be on the accused to show that he was not attempting to commit any of these offences (s. 1(2)).

Killing, taking or injuring a badger, or attempting any of these things, is not an offence if the action is necessary to prevent serious damage to land, crops, poultry or other property (s. 7(1)). However, if it was known that the action would need to be taken and no application was made for a licence as soon as reasonably practicable after this became obvious, this defence does not apply (s. 7(2)(a)).

Section s. 1(3) does not apply if the badger had not been killed or had been killed lawfully (s. 1(4)(a)) or if the badger (or other thing) had been sold and the purchaser at the time had no reason to believe that the badger had been unlawfully killed (s. 1(4)(b)).

Protection of Badgers Act 1992 s. 2

(1) A person is guilty of an offence if –
　　(a) he cruelly ill-treats a badger;
　　(b) he uses any badger tongs in the course of killing or taking, or attempting to kill or take, a badger;
　　(c) ... he digs for a badger.

Where there is evidence that the accused was digging for a badger, the onus is on the accused to show he was not digging for a badger (s. 2(2)).

It is also an offence under s. 2(1)(d) to use particular types of firearms for the purpose of killing or taking a badger.

Protection of Badgers Act 1992 s. 3

A person is guilty of an offence if he interferes with a badger sett by doing any of the following things –
(a) damaging a badger sett or any part of it;
(b) destroying a badger sett;
(c) obstructing access to, or any entrance of, a badger sett;
(d) causing a dog to enter any badger sett; or
(e) disturbing a badger when it is occupying a badger sett,
intending to do any of those things or being reckless as to whether his actions would have any of those consequences.

A badger sett is defined under s. 14 as 'any structure or place which displays signs indicating current use by a badger'. Offences therefore only apply to badger setts in current use by badgers. Even if planning permission exists, a licence is still required (under s. 10) if an occupied badger sett is to be damaged or disturbed.

Section 3 offences do not apply if the action was necessary to prevent serious damage to land, poultry, crops or other property. However, if it was known that the action would need to be taken and no application was made for a licence as soon as reasonably practicable after this became obvious, this defence does not apply (s. 8(2)).

Offences under s. 3(a),(c) and (e) do not apply when the action was the incidental result of a lawful operation and could not reasonably have been avoided (s. 8(3)) or was temporary and done while fox hunting with hounds (subject to certain restrictions) (s. 8(4–9)) (see Section 4.9.2).

Protection of Badgers Act 1992 s. 4

A person is guilty of an offence if ... he sells a live badger or offers one for sale or has a live badger in his possession or under his control.

The offence of possessing or controlling a live badger does not apply to persons while working as carriers or to persons who are tending injured badgers, provided the badger was not injured by their actions (s. 9).

Protection of Badgers Act 1992 s. 5

A person is guilty of an offence if, except as authorised by a licence under section 10 …, he marks, or attaches any ring, tag or other marking device to, a badger other than one which is lawfully in his possession by virtue of such a licence.

No offence is committed under this Act where an injured badger is taken (or an attempt is made to take it) for the sole purpose of tending it (s. 6(a)). This exception only applies if the badger was not injured by the actions of the person taking the badger. Other exceptions include killing or attempting to kill a seriously injured badger as an act of mercy, and unavoidably killing or injuring a badger as the incidental result of a lawful action.

Under s. 10(1) of the Act licences may be issued by the appropriate Conservancy Council:

(a) For scientific or educational purposes or the conservation of badgers
   (i) to kill, take, sell or possess badgers;
   (ii) to interfere with a badger sett.
(b) For the purposes of any zoo or collection, to take, sell or possess badgers.
(c) To take badgers for marking.
(d) To interfere with a badger sett for the purposes of development as defined by s. 55(1) of the Town and Country Planning Act 1990 or s. 26 of the Town and Country Planning (Scotland) Act 1997.
(e) To interfere with a badger sett for the purpose of the preservation or archaeological investigation of a monument scheduled under s. 1 of the Ancient Monuments and Archaeological Areas Act 1979.
(f) To interfere with a badger sett for the purposes of investigating possible offences or collecting evidence in connection with court proceedings.

Under s. 10(2) of the Act the appropriate minister may grant a licence:

(a) To kill or take badgers or interfere with a badger sett to prevent the spread of disease.
(b) To kill or take badgers or interfere with a badger sett to prevent serious damage to land, crops, poultry or any other property.

(c) To interfere with a badger sett for the purposes of any agricultural or forestry operation.
(d) To interfere with a badger sett for the purpose of maintaining or improving land drainage, or defence against the sea or tidal water.

These licences may include conditions. Failure to comply with any such conditions is an offence. Licences may be revoked at any time by the issuing authority.

Where more than one badger is involved in an offence, for the purposes of imposing fines, a separate offence arises with respect to each badger (s. 12(2)). The court may impose fines or a prison term or both for offences against badgers (s. 12(1)), and it may also seize any weapon or article used in the commission of the offence (s. 12(4)) or any dog which was used in or present at the commission of the offence (s. 13(1)).

Badger setts and foxes

Under s. 10(3) of the Act the appropriate statutory nature conservation agency or the appropriate minister may issue a licence to interfere with a badger sett for the purposes of controlling foxes in order to protect livestock, game or wildlife.

The Protection of Badgers Act 1992 permits fox hunts to obstruct the entrances to setts to prevent the entry of foxes. This action does not require a licence but the manner in which setts may be obstructed is carefully prescribed (s. 8(4)).

Case law

David Green, Ian Peter Reynolds, David Rowbotham and Martin George Trench v. *Stipendary Magistrate for the County of Lincolnshire* (2000)
This case examined the definition of a badger sett and concluded that a sett did not include the area up to and including the surface area above the systems of tunnels and chambers.

4.4.4   Mitigation measures

Sett capping

Some badger groups protect badgers by sett capping. Badger diggers locate badgers by sending terriers wearing radio collars into setts. Diggers monitor the dog's position from above ground by using receiving equipment. Once a dog has cornered a badger the diggers receive a static signal that tells them where to dig in order to retrieve the terrier and take the trapped badger. Sett capping involves using several layers of high tensile steel and concrete to cap the sett at the surface, preventing diggers from digging down into the tunnel system. Some setts are also protected by special alarm systems. Unfortunately, badger diggers are increasingly turning to lamping: using powerful lamps to startle badgers caught in the open which are then attacked by dogs, usually lurchers.

Fences, tunnels and badger gates

Fences used to keep rabbits out of woodland plantations should include a badger gate wherever a badger path crosses the fence line. A suitable gate was designed by King (1964) and consists of a wooden door suspended from the top of a wooden frame (Figure 4.15). The door should be capable of swinging in both directions and of heavy construction so that rabbits are unable to pass through.

Where a badger path is crossed by a new road, provision should be made for the badgers to pass under the road through a tunnel (see Section 8.2.2). Where badger setts are destroyed as the result of development the badgers may be captured and translocated to an artificial sett. This technique has been used at Manchester International

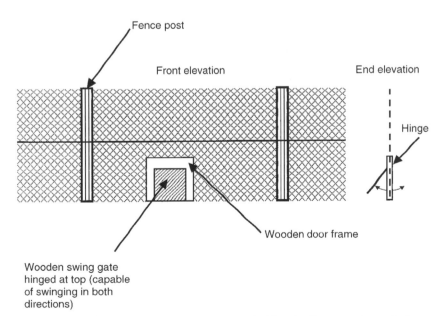

Figure 4.15 A badger swing gate recommended by the Forestry Commission. (Based on a photograph in Neal and Cheeseman (1996) of a gate designed by King (1964).)

Airport where badgers were moved to make way for a second runway (see Section 9.3.1).

4.4.5   Case studies

Prison sentence for badger-snaring gamekeeper

Local badger groups are extremely important in protecting badgers. On the 11 June 1999 the head gamekeeper on the Holker Estate in Cumbria was sentenced to three months' imprisonment at Kendal Magistrates' Court for offences against badgers. He was found guilty of taking a badger, killing two badgers, cruelly ill-treating all three of these badgers, 20 charges of using an illegal self-locking snare and 20 charges of using a snare to take or kill a badger. Evidence against the defendant was collected by the South Lakes Badger Group, in cooperation with the police, the RSPB and the RSPCA. The case was prosecuted by the RSPCA (Anon., 1999c).

Interfering with badger setts

On 19 September 2000 a man was found guilty of interfering with a badger sett after obstructing entrances in his garden in Hockley, Essex. The offence was investigated by MAFF's Farming and Rural Conservation Agency (FRCA), supported by the police. The defendant refused to be interviewed by police and refused to allow them to examine the sett (claiming in court that he believed they might have been impostors). Southend Magistrates' Court imposed a fine of £500 and £250 costs (Anon., 2000b).

On 28 February 2001, two men were fined £150 each and £50 costs after being found guilty of interfering with a badger sett by recklessly allowing their dog to enter it. The incident occurred at Ladywalk Nature Reserve, Hams Hall in Warwickshire. The two men denied digging for badgers but a hair from the dog which had entered the sett was found on a spade discovered hidden close to the scene of the crime. The dog was wearing a 'ferret finder' locating collar used for locating the dog while underground (Anon., 2001e).

On 26 April 2001 the Master of the Essex and Farmers Union Hunt was found guilty of aiding, abetting, counselling and procuring interference with a badger sett at the Othona Community, Bradwell-on-Sea, Essex. She was fined £250 plus £500 costs by Epping Magistrates' Court. The hunt had not sought permission to block the badger setts and the method of blocking had been unlawful. Soil had been packed hard into the sett entrances and the hunt had cut into the tops and sides of the entrances (Anon., 2001f).

Figure 4.16 A sign displayed by police on moorland in Lancashire requesting information on badger diggers from the public.

Pennine Link Badger Monitor

Pennine Link Badger Monitor collects information on badger sightings and the location of badger setts in an attempt to improve the protection of badgers in the areas of Oldham, Tameside and Rochdale. The organisation is made up of individuals from Oldham Countryside Service, Oldham Museum, Saddleworth Conservation Action Group, Rochdale Countryside Service and has the support of Greater Manchester Police.

Badgers close road

In 1998 a country road in Yorkshire had to be closed after badgers had dug eight tunnels which left it close to collapse. The Ministry of Agriculture would not allow North Yorkshire County Council to remove the badgers in case there were young in the setts (Anon., 1998b).

4.5   THE PROTECTION OF DEER

4.5.1   Introduction

The UK has just two native species of deer but four other species have been introduced in recent times. Two of these introduced species, muntjac and sika, have undergone population explosions in recent years.

4.5.2   Natural history

Classification

Class: Mammalia

Order: Artiodactyla

Family: Cervidae

Species: *Cervus elephus*   Red deer   Native
            C. nippon   Sika deer   Introduced
            Capreolus capreolus   Roe deer   Native
            Dama dama   Fallow deer   Introduced
            Muntiacus reevesi   Muntjac deer   Introduced
            Hydropotes inermis   Chinese water deer   Introduced

Distribution

The red deer is Britain's largest native land mammal. It inhabits moorland, open woodland and farmland copses. The red deer is widespread

Figure 4.17 A fallow deer buck (*Dama dama*).

and common in Scotland and occurs in more isolated populations in other areas of Britain and Ireland. The roe deer is the only other native deer species. It is widespread in Scotland and southern England, and is dispersing to other areas, but is absent from Ireland. It inhabits woodlands with dense undergrowth, copses and farmland with thick hedgerows.

Fallow deer and muntjac were both introduced. Both are widespread in the south of England and fallow deer also occur in isolated populations in Scotland and Ireland. Sika deer were introduced from Japan in 1860 and are now naturalised in many localities in Britain and Ireland. All three species are found in woodland, and while fallow deer and sika may also frequent farmland, muntjac are increasingly seen in shrubby gardens. The Chinese water deer occurs in eastern England.

Behaviour

Social organisation varies between species. In red, fallow and sika deer the females live in groups with their young while the males live alone or in small groups. Muntjac are generally solitary while roe deer are organised into loose social groups. Deer feed on grasses, fruits, herbs and bark.

Signs

Footprints and droppings are the most likely signs of the presence of deer, but tree damage may also be apparent. Antlers which have been shed may also be found.

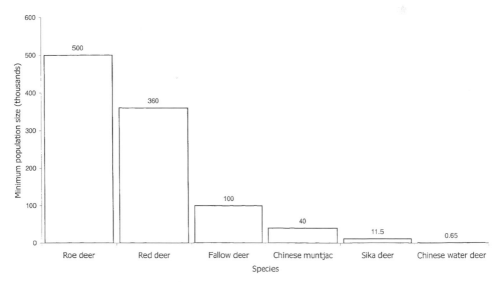

Figure 4.18 Population sizes of deer species. (Source: based on data in Harris *et al.*, 1995.)

Breeding

The mating period of deer is known as the 'rutting season', during which the stags bellow or roar. In red deer this lasts from mid-September to late October, but the precise period varies with the species. Muntjac produce a single fawn at any time of the year. The other species generally produce a single fawn between May and July, fallow and roe deer rarely producing two fawns.

4.5.3   Deer and the law

Wild deer are a quarry species and are also pests of significant economic importance. The killing and taking of deer are regulated under the Wildlife (Northern Ireland) Order 1985, the Deer Act 1991 and the Deer (Scotland) Act 1996. The Wild Mammals (Protection) Act 1996 and certain provisions of the Wildlife and Countryside Act 1981 also apply to deer.

The Deer Act 1991 applies to England and Wales only. For the purposes of this Act deer are defined as 'deer of any species and includes the carcase of any deer or any part thereof' (s. 16).

Deer Act 1991 s. 1

(1) ... if any person enters any land without the consent of the owner or occupier or other lawful authority in search or pursuit of any deer with the intention of taking, killing or injuring it, he shall be guilty of an offence.

(2) ... if any person while on any land –
  (a) intentionally takes, kills or injures, or attempts to take, kill or injure, any deer,
  (b) searches for or pursues any deer with the intention of taking, killing or injuring it, or
  (c) removes the carcase of any deer,
  without the consent of the owner or occupier of the land or

---

No offence is committed under subsection (1) or subsection (2) where the accused believes that he would have had the consent of the owner or occupier if he knew he was doing it (s. 1(3)(a)) or if he has other lawful authority. Under s. 1(4) it is an offence for anyone suspected of committing an offence under subsection (1) or subsection (2) to fail to provide his name and address to the owner or occupier of the land or any person authorised by him, and to quit the land.

Deer Act 1991 s. 2

(1) ... if any person takes or intentionally kills any deer of a species and description mentioned in Schedule 1 to this Act during the prescribed close season, he shall be guilty of an offence.

Section 2(1) does not apply to deer which are clearly marked and, by way of business, are kept on enclosed land for the production of food, skins, other by-products or as breeding stock (s. 2(3)).

Deer may not be taken or killed at night (s. 3) or by the use of certain types of weapons and other articles. The use of the following for taking or killing deer are prohibited under s. 4:

- trap, snare, poisoned or stupefying bait, or net;
- certain types of firearms and ammunition (listed in Schedule 2);
- arrow, spear or similar missile;
- missile carrying or containing any poison, stupefying drug or muscle-relaxing agent;
- mechanically propelled vehicle (for firing at or driving deer).

An offence under this section is not committed if the act is done by, or with the written authority of, the occupier of any enclosed land where deer are usually kept and in relation to any deer on that land (s. 4(5)). Section 5 creates offences of attempting to commit an offence under sections 2 to 4.

Exceptions to the offences created by section 2 and 3 are listed in section 6. They include anything done in pursuance of any requirement under s. 98 of the Agriculture Act 1947; anything done to prevent the suffering of an injured or diseased deer (including the use of traps or nets under s. 4(1)(a) and (b)); and the use of a smooth-bore gun for humane reasons or the use of certain smooth-bore guns for slaughtering).

Section 7 makes provisions for the taking or killing of deer by the occupier of the land, and certain other persons, in order to protect crops, fruit, vegetables, growing timber and any other form of property on the land.

Where more than one deer is involved in an offence under any of the aforementioned provisions, for the purposes of imposing fines, a separate offence arises with respect to each animal (s. 9(2)).

Sections 10 and 11 of the Act create offences relating to the sale of venison by persons who are not licensed game dealers and requires such dealers to keep detailed records of transactions.

PANEL 4.10

Red deer   *Cervus elaphus*

Identification/size
Head and body 1.65–2.5 m. Male larger than female. Coat red-brown in summer and grey-brown in winter. Males possess large branching antlers.

Habitat
Deciduous and mixed woodland, montane forests and beyond the tree line, parkland, moorland and river valleys.

Distribution
Native to Scottish Highlands, the Lake District and Exmoor; introduced or escaped species elsewhere. Kept on some private estates and National Trust property.

Ecology and behaviour
Mainly active at dawn, dusk and at night. Runs and swims well. Lives in herds. Males are solitary or form bachelor groups outside breeding season. Feeds on grass, herbs, leaves, bark, fruit and tree shoots. Uses fixed tracks. Males fight during rutting season (September–October). Hind usually produces a single calf.

Any person convicted of any offence under this Act may be ordered to forfeit any deer or venison in respect of which the offence was committed, and any vehicle, animal, weapon or other thing which was used to commit the offence or capable of being so used (s. 13(1)).

Corporate bodies and their directors, managers and other officers may commit offences under sections 1, 10 and 11 of the Act, through their consent, connivance or neglect (s. 14).

The following Schedules form part of the Act:

■ Schedule 1   Close seasons

- Schedule 2   Prohibited firearms and ammunition
- Schedule 3   Form of record to be kept by licensed game dealers

In Scotland, the Deer Commission has wide powers to cull deer where they present a danger to enclosed woodland, public safety or the natural heritage.

4.5.4   Mitigation measures

Measures which may be taken to reduce road deaths in deer are discussed in Section 8.2.2.

4.6   THE PROTECTION OF SEALS

4.6.1   Introduction

Around 35 000 common seals and 93 500 grey seals live around British coasts. Seals are protected under the Conservation of Seals Act 1970, which applies in England, Wales and Scotland and in the adjacent territorial waters.

4.6.2   Natural history

Classification

Class:	Mammalia	
Order:	Pinnipedia	
Family:	Phocidae	
Species:	*Phoca vitulina*	Common seal
	Halichoerus grypus	Grey seal

Distribution

The common seal prefers sheltered coastlines and is found along North Sea coasts and the sheltered parts of north and west Scotland and Ireland. It usually hauls out on sandbanks or in sea lochs. The grey seal is found in open seas and along rocky coasts and estuaries. It breeds on remote rocky shores and is widespread around the coasts of Britain and Ireland, except for south-east England.

Behaviour

Seals are generally solitary and only gather into groups in the breeding season. They feed predominantly on fish, crustacea and molluscs, but the grey seal also occasionally takes birds.

Signs

Seal tracks may be visible in sand or mud where seals haul out of the water. Flipper marks are made on either side of a central body drag. Moulted fur may be found on land. The moulting period is between July and September.

Breeding

The common seal produces a single pup in June–July. The grey seal produces a single pup in October–November.

4.6.3   Seals and the law

The Conservation of Seals Act 1970 prohibits certain methods of killing seals, imposes a close season for seals, and makes provision for the making of orders prohibiting the killing of seals for the purposes of

---

PANEL 4.11

Grey seal   *Halichoerus grypus*

Identification/size
2.5–3 metres long. Grey to brownish in colour with patchy appearance. Elongated head with heavy muzzle and small external ears.

Habitat
Open sea, estuaries and rocky coasts. May remain at sea for long periods.

Distribution
Widespread around coasts of the British Isles, apart from south-east England.

Ecology and behaviour
Active during the day. Rarely comes onto land. Dives well. Feeds on fish and squid. In breeding season male accompanies a harem of up to 10 females. Produces a single pup between October and November.

conservation. It also makes provision for the Secretary of State to take action to control seal populations to protect fisheries. The Act does not cover anything done outside the seaward limits of the territorial waters adjacent to Great Britain.

Conservation of Seals Act 1970 s. 1

(1) ... if any person–
  (a) uses for the purposes of killing or taking any seal any poisonous substance; or
  (b) uses for the purpose of killing, injuring or taking any seal any firearm other than a rifle using ammunition having a muzzle energy of not less than 600 footpounds and a bullet weighing not less than 45 grains,
he shall be guilty of an offence.

The Secretary of State may by order alter the description of the prohibited firearm and ammunition mentioned in paragraph (b) of subsection (1) of the Act (s. 1(2)). The Act imposes a close season in relation to both seal species.

Conservation of Seals Act s. 2

(1) There shall be an annual close season for grey seals ... *Halichoerus grypus* ... and an annual close season for common seals ... *Phoca vitulina*...
(2) ... if any person wilfully kills, injures or takes a seal during the close season ... he shall be guilty of an offence.

The Secretary of State may make orders to protect seals for conservation purposes in specified areas.

Any person who attempts to commit any offence under this Act shall be guilty of an offence (s. 8(1)). The possession of any prohibited poison, firearm or ammunition for the purpose of committing an offence is also itself an offence (s. 8(2)). The court may order the forfeiture of any seal or seal skin unlawfully taken, and any firearm, ammunition or poison used in the commission of an offence (s. 6).

Conservation of Seals Act 1970 s. 3

---

(1) Where, after consultation with the Council,* it appears to the Secretary of State necessary for the proper conservation of seals he may by order prohibit with respect to any area specified in the order the killing, injuring or taking of the seals of both or either of the species mentioned in section 2 of this Act.

(2) ... if any person wilfully kills, injures or takes a seal in contravention of an order made under subsection (1) of this section he shall be guilty of an offence.

*Natural Environment Research Council

No offence is committed under section 2 or 3 where a disabled seal is taken solely for the purposes of tending it (s. 9(1)(a)) or where a seal is unavoidably killed or injured as an incidental result of a lawful action (s. 9(1)(b)). A seal may also be killed to prevent it from causing damage to a fishing net or fishing tackle or to any fish in a fishing net, provided that the seal was in the vicinity of the net or tackle (s. 9(1)(c)). A seal may be lawfully killed if it is so seriously disabled that there is no reasonable chance of recovery (s. 9(2)).

Seals may be taken or killed under a licence granted under s. 10 of the Act. Such a licence may be issued for:

- scientific or educational purposes;
- zoological gardens or other collections;
- the prevention of damage to fisheries;
- seal population management;
- the protection of flora or fauna in a certain statutory nature reserves (s. 10(4)).

The Secretary of State is required to consult the Natural Environment Research Council before granting a licence under s. 10. He or she is further required to consult the Nature Conservancy Council before granting a licence with respect to any seals within certain statutory nature reserves (defined under s. 10(4)), except in relation to the prevention of damage to fisheries (s. 10(3)). Seals may not be killed under licence by the use of strychnine (s. 10(1)(a)).

Section 11 of the Act gives the Secretary of State, after consulting NERC, a power to authorise persons to enter upon land for the purpose of obtaining information relating to seals, or for the purpose of killing or taking seals to protect fisheries. NERC is required under

s. 13 to provide the Secretary of State with scientific advice on seal management.

4.7   PROTECTION OF WILD PLANTS

4.7.1   Introduction

This section is concerned with the protection of wild plants under the Wildlife and Countryside Act 1981. The additional protection given to trees and hedgerows is described in Chapter 6.

Some species are under particular threat from organised criminal activity. 'Bulb rustlers' pose a significant threat to species like the snowdrop (*Galanthus nivalis*). Such species are illegally removed from woodland and other habitats and then sold.

PANEL 4.12

Fungus   *Mycena maculata*

Identification/size
Small with conical or bell-shaped cap. Grey-buff, but fruit body has red-brown spots when old. Cap 1–4 cm diameter, stem 20–60 cm long.

Habitat
Occurs in broad-leaved woodland in August–November.

Distribution
Widespread.

Ecology
Often occurs in small clusters on leaf litter and rotten wood, particularly on stumps or logs of beech.

4.7.2   Natural history

Classification

Class: Lycopodiopsida	Club mosses and quillworts	
Class: Equisetopsida	Horsetails	
Class: Pteropsida	Ferns	
Class: Pinopsida	Gymnosperms, conifers	
Class: Magnoliopsida	Angiosperms, flowering plants	
Subclass:	Magnoliidae (Dicotyledonidae)	Dicotyledons
Superorder:	Magnoliiflorae	Buttercups, poppies, water-lilies etc.
	Hamameliflorae	Elms, birches, mulberrys, nettles etc.
	Caryophylliflorae	Thrifts, knotweeds, glassworts etc.
	Dilleniiflorae	Limes, violets, willows, heather etc.
	Rosiflorae	Saxifrages, roses, gooseberries etc.
	Asteriflorae	Periwinkles, dodders, mints etc.
Subclass:	Liliidae (Monocotyledonidae)	Monocotyledons
Superorder:	Alismatiflorae	Arrowheads, pondweeds, eelgrasses etc.
	Areciflorae	Duckweeds etc.
	Commeliniflorae	Rushes, sedges, grasses, bulrushes etc.
	Zingiberiflorae	*Rhodostachys*
	Liliiflorae	Lilies, orchids, daffodils, snowdrops etc.

The classification of plants listed above is based on that used by Stace (1997) in the *New Flora of the British Isles*.

Distribution

Some plant species have a widespread distribution within the UK while others are rare and occur only locally. Identification is often difficult, especially in winter when flowers are absent. Grasses cause a particular problem. To the casual observer grass is grass, but worldwide there are some 10 000 species. Around 160 species of grass are indigenous or naturalised in the British Isles. There is disagreement over the exact number because some types are treated as a species by one authority but as a subspecies by another. Some species are wide-spread while others are extremely rare, like Somerset grass (*Koeleria vallesiana*). This species is restricted to a few limestone hills in Weston-super-Mare, in the south-west of England.

It would be meaningless to attempt to summarise the distribution of plants in the UK in a work of this nature. Stace (1997) provides useful summaries of both the degree of rarity and the distribution of plant species.

4.7.3   Plants and the law

The WCA 1981 (s. 27(1)) defines a wild plant as 'any plant which is or (before it was picked, uprooted or destroyed) was growing wild and is of a kind which ordinarily grows in Great Britain in a wild state'.

Wildlife and Countryside Act 1981 s. 13

---

(1) ... if any person –
   (a) intentionally picks, uproots or destroys any wild plant included in Schedule 8; or
   (b) not being an authorised person,* intentionally uproots** any wild plant not included in that Schedule,
   he shall be guilty of an offence.

*the owner or occupier, or any person authorised by the owner or occupier, or other authorised person (s. 27(1)).

**does not apply to 'picking'.

As with other living things, it is an offence to sell, possess, transport or offer for sale certain protected species.

Wildlife and Countryside Act 1981 s. 13

---

(2) ... if any person –
   (a) sells, offers or exposes for sale, or has in his possession or transports for the purpose of sale, any live or dead wild plant included in Schedule 8, or any part of, or anything derived from, such a plant; or
   (b) publishes or causes to be published any advertisement likely to be understood as conveying that he buys or sells, or intends to buy or sell, any of those things,
   he shall be guilty of an offence.

Section 13(1) offences do not apply to anything done as an unintentional and unavoidable consequence of a lawful operation (s. 13(3)). This would include construction activities which have planning permission.

Any plant which is the subject of any proceedings under subsection (2)(a), shall be presumed to have been a wild plant unless the contrary is shown (s. 13(4)). Anything done in accordance with the terms of a licence granted by the appropriate authority cannot be unlawful under s. 13(2) (see Section 4.2.3).

Wild mushrooms may be picked at any time and by anyone, including on private land. Protected plant species are listed in WCA 1981 Schedule 8 (Plants which are protected) (see Appendix 4). Plants are not property for the purposes of the Theft Act 1968 (although cultivated mushrooms are property and may be stolen).

Some plant species are protected by the Habitats Directive whose provisions have been implemented in domestic legislation by the Conservation (Natural Habitats etc.) Regulations 1994 (see Chapter 7 and Appendix 4).

4.7.4   Mitigation measures

Large plants, including trees (see Chapter 6), may be transplanted. Wild flower seeds are widely available and many restoration schemes involve the re-creation of meadows, woodlands, moorland or other habitats. However, care must be taken to use individual plants or seeds which are genetically as similar as possible to those that have been lost or removed. The National Wildflower Centre in Merseyside promotes the planting of native species. A number of examples of habitat restoration projects using native plants are described in Chapters 8 and 9.

4.7.5   Case study

Rare plants on the line

The Deptford pink (*Dianthus aemeria*), a wild relative of the carnation, is a native species of southern Europe but has never been common in Britain. The recent discovery of the species on a 100-yard stretch of railway test track halted the testing of new railway carriages worth more than £700 million (Leathley, 2000). In order to test the new rolling stock a new multimillion-pound electrification of a stretch of track between the villages of Edwalton, south of Nottingham, and Asfordby, near Melton Mowbray, Leicestershire, had been planned. However, English Nature intervened to prevent a four-mile stretch of new track being laid alongside existing track because the plant would be destroyed. The Deptford pink is an annual plant, which germinates, flowers and dies

PANEL 4.13

Military orchid   *Orchis militaris*

Identification/size
Stems generally up to 45 cm long. Leaves elliptic-oblong, unspotted. Flowers pinkish- to reddish-purple.

Habitat
Chalk grassland and old chalk pits.

Distribution
One site in Buckinghamshire and one in West Sussex. Two sites sporadically in Oxfordshire.

Ecology
Very rare. Only occurs on chalk habitats.

in the same year. Although it is found in 30 sites in Britain it grows strongly in just ten, including the test track. While the track lay unused by British Rail the rare plant flourished, but English Nature feared that the thousands of hours needed to test the new carriages would quickly destroy it. In order to avoid breaking the law, the train manufacturer was advised that it might have to wait ten months to see if the plant would regenerate.

4.8   INTRODUCTION OF NEW SPECIES

4.8.1   Introduction

The UK is home to a large number of naturalised species of animals and plants. Some exist as small, harmless populations in localised areas. Others have spread widely across the country. Some exotic animal species have escaped from zoos or fur farms, while others were unwanted pets.

Grey squirrels were introduced from North America and have established themselves as a pest species, forcing the native red squirrel into a small number of safe havens. Zoos are now breeding this species in captivity and re-introducing it into appropriate woodlands. Coypu have escaped from farms in East Anglia. These large rodents have done considerable damage to the drainage ditches in this region. They are now considered to be extinct.

Several bird species have been re-introduced into their former ranges in the UK, including the red kite and the osprey. Scottish Natural Heritage is planning to re-introduce beavers into Scotland and scientists have also considered the feasibility of re-introducing the wild boar.

Introduced plants may completely change the nature of the countryside, altering both the landscape and the ecology. There are around 100 introduced plant species which are considered invasive in Britain. The rhododendron is one of the most ecologically damaging of these plants. *Rhododendron ponticum* (Figure 4.19) was introduced into Britain from Spain and Portugal in 1763. It may grow to a height of five metres and a mature bush may produce more than a million seeds. Since it has no natural enemies in Britain it has been able to spread throughout the country and as an evergreen it shades out other species forming a

Figure 4.19 Rhododendron (*Rhododendron ponticum*) is an exotic plant which has covered large areas of the UK and is extremely difficult to eradicate.

monoculture over large areas. It has been estimated that 1.5 per cent of
Snowdonia National Park is infested with this species and it would
cost £40 million to control it.

4.8.2 Introduced species and the law

The introduction of exotic animal species into the wild is prohibited
under the Wildlife and Countryside Act 1981. In addition, certain listed
animal and plant species may not be released.

Wildlife and Countryside Act 1981 s. 14

(1) ... if any person releases or allows to escape into the wild any
    animal* which:
    (a) is of a kind not ordinarily resident in and is not a regular
        visitor to Great Britain in a wild state; or
    (b) is included in Part I of Schedule 9,
    he shall be guilty of an offence.
(2) ... if any person plants or otherwise causes to grow in the
    wild any plant which is included in Part II of Schedule 9, he
    shall be guilty of an offence.

*Defined by s. 27(3) as:
... any animal of any kind ... an egg, larva, pupa, or other
immature stage ...

Part I of Schedule 9 lists animals which are established in the wild
and may not be released except under licence including:

- Coypu   *Myocaster coypus*
- Red-necked wallaby   *Macropus rufogriseus*
- American mink   *Mustela vison*
- Black rat   *Rattus rattus*
- Grey squirrel   *Sciurus carolinensis*
- Golden pheasant   *Chrysolophus pictus*
- African clawed toad   *Xenopus laevis*

(See Appendix 4.)

Birds listed in Schedule 9 Part I include the Canada goose (*Branta canadensis*), which was introduced from North America about 280 years
ago and is now commonly seen in Britain (Figure 4.20). The little owl
(*Athene noctua*) is not a native species but was introduced from the
continent in 1889–96. It is now widespread in Britain, but is not listed
in Schedule 9. Barn owl populations have been in decline for many

years but readily breed in captivity. The barn owl was added to Schedule 9 in 1992 in order to control the release of captive bred birds.

Many people keep exotic species and it would be unreasonable to prosecute members of the public every time a pet animal accidentally escapes into the wild. Section 14(3) of the Act provides a defence where an introduction has been caused unintentionally, provided that all reasonable steps were taken and all due diligence exercised to avoid committing an offence.

Part II of Schedule 9 lists plants which may not be introduced (under s. 14) including:

- Giant hogweed   *Heracleum mantegazzianum*
- Giant kelp   *Macrocystis pyrifera*
- Japanese knotweed   *Polygonum cuspidatum*

(See Appendix 4.)

Section 14 does not apply to anything done under and in accordance with the terms of a licence granted by the appropriate authority (s. 16(4)(c)).

Figure 4.20 The Canada goose (*Branta canadensis*) is an introduced species which is listed in Schedule 9 of the Wildlife and Countryside Act 1981.

4.8.3   A legal obligation to restore wildlife and habitats

Since 1979 there has been a clear obligation under international law for European States to re-introduce native species (Berne Convention, Art. 11(2)). This obligation was extended to the parties to the Biodiversity Treaty in 1992 and there is some evidence that national laws are beginning to reflect these obligations (Rees, 2000) The Convention on the Conservation of European Wildlife and Natural Habitats, 1979 (the Berne Convention), was the first wildlife treaty to encourage its parties to re-introduce native species as a method of conservation.

Under Article 11(2) of the convention the contracting parties undertake,

a) to encourage the re-introduction of native species of wild flora and fauna when this would contribute to the conservation of an endangered species, provided that a study is first made in the light of the experiences of other Contracting Parties to establish that such re-introductions would be effective and acceptable; …

More recently, the Convention on Biological Diversity, 1992, has re-affirmed an international commitment to the recovery of species. Article 8(d) creates an obligation upon contracting parties to:

Promote the protection of ecosystems, natural habitats and the maintenance of viable populations of species in natural surroundings;

and Article 8(f) creates a further obligation to:

… promote the recovery of threatened species …

Article 9(c) creates an obligation to re-introduce threatened species, requiring that:

Each Contracting Party shall, as far as possible and as appropriate, and predominantly for the purpose of complementing *in-situ* measures:
… (c) Adopt measures for the recovery and rehabilitation of threatened species and for their re-introduction into their natural habitats under appropriate conditions.

The Countryside and Rights of Way Act 2000 (s. 74) has imposed a duty upon the government to have regard to the purpose of conserving biological diversity in the carrying out of its functions, in accordance with the UN Convention on Biological Diversity. The UK Biodiversity Action Plan, which was implemented as a matter of policy, now has a statutory basis. The duty includes an obligation to restore wildlife populations and habitats. Although some existing English law could have been used to require the restoration of protected trees, hedgerows and some protected sites when damaged, this is the first attempt to impose a general duty to restore a wide range of listed species and habitats (Rees, 2001)

4.8.4   Case studies

Red kites

The red kite died out in England in the 1880s and had been lost from Scotland by the 1890s. The entire kite population of Wales appears to have descended from just two birds. From a secret site in the Chilterns 93 young kites from Spain, Sweden and Wales have been introduced into England. A similar programme operated in Scotland until 1993 and another release site has been used in the Midlands. The English and Scottish kite populations now number 120 and 70 birds respectively. Illegal poisoning remains a threat and a Herefordshire gamekeeper who stored massive quantities of the toxin Endrin was prosecuted for poisoning kites in 1989. Kites are slowly gaining acceptance among farmers and landowners and in May 1995 the Royal Society for the Protection of Birds launched the 'Kite Country' tourism scheme which allows visitors to central Wales to observe nesting kites by video links. Deaths of kites still occur in Wales.

Red squirrels

The Welsh Mountain Zoo in North Wales is currently engaged in a re-introduction programme for red squirrels called the 'Reds Return Project'. The zoo has constructed a series of large cages that are connected to adjacent trees by narrow wire walkways. The intention is to captive breed as many squirrels as possible in the cages and then to release them into the pine woodland within the grounds of the zoo.

In north-east England 25 refuges for red squirrels are to be established by Durham and Northumberland Wildlife Trusts. There are only about 160 000 red squirrels in the UK. Most of them are in Scotland with only about 30 000 in England, where the north-east is their major stronghold.

Experimental tightropes have been installed high above the traffic in West Wight, Isle of Wight, in an attempt to reduce the number of road casualties suffered by red squirrels. The experiment was carried out by 'NPI Red Alert' (Anon., 1998c).

Re-introduction of beavers

Beavers have not been seen in England since the twelfth century and the last beaver sighting in Scotland was in 1527 near Loch Ness

(Macdonald, 1995). Now there are plans for its return.

Those involved with re-introduction programmes often give unscientific and emotive reasons for their actions. They are rarely based on scientific or ecological reasoning. The Mammal Society has recently given its support to the proposed re-introduction of beavers in Scotland. Gorman and Kruuk (1998) have argued that the beaver should be re-introduced into Scotland because, *inter alia*, Britain is one of the few countries in the original geographic range of the species in Europe which is currently without beavers. But do the people of Scotland want beavers?

Scottish Natural Heritage has undertaken an extensive public consultation exercise in order to establish public opinion in relation to the re-introduction of the European beaver to the wild in Scotland (Anon., 1998d). Almost two thirds (63 per cent) of the 2141 members of the passive public sample consulted were in favour of re-introduction with just 12 per cent against (although many of these gave 'lack of interest' as their reason). A further 1944 responses were received from a 'pro-active public' sample consisting of academics, ecologists, zoologists, conservationists, hillwalkers and others with similar interests including land managers. Overall 86 per cent of this sample favoured re-introduction but the survey identified a clear lack of support from those with a fishing or agricultural interest.

Release of mink

On the 8 August 1998 around 6000 mink (*Mustela vison*) were released from a fur farm near Ringwood in Hampshire, by animal rights activists. Within hours there were reports of the animals attacking a variety of small animals from pet rabbits to small dogs. Whether or not the presence of these mink had a significant effect upon the local wildlife is difficult to assess in the absence of detailed population data for potential prey species and potential competitors. However, even if the ecological effects are negligible the human response to their release was negative, particularly since there was damage to domestic and farm animals (and birds of prey in a local sanctuary).

Mink are not native to Britain but the species is established in the wild, and this incident illustrates the type of public response that could be expected from a mass re-introduction of predators. It is admittedly a poor example because no sensible ecologist would advise that predators should be re-introduced into an ecosystem in such large numbers over such a short period of time. This particular release was unlawful and in contravention of s. 14(1) of the Wildlife and Countryside Act 1981. In October 2001, more than 500 mink were deliberately released into the New Forest, probably by animal rights activists (Anon., 2001g).

The Hebridean Mink Project is a five-year operation to remove mink from North and South Uist which will cost an estimated £1.65 million. In the Outer Hebrides mink threaten bird populations, wild and farmed fish populations, crustacean colonies, and the chickens kept by crofters (Morgan, 2001). At the beginning of the project 12 trappers were employed using live traps to capture the animals. The initial aim is to eradicate mink from Uist and substantially to reduce the population on Harris. The long-term aim of the project is to eradicate mink from the Western Isles.

Wolves

Some zoologists believe that we should re-introduce wolves into Britain. Wolves could help to control the deer population in Scotland, and allow the regeneration of the native Caledonian forest. Unfortunately, a proposal to place a wolf pack on Rhum was shelved

---

PANEL 4.14

American mink   *Mustela vison*

Identification/size
Usually glossy dark brown or black. Underside mostly white. Feet slightly webbed. Slightly bushy tail. Head, body and tail 42–65 cm.

Habitat
Lake margins, well-vegetated river banks, ditches, estuaries and rocky shores.

Distribution
Widespread in British Isles, including offshore islands.

Ecology and behaviour
Introduced. Very destructive to native wildlife. Feeds on small mammals and birds, fish, amphibians, molluscs and large insects. Food may be stored in caches on river banks. Partially aquatic.

by Scottish Natural Heritage when concern was expressed that they might take sheep and even attack people (Anon., 1999c).

Wolf attacks on man are extremely rare and most of the recorded cases have been attributed to rabid animals (Macdonald, 1980). Nevertheless, the general public are still wary of wolves. In April 2001 a pet timber wolf slipped her leash after a confrontation with a dog in Broxbourne Wood, Hertfordshire. She was eventually coaxed back to her owner with food after a search of several hours involving armed police and a helicopter (Anon., 2001h).

4.9   FISH, GAME AND QUARRY SPECIES

4.9.1   Fish, game and quarry species and the law

A wide range of legislation controls the taking of game and quarry species. The following list is intended only as a guide.

- Game (Scotland) Act 1772
- Night Poaching Act 1828
- Game Act 1831
- Game (Scotland) Act 1832
- Night Poaching Act 1844
- Hares Act 1848
- Hares (Scotland) Act 1848
- Game Licences Act 1860
- Poaching Prevention Act 1862
- Ground Game Act 1880
- Hares Preservation Act 1892
- Ground Game (Amendment) Act 1906
- Game Preservation Act (Northern Ireland) 1928
- Agriculture Act 1947
- Agriculture (Scotland) Act 1948
- Pests Act 1954
- Protection of Animals Act 1911
- Protection of Animals (Scotland) Act 1912
- Deer Act 1991
- Deer (Scotland) Act 1996
- Spring Traps (Approval) Order 1995
- Spring Traps (Scotland) (Approval) Order 1996
- Wild Mammals (Protection) Act 1996
- Welfare of Animals (Transport) Order 1997
- The Game Birds Preservation Order (Northern Ireland) 1999

This legislation variously imposes close seasons on the hunting of particular species, restricts the methods of capture, makes certain cruel acts unlawful, and requires the licensing of certain activities. A detailed discussion of game laws is to be found in Parkes and Thornley (1997).

Various provisions of the Wildlife and Countryside Act 1981 and the Wildlife (Northern Ireland) Order 1985 also apply to game and quarry species, such as methods of taking animals.

Provisions relating to freshwater fish occur in a number of Acts:

- Salmon and Freshwater Fisheries (Protection) (Scotland) Act 1951
- Salmon and Freshwater Fisheries Act 1975
- Salmon Act 1986
- Salmon Conservation (Scotland) Act 2001

A detailed history of the development of freshwater fishery law is provided by Howarth (1987).

The Wildlife and Countryside Act 1981 includes some fish in Schedule 5, but no fish species are scheduled under the Wildlife (Northern Ireland) Order 1985. In England and Wales, fishing licences are issued by the Environment Agency. The Agency also enforces legislation aimed at controlling and preventing pollution which may result in fish kills. It is an offence to poison fish by polluting water (see Section 5.10).

4.9.2   Foxes

Foxes are an exception to the general rule that urbanisation has reduced the carrying capacities for mammals. This species has adapted well to urban living (Harris, 1977, 1978 and 1986; Macdonald, 1977 and 1987) and has penetrated deep into the heart of our largest cities, feeding on food waste taken from refuse bins.

Harris (1986) has shown that while fox density in lowland Britain is one per km^2, in towns like Bristol it may reach four per km^2. Death rates in urban foxes are high, with up to 60 per cent of the population dying each year. About half of these deaths are due to road accidents and a quarter are due to deliberate killing by man. This high death rate prevents the normal large family groups found in the countryside from developing.

Analysis of the stomach contents of 571 urban foxes living in London has shown that scavenged food formed an important part of the diet, but they also ate wild mammals and birds, earthworms and even pets. Inner city animals consumed a higher proportion of scavenged food than those living in outer London (Harris, 1986).

The home range of urban foxes varies from 25 to 40 ha in privately owned, interwar suburbs where supplementary feeding is available, to over 100 ha in less favourable habitats, such as industrial areas and local authority housing estates. In his study of suburban foxes around

Oxford, Macdonald (1977) recorded an average territory size of about 45 ha but in a sheep farming area in the north of England territories reached 1300 ha. This illustrates the importance of urban areas as possible strongholds of fox populations.

Foxes have been recorded from almost every urban habitat in Oxford, from terraced housing to the university's grounds. Macdonald (1980) describes one earth that had been excavated under the university's Astrophysics Department and into the housing of a seismograph. Another family spent most of its time within a large industrial warehouse.

Foxes and the law

Special provisions are made within the Protection of Badgers Act 1992 for the temporary blocking of badger sett entrances to prevent foxes from seeking refuge. The Wildlife and Countryside Act 1981 protects animals, including foxes, from being trapped or killed in certain ways (see Section 4.2.3). Other legislation prohibits the use of certain poisons, pesticides and certain types of trap against foxes. The Wild Mammals (Protection) Act 1996 specifically excludes cruelty to mammals caused by hunting with dogs and as such does not make any cruelty caused by fox hunting an offence. Section 2(b) allows the killing of mammals in a 'reasonably swift and humane manner' if taken in the course of 'lawful shooting, hunting, coursing, or pest control activity' (see Panel 1.2).

Hunting with dogs in Scotland
Progress has been made on the eradication of fox hunting in Scotland by the passing of the Protection of Wild Mammals (Scotland) Act 2002, by the Scottish Parliament. The Act protects wild mammals from being hunted with dogs and came into force in August 2002.

Protection of Wild Mammals (Scotland) Act 2002 s. 1

(1) A person who deliberately hunts a wild mammal with a dog commits an offence.
(2) It is an offence for an owner or occupier of land knowingly to permit another person to enter or use it to commit an offence under subsection (1).
(3) It is an offence for an owner of, or person having responsibility for, a dog knowingly to permit another person to use it to commit an offence under subsection (1).

Exceptions are listed in s. 2 and include using a dog under control (with the permission of the owner or lawful occupier of the land) to stalk a wild mammal or flush it from cover for the purpose of:

- protecting livestock, ground-nesting birds, timber, fowl (including wildfowl), game birds or crops from attack by wild mammals;
- providing food;
- protecting human health;
- preventing the spread of disease;
- controlling pest numbers;
- controlling the numbers of particular species to safeguard the welfare of that species.

The Act also allows the use of dogs to stalk or flush wild mammals which may be shot for sport or killed using a bird of prey used in connection with falconry (s. 3). Section 4 allows authorised persons (a local authority officer, someone authorised by him, or a constable) to use a dog to search for or catch a wild mammal provided he does not intend to harm the wild mammal. Under s. 5, a dog may be used to:

- retrieve a hare which has been shot (s. 5(1)(a));
- locate a wild mammal that has escaped from or been released from captivity (provided it is captured or shot once located) (s. 5(1)(b));
- retrieve or locate an injured or orphaned wild mammal (provided that once located it is treated or killed humanely) (s. 5(1)(c)).

Subsection (1)(b) does not apply to foxes or hares, and only applies to deer, boar or mink if they have escaped from a farm or zoo. A dog may not be used to locate a wild mammal that has been released after having been specifically raised for the purpose of being hunted (s. 5(2)(c)). Subsections 4–9 of Section 8 of the Protection of Badgers Act 1992, which allow the temporary blocking of setts to exclude foxes, are repealed in Scotland.

4.10   ACCESS TO LAND FOR CONSERVATION PURPOSES

The Countryside and Rights of Way Act 2000 (CRoW) gives a right of access (s. 40) to the countryside agencies for the purpose of determining whether or not to restrict or exclude access to land where there is a threat to any conservation interest, any scheduled monument or other important site (s. 26).

4.10.1   Case study

Monitoring hen harrier nesting sites

English Nature intends to monitor hen harrier nesting sites over the

---

Countryside and Rights of Way Act 2000 s. 26

---

(1) The relevant authority* may by direction exclude or restrict access by virtue of 2(1) to any land during any period if they are satisfied that … [it] is necessary for either of the purposes specified in subsection (3).

(3) The purposes referred to in susbsection (1) are –
   (a) the purpose of conserving flora, fauna or geological or physiographical features of the land in question;
   (b) the purpose of preserving –
      (i) any scheduled monument as defined by section 1(11) of the Ancient Monuments and Archaeological Areas Act 1979, or
      (ii) any other structure, work, site, garden or area which is of historic, architectural, traditional, artistic or archaeological interest.

*The relevant authority means the Countryside Agency (in England), the Countryside Council for Wales (in Wales), the National Park Authority (in any National Park) or in relation to some woodlands, the Forestry Commissioners (s. 21(5)).

---

next few years in the hope of improving breeding success on grouse moors. The number of breeding pairs in England fell to just five in 2000. In these areas this species has been suffering from illegal poisoning, trapping and shooting because it preys on grouse. Previously raptor workers have been unable to enter land without the owner's permission. The new powers granted under the CRoW 2000 will allow them to enter private land for monitoring purposes (Theobald, 2001).

4.11   CRUELTY TO WILDLIFE

4.11.1   Introduction

The law prohibiting cruelty to animals is largely concerned with domestic and farm animals. However, the Wild Mammals (Protection) Act 1996 created cruelty offences relating to wild mammals.

4.11.2   The law and cruelty to wild mammals

The Wild Mammals (Protection) Act 1996 is concerned with preventing certain types of cruelty to wild mammals (see Panel 1.2). It is not a conservation law *per se* but, nevertheless, affords wild mammals protection against cruel acts that was previously available only to domestic and farm animals.

Wild Mammals (Protection) Act 1996 s. 1

If ... any person mutilates, kicks, beats, nails or otherwise impales, stabs, burns, stones, crushes, drowns, drags or asphyxiates any wild mammal with intent to inflict unnecessary suffering he shall be guilty of an offence.

No offence is committed under s. 1 as a result of the attempted killing of a seriously disabled wild mammal as an act of mercy (provided that the person concerned was not responsible for the original injury) (s. 2(a)). The Act also provides other exceptions where a wild mammal has been injured or taken in the course of lawful shooting, hunting, coursing or pest control activity; the activity is authorised by or under any enactment; the lawful use of any snare, trap, dog or bird used for taking or killing any wild mammal; and the lawful use of poisons (see Panel 1.2 for the exact wording of the Act). The exceptions were carefully worded so that the Act could not be used to ban fox hunting since it specifically excludes any act done by dogs lawfully used for killing or taking any wild mammal (s. 2(d)) (see Panel 4.15).

For the purposes of determining the maximum fine the courts will consider each mammal that has been cruelly treated as the subject

---

PANEL 4.15   CRUELTY TO WILD MAMMALS

Until recently, only captive mammals received any general legal protection in the United Kingdom. Increased awareness of cruelty to wild mammals led to the passing of the Wild Mammals (Protection) Act in 1996.
  Such cruelty has included:
  ■ Maiming of foxes by cutting off their ears or tails.
  ■ A vixen found alive, nailed to a tree, her cubs left to die.
  ■ A squirrel nailed alive to tree.
  ■ A hedgehog deliberately impaled on a piece of wood.
  ■ Hedgehogs beaten to death, and severe injuries caused by kicking and striking with a golf club.
Prior to 1996 none of these action was illegal under British law. The Protection of Animals Act (1911) applies only to animals kept in captivity. Some wild mammals are protected from cruelty by other specific legislation, e.g. badgers.
Source: Ferris, F. (1995) The crime of being wild and free. *BBC Wildlife*, 13(4), 52–53.

of a separate offence (s. 5(2)). Under s. 6 the court may order the confiscation of any equipment or vehicle used in the commission of an offence. The Protection of Wild Mammals (Scotland) Act 2002 has amended this Act in relation to Scotland.

Case law

A boy aged 14 who sang the theme tune to *Match of the Day* while kicking a hedgehog around like a football was found guilty by Guildford Youth Court of causing unnecessary suffering (Anon., 2001i). The hedgehog was eventually killed and ended up impaled by its spikes on the boy's trainers.

Stuart Bandeira and Darren Brannigan v. *Royal Society for the Prevention of Cruelty to Animals* (2000)
The appellants had deliberately put a dog into a badger sett. The badger could not escape and caused serious injury to the dog. The appellants were guilty of ill-treatment to an animal under s. 1(1)(a) Protection of Animals Act 1911.

4.11.3   Cruelty to other species

The law restricts the method of taking and killing many animal species but does not prohibit general cruelty to wild non-mammalian species.

In April 2001 the RSPCA was called to an incident at two ponds in Lincoln, where they discovered 100 mutilated frogs. Some had their back legs knotted together, others had been stood on, burnt, disembowelled or cut in half (Anon., 2001j).

4.11.4   Illegal poisoning of wildlife

In 1997 the Wildlife Incident Investigation Scheme registered 607 suspected cases of wildlife poisoning in the UK. The cause was determined in 300 cases, of which 185 were attributed to pesticides. All but six of these incidents arose from the misuse of pesticides. Pesticides were deliberately used to poison animals in 125 cases (Anon., 2000c).

REFERENCES

Anon. (1998a) Graveyard for whales. Mammal Briefs. *Mammal News*, No. 114 (Summer 1998). The Mammal Society, London, p. 12.
Anon. (1998b) Badger tunnels. Mammal Briefs. *Mammal News*, No. 114 (Summer 1998). The Mammal Society, London, p. 12.
Anon. (1998c) Red squirrel crossings. Mammal Briefs. *Mammal News*, No. 114 (Summer 1998). The Mammal Society, London, p. 12.

Anon. (1998d) *Re-introduction of the European Beaver to Scotland: Results of a Public Consultation*. Scottish Natural Heritage Research Survey and Monitoring Series No. 121, Scott Porter Research and Marketing.

Anon. (1999a) *Scottish Natural Heritage Annual Report 1999/2000*. Scottish Natural Heritage, Edinburgh.

Anon. (1999b) Prison sentence for badger-snaring gamekeeper. News release, 12 July 1999. National Federation of Badger Groups. *www. badgers. org.uk/nfbg/news/990712.htm*, accessed 23 April 2001.

Anon. (1999c) 14,000 bears in ancient Britain. *The Times*, 18 September 1999, p. 18.

Anon. (2000a) *State of the UK's Birds*. Royal Society for the Protection of Birds, The Lodge, Sandy, Bedfordshire.

Anon. (2000b) Fine for interfering with a badger sett. News release, 26 September 2000. National Federation of Badger Groups. *www.badgers.org. uk/nfbg/news/000926.htm*, accessed 23 April 2001.

Anon. (2000c) *The State of the Environment of England and Wales: The Land*. Environment Agency, The Stationery Office Ltd, London.

Anon. (2001a) Watch the birdie. *The Times*, 7 November 2001, p. 14.

Anon. (2001b) One amazing prey. *The Times*, 26 June 2001, p. 6.

Anon. (2001c) Osprey killed. *The Times*, 7 May 2001, p. 6.

Anon. (2001d) Bats in the Dales. *The Times*, 25 October 2001, p. 14.

Anon. (2001e) Fine for interfering with a badger sett. News release, 28 March 2001. National Federation of Badger Groups. *www.badgers.org. uk/nfbg/news/010328.htm*, accessed 27 April 2001.

Anon. (2001f) Hunt fined for illegally blocking a badger sett. News release, 26 April 2001. National Federation of Badger Groups. *www.badgers.org. uk/nfbg/news/010426.htm*, accessed 27 April 2001.

Anon. (2001g) Mink set free. *The Times*, 25 October 2001, p. 14.

Anon. (2001h) Pet wolf recaptured in woods. *The Times*, 17 April 2001, p. 10.

Anon. (2001i) Hedgehog cruelty. *The Times*, 26 January 2001, p. 4.

Anon. (2001j) Frogs mutilated. *The Times*, 11 April 2001, p. 9.

Anon. (2002) Recent prosecutions. Partnership for Action Against Wildlife Crime, DEFRA. *www.defra.gov.uk/paw/prosecutions*, accessed 26 April 2002.

Armstrong, E.A. (1970) *The Folklore of Birds. An Enquiry into the Origin and Distribution of Some Magico-Religious Traditions*, 2nd edn, Dover Publications, New York.

Boland, L. (1987) *Nest Boxes for the Birds of Britain and Europe*. Sainsbury Publishing Ltd, Nottinghamshire, England.

Carwardine, M. (1995) *The Guiness Book of Animal Records*. Guinness Publishing Ltd, Enfield, Middlesex.

Cramp, S., Parrinder, E.R. and Richards, B.A. (1964) Roosts and fly-lines. In *The Birds of the London Area*. Rupert Hart-Davis, London, pp. 106–117.

English, S. (2000) Turbine deadline lets birds nest in peace. *The Times*, 8 April 2000, p. 7.

Fisher, J. (1970) *Wildlife Crisis*. Hamish Hamilton Ltd, London.

Gorman, M. and Kruuk, H. (1998) Re-introduction of the European beaver in Scotland. The Mammal Society Statement. *Mammal News*. Mammal Society, Autumn 1998, No. 115, p. 6.

Harris, G. (2001) £2000 fine for man who shot rare hawk. *The Times*, 26 May 2001, p. 6.

Harris, S. (1977) Distribution, habitat utilisation and age structure of a suburban fox (*Vulpes vulpes*) population. *Mammal Rev.*, 7: 25–39.

Harris, S. (1978) Injuries to foxes (*Vulpes vulpes*) living in suburban London. *J. Zool. Lond.*, 186: 567–72

Harris, S. (1986) *Urban Foxes*. Whittet Books, London.

Harris, S., Morris, P., Wray, S. and Yalden, D. (1995) *A review of British mammals: population estimates and conservation status, other than cetaceans*. JNCC, Peterborough.

Hawkes, N. (1999) Duck cull prompts call of ruddy racism. *The Times*, 2 February 1999, p. 3.

Hilton-Brown, D. and Oldham, R.S. (1991) The status of the widespread amphibians and reptiles in Britain, 1990, and changes during the 1980s. *Focus on Nature Conservation*, No. 131. Nature Conservancy Council.

Howarth, W. (1987) *Freshwater Fishery Law*. Blackstone Press, London.

Hudson, W.H. (1898) *Birds in London*. Longmans, Green and Co., London.

King, R.J. (1964) The badger gate. *Q.J. Forestry*, 58: 505–506.

Leathley, A. (2000) Flower power delays tests on £700 m trains. *The Times*, 16 October 2000, p. 11.

Leeson, R.C. and Mills, B.M.C. (1977) *Survey of Excavated Badger Setts in the County of Avon*. ADAS, MAFF.

Macdonald, D.W. (1977) The behavioural ecology of the red fox (*Vulpes vulpes*): a study of social organisation resource exploitation. D. Phil. thesis, University of Oxford.

Macdonald, D.W. (1980) *Rabies and Wildlife. A Biologist's Perspective*. Oxford University Press, Oxford.

Macdonald, D.W. (1987) *Running with the Fox*. Unwin Hyman, London and Sydney.

Macdonald, D. (1995) *European Mammals. Evolution and Behaviour*. Harper Collins Publishers Ltd, London.

Morgan, A. (2001) Unnatural born killers run riot. *The Times, Weekend Supplement*, 7 April 2001, p. 13.

Moss, S. (2001) The fall of the sparrow. *BBC Wildife*, 19(11): 45.

Neal, E. and Cheeseman, C. (1996) *Badgers*. T. & A.D. Poysner Ltd, London.

Owen, G. (1999) 'Bombsite bird' threatens to stop work on the Dome. *The Times*, 5 December 1999, p. 1.

Parkes, C. and Thornley, J. (1997) *Fair Game. The Law of Country Sports and the Protection of Wildlife*. Pelham Books, London.

Pigott, C.D. (1975) Natural history. In *Peak District. National Park Guide No. 3*, 2nd edn, P. Monkhouse, (ed.). The Stationery Office Ltd, London, pp. 13–20.

Potts, G.R. (1967) Urban starling roosts in the British Isles. *Bird Study*, 14: 25–42.

Rees, P.A. (2000) Is there a legal obligation to reintroduce animal species into their former habitats? *Oryx*, 35(3): 216–223.

Rees, P.A. (2001) The emergence of a legal duty to restore wildlife and wildlife habitats in England and Wales. *Journal of Practical Ecology and Conservation*, 4(2): 32–38.

Roper, T.J., Tait, A.I. and Christian, S. (1991) Internal structure and contents of three badger (*Meles meles*) setts. *J. Zool. Lond.*, 225: 115–124.

Spencer, K.G. (1966) Some notes on the roosting behaviour of starlings. *Naturalist*, 898: 73–80.

Stace, C. (1997) *New Flora of the British Isles,* 2nd edn. Cambridge University Press, Cambridge.

Thearle, R.J.P. (1968) Urban bird problems. In *The Problems of Birds as Pests,* R.K. Murton and E.N. Wright (eds). Institute of Biology Symposia Number 17. Academic Press, London and New York, pp. 181–197.

Theobald, J. (2001) New powers put to the test. *BBC Wildlife,* 19(6): 37.

5

Habitat and landscape protection under UK law

5.1 INTRODUCTION

A wide range of designations protects the landscapes and individual habitats in the UK. Some are intended to conserve large areas of high landscape value, while others protect the habitats of particular rare species. In many cases, habitat protection alone will not result in conservation, and specific management will be essential.

In 1932 botanists purchased a site of just 290 square metres in western England. It was one of just two locations in England where the adders-tongue spearwort (*Ranunculus ophioglossifolius*) could be found. In 1933 they erected a stout fence around the site, unwittingly excluding the very cattle whose grazing and trampling had maintained the habitat required by the rare plant for centuries. The spearwort went into a rapid decline, and some years later there were plants outside the reserve but none inside. In 1962 the botanists decided that the site needed to be managed and began cutting back the vegetation over part of the reserve. The spearwort quickly reappeared as seed that had remained dormant in the soil for years began to germinate (Noel, 1993).

The way to protect wild species is to protect habitats. Once the habitat is protected the animals and plants will generally look after themselves. But often it is not enough simply to declare that a site is protected. Without careful management habitats change, and this change may mean that some species will be lost and others gained. Recent changes to the law have put increased emphasis on the need for ecological management (see Section 5.4.4).

A number of statutory designations are used in the UK to protect areas of high landscape value. These include National Parks, Areas of

Outstanding Natural Beauty (AONBs) and National Scenic Areas (NSAs). Other, smaller areas are protected by virtue of the presence of rare species or habitats as, for example, Sites of Special Scientific Interest (SSSIs), National Nature Reserves (NNRs), or Local Nature Reserves (LNRs).

5.2   MULTIPLE DESIGNATION OF SITES

The UK system of habitat protection is extremely complex. It includes a diverse array of designations and yet none is capable of completely protecting any particular site from development or damage. In some cases, a site may be notified under several UK designations as well as being protected under European and international law. Outside these systems there exists a large number of locally important sites and protected sites that are privately owned.

Brownsea Island is owned by the National Trust but part of the island is leased as a nature reserve to the Dorset Wildlife Trust. The National Trust owns many SSSIs and also areas within National Parks. Marsden Moor Estate in West Yorkshire is 2300 hectares of unenclosed moorland owned by the Trust. It is an important breeding area for moorland birds and has been designated as a SSSI. It also forms part of a Special Protection Area (SPA) designated under the Wild Birds Directive.

Watergrove is a catchment in the South Pennines owned by North West Water (part of United Utilities). It is managed in partnership with Rochdale Metropolitan Borough Council who jointly fund the Ranger Service. Watergrove reservoir is located within a buffer zone around a SSSI and a SPA, and a nature reserve has been fenced off around the edge of part of the reservoir, to protect breeding birds

The whole of the Ribble Estuary is protected under European law as a Special Protection Area (under the Wild Birds Directive) and its international importance as a wintering site for wildfowl is recognised by its designation as a Ramsar site (under the Convention on Wetlands of International Importance Especially as Waterfowl Habitat). In winter, over 100 000 wigeon (*Anas penelope*) may be present on the estuary, making it the most important site in Britain for this species. Much of the estuary is also protected within the Ribble National Nature Reserve (designated under the Wildlife and Countryside Act 1981) and managed by English Nature. Adjacent to the NNR is Marshside Nature Reserve, which is managed by the RSPB (but included in the SPA and the Ramsar site) and leased from Sefton Metropolitan Borough Council.

The Ministry of Defence owns or leases many sites which have a high conservation value, including SSSIs (over 250), AONBs, NNRs, Ramsar sites, SPAs and SACs. Some MOD land is within National Parks (Panel 5.1).

PANEL 5.1   BIRD CONSERVATION ON THE MINISTRY OF DEFENCE ESTATE

The Ministry of Defence is the second largest landowner in the UK. The Salisbury Plain Training Area (SPTA) covers an area of 40 000 hectares and is the largest and most heavily used military training area in the UK. Approximately half of the Plain is protected as a:

- Site of Special Scientific Interest (SSSI)
- Special Protection Area (SPA), or
- Candidate Special Area of Conservation (cSAC)

A survey conducted by the RSPB and the Defence Estate on the SPTA identified an estimated 48 000 pairs or territories of breeding birds.

Species	Breeding pairs or territories	Percentage of national population
Whinchat	586	2.09–4.19
Grasshopper warbler	264	2.51
Corn bunting	391	1.70–2.44
Stonechat	223	1.01–2.62
Skylark	14 612	1.46
Whitethroat	4 008	0.61
Meadow pipit	8 869	0.47
Linnet	2 294	0.44
Grey partridge	261	0.17–0.19
Reed bunting	267	0.08–0.23

Source: Wynne, G. (2001) working together for wildlife. The partnership between the MOD and the RSPB. *Sanctuary*, 30: 66–69.

5.3   LANDSCAPE DESIGNATIONS

5.3.1   National Parks

National Parks were first established in Britain by the National Parks and Access to the Countryside Act 1949. This legislation set up a

National Parks Commission that was later replaced by the Countryside Commission, and has since been absorbed into the Countryside Council for Wales and Scottish Natural Heritage, and more recently, the Countryside Agency in England. The original purpose of National Parks was defined in s. 5 of the Act but this has since been amended by s. 61 of the Environment Act 1995.

National Parks and Access to the Countryside Act 1949 s. 5

(1) The provisions of this Part of this Act* shall have effect for the purpose –
   (a) of conserving and enhancing the natural beauty, wildlife and cultural heritage of the areas specified ... ; and
   (b) of promoting opportunities for the understanding and enjoyment of the special qualities of those areas by the public.

*This relates to the establishment of National Parks.

There are currently seven National Parks in England and three in Wales, each administered by a National Park Authority. In England the National Parks are in:

- Dartmoor
- Exmoor
- Lake District
- Northumberland
- North York Moors
- Peak District
- Yorkshire Dales

In addition, the Broads have equivalent status and are administered by the Broads Authority. The Countryside Agency is in the process of designating two new National Parks: the New Forest and the South Downs (East Sussex).

The National Parks in Wales are:

- Brecon Beacons
- Pembrokeshire Coast
- Snowdonia

Loch Lomond and the Trossachs is Scotland's first National Park and the Cairngorms will become the second in 2003 (National Parks (Scotland) Act 2000). There are no National Parks in Northern Ireland.

Section 11A(2) of the 1949 Act requires all public bodies and statutory undertakers to have regard to the new purposes of National Parks when exercising or performing any of their functions. It also requires that purpose (a) is given greater weight than purpose (b) in s. 5(1) where any conflict arises between these two purposes.

Section 67 of the Environment Act 1995 established the National Park Authority as the sole planning authority for areas within any National Park. However, the district council retains concurrent jurisdiction in relation to tree preservation orders (see Chapter 6).

Under s. 43 WCA 1981 (as amended by the WCA (Amendment) Act 1985), every National Park Authority is under a duty to prepare a map of any area of mountain, moor, heath, woodland, down, cliff or foreshore within the National Park whose natural beauty it is important to conserve. These maps will provide a means of monitoring changes to the landscape.

5.3.2   Areas of Outstanding Natural Beauty (AONBs), National Scenic Areas (NSAs) and Natural Heritage Areas (NHAs)

Areas of Outstanding Natural Beauty were originally established under the National Parks and the Countryside Act 1949, along with National Parks. A total of 50 AONBs have been designated: 37 in England and 5 in Wales (with the Wye Valley AONB extending across the border) and 9 in Northern Ireland (Tables 5.1 and 5.2).

AONBs are designated by the Countryside Agency in England and the Countryside Council for Wales, in Wales (Countryside and Rights of Way Act 2000 s. 82). The CRoW 2000 has improved protection by introducing a requirement for management plans for these areas that must be incorporated into local authority development plans. The Act also gives the Secretary of State (with respect to areas in England) and the National Assembly for Wales (with respect to areas in Wales) a power to establish conservation boards to promote the conservation and management of AONBs (s. 86) and imposes obligations on planning authorities to consider the effects of planning applications upon these areas (s. 84(4)).

Within two years after a conservation board has been established the board must produce a management plan. In the exercise of its functions the conservation board must have regard to:

- the conservation and enhancement of the natural beauty of the AONB, and
- increasing the understanding and enjoyment by the public of the special qualities of the AONB.

If there is a conflict between these two purposes, greater weight must be given to the first.

Table 5.1 Areas of Outstanding Natural Beauty (England and Wales)

England		England	
1	Arnside and Silverdale	23	Norfolk Coast
2	Blackdown Hills	24	North Devon
3	Cannock Chase	25	North Pennines
4	Chichester Harbour	26	Northumberland Coast
5	Chilterns	27	North Wessex Downs
6	Cornwall	28	Quantock Hills
7	Cotswolds	29	Shropshire Hills
8	Cranborne Chase and West Wiltshire Downs	30	Solway Coast
		31	South Devon
9	Dedham Vale	32	South Hampshire Coast
10	Dorset	33	Suffolk Coast and Heaths
11	East Devon	34	Surrey Hills
12	East Hampshire	35	Sussex Downs
13	Forest of Bowland	36	Tamar Valley
14	High Weald	37	Wye Valley (England and Wales)
15	Howardian Hills		
16	Isle of Wight	*Wales*	
17	Isles of Scilly	37	Wye Valley (England and Wales)
18	Kent Downs	38	Anglesey
19	Lincolnshire Wolds	39	Clwydian Range
20	Malvern Hills	40	Gower
21	Mendip Hills	41	Lleyn
22	Nidderdale		

Table 5.2 Areas of Outstanding Natural Beauty (Northern Ireland)

1	Antrim Coast and Glens
2	Causeway Coast
3	Lagan Valley
4	Lecale Coast
5	Mourne
6	North Derry
7	Ring of Gullion
8	Sperrin, Co. Tyrone
9	Strangford Lough

National Scenic Areas (NSAs) are the Scottish equivalent of AONBs and are designated by Scottish Natural Heritage. They are nationally important areas of outstanding natural beauty and were originally designated by the Countryside Council for Scotland under town and country planning legislation in 1980. There are 39 National Scenic Areas (Table 5.3).

Table 5.3 National Scenic Areas (Scotland)

1	Assynt-Coigach	21	Loch Rannoch and Glen Lyon
2	Ben Nevis and Glen Coe	22	Loch Tummel
3	Cairngorm Mountains	23	Lynn of Lorn
4	Cuillin Hills	24	Morar, Moidart and Ardnamurchan
5	Deeside and Lochnagar	25	North-West Sutherland
6	Dornoch Firth	26	Nith Estuary
7	East Stewartry Coast	27	North Arran
8	Eildon and Leaderfoot	28	River Earn
9	Fleet Valley	29	River Tay
10	Glen Affric	30	St Kilda
11	Glen Strathfarrar	31	Scarba, Lunga and the Garvellachs
12	Hoy and West Mainland	32	Shetland
13	Jura	33	Small Isles
14	Kintail	34	South Lewis, Harris and North Uist
15	Knapdale	35	South Uist Machair
16	Knoydart	36	The Trossachs
17	Kyle of Tongue	37	Trotternish
18	Kyles of Bute	38	Upper Tweeddale
19	Loch Na Keal, Mull	39	Wester Ross
20	Loch Lomond		

The Natural Heritage (Scotland) Act 1991 gave Scottish Natural Heriatge a power to designate Natural Heritage Areas (NHAs). These are intended to be large, discrete areas of outstanding natural heritage value, containing a wide range of landscape and nature conservation interests. Integrated management will be encouraged in NHAs taking account of recreational use of the land and socio-economic activities. However, to date none has been designated.

5.3.3   Heritage Coasts and coastal erosion

The coastline is threatened by a wide range of problems including, urban development, leisure facilities, changing agricultural practices, industrial use and the extraction of minerals. Some sections of the coastline fall within protected areas while others are owned and managed by the National Trust and other organisations. The National Trust protects nearly 600 miles of coastline in England, Wales and Northern Ireland, some 19 per cent of the total. It protects Robin Hood's Bay on the North Yorkshire coast, sections of the White Cliffs of Dover, more than 40 per cent of the Cornish coast, including Lizard Point, and over 138 miles of the Welsh coast.

Heritage Coasts (Table 5.4) are designated by agreement between local authorities and the Countryside Agency (in England) or the Countryside Council for Wales (in Wales) for the purpose of conserving undeveloped coasts for public enjoyment, but they have no formal

Table 5.4 Heritage Coasts

	Location	Part of the AONB at...
1	Dover–Folkestone	
2	East Devon	
3	Exmoor	
4	Flamborough Headland	
5	Godrevy–Portreath	
6	Gribbin Head–Polperro	South Devon
7	Hamstead	
8	Hartland (Cornwall only)	
9	Hartland (Devon)	North Devon
10	Isles of Scilly	Isles of Scilly
11	Lundy	
12	North Devon	North Devon
13	North Norfolk	Norfolk Coast
14	North Northumberland	
15	North Yorkshire and Cleveland	
16	Pentire Point–Widemouth	
17	Penwith	
18	Purbeck	
19	Rame Head	South Devon
20	St Agnes	
21	St Bees Head	
22	South Devon	South Devon
23	South Foreland	
24	Spurn	
25	Suffolk	
26	Sussex Downs	Sussex Downs
27	Tennyson	
28	The Lizard	
29	The Roseland	South Devon
30	Trevose Head	
31	West Dorset	

statutory basis. There are around 45 Heritage Coasts in England and Wales, covering more than 1500 km of coastline. Protection is afforded through policies in local and structure plans. However, significant stretches of Heritage Coast lie within National Parks and AONBs, and most contain at least one SSSI. Planning Policy Guidance Note No. 20 (Coastal Planning) establishes a number of restraint policies.

Attempts at preventing coastal erosion have generally been unsuccessful and there have been a number of significant cliff falls in recent years. The cliff-top footpath at Robin Hood's Bay has been diverted inland as a result of loss of the original route to erosion and a cliff-top hotel in Scarborough was lost to the sea a few years ago. In 1993 the Holbeck Hall Hotel fell 50 metres into the sea as a result of a landslip. The owners won £2 million in compensation from Scarborough

Council in 1997, but this decision was reversed by the Appeal Court, which ruled that the Council had not breached its duty to maintain the undercliff (*Holbeck Hall Hotel* v. *Scarborough Borough Council* (2000)).

In 1999, a large section of the cliff at Beachy Head crashed into the sea, effectively joining the unmanned Beachy Head lighthouse to the Sussex coast by many thousands of tonnes of chalk (Anon., 1999).

In many areas the government has acknowledged the futility of trying to prevent losses of the coastline using engineering solutions and has a general policy of 'managed retreat'. However, at some seaside resorts, such as Blackpool in Lancashire where the sea regularly floods the promenade and coast roads, substantial new sea defences have been constructed (Figure 5.1).

5.3.4   Natural Areas

English Nature has developed a 'Natural Areas' approach to conservation as a key component of its *Strategy for the 1990s*. Some 120 Natural Areas have been defined, each identified by its local distinctiveness, based on its characteristic wildlife and natural features, for example,

Figure 5.1 The new coastal defences at Blackpool, Lancashire.

Dartmoor, the Lancashire Plains and Valleys, and the Urban Mersey Basin. The boundaries of theses areas are based on the distribution of wildlife and natural features, and on land-use pattern and human history. Such a system offers a better framework for planning and nature conservation than does one based on administrative boundaries, and it recognises that not all wildlife is restricted to designated and protected sites.

Natural Areas are not nature conservation designations, but for each area English Nature has produced a *Natural Area Profile* which contains details of its boundaries, the key species and wildlife habitats which require conservation attention, and important geological and landform features.

The Urban Mersey Basin Natural Area

The Urban Mersey Basin Natural Area extends from Southport, Liverpool and the Wirral peninsula across parts of Cheshire and Lancashire to Rochdale, Glossop and Stockport, including Warrington, Wigan, Bolton and Greater Manchester (Hill and Fleming, 1999). This Natural Area encompasses some of the most densely populated urban areas in Britain and includes an area of approximately 2000 square kilometres; its character and landscape are largely a result of industrial activity in the area, particularly coal mining.

The habitats which are most characteristic of the Urban Mersey Basin Natural Area are:

- peat mosslands;
- wooded cloughs;
- wood-pasture;
- hay meadows; and
- field ponds.

Other habitat types owe their existence to the industrial past of the region and include:

- reservoirs;
- canals;
- mining subsidence 'flashes'; and
- industrial spoil.

The Natural Area includes 30 SSSIs (8 of which protect important geological features), part of the Ribble Marshes NNR, and 18 Local Nature Reserves (designated under the National Parks and Access to the Countryside Act 1949). In addition there are some 750 non-statutory sites which have been identified by local authorities as sites of importance to nature conservation (the precise designation varying from one authority to another).

5.3.5   National Trails

The Countryside Agency has designated 13 long distance routes as National Trails (Table 5.5). The first of these was the 412 km Pennine Way which was opened in 1965. The latest is the Thames Path which is 288 km long and follows the River Thames from its source in Gloucestershire to the Thames Barrier in London.

Table 5.5 National Trails

1	Cleveland Way
2	North Downs Way
3	Offa's Dyke Path
4	Peddars Way/Norfolk Coast Path
5	Pennine Way
6	Ridgeway
7	South Downs Way
8	South West Coast Path
9	Thames Path
10	Wolds Way
11	Cotswold Way
12	Hadrian's Wall Path
13	Pennine Bridleway

5.4   SITES OF SPECIAL SCIENTIFIC INTEREST

5.4.1   Introduction

In 1945 the Wildlife Conservation Special Committee chaired by the eminent zoologist J. S. Huxley was asked to examine the efficacy of legislation on wildlife in England and Wales. During a visit to the Suffolk coast in 1946 several members of the Committee discovered plans to excavate more shingle from Benacre Ness and to transform a naval camp at Covehithe Cliffs into a holiday resort. In order to draw attention to the scientific value of these sites the Huxley Committee (Huxley, 1947) recommended that they should be scheduled as Sites of Special Scientific Interest and that lists of such places should be sent to the then Ministry of Town and Country Planning, the local planning authorities and landowners and made available for public inspection. Although the Huxley Committee's proposal was adopted in principle and incorporated into the legislation of the time, it was over thirty years before the legislation required the degree of public awareness of the existence of the designated sites that the Committee had originally envisaged and which was essential to (but did not guarantee) their survival.

Sites of Special Scientific Interest (SSSIs) are areas of special interest by reason of any of their flora, fauna, or geological or physiographical features. SSSIs were first defined under section 23 of the National Parks and Access to the Countryside Act 1949. Under this Act such sites were designated under the direction of the Natural Environment Research Council (NERC). All that was required was that NERC inform the relevant planning authority. This resulted in many landowners not being aware of the designation and, since they were outside of any planning control, agricultural and forestry operations could take place. The equivalent designation in Northern Ireland is Area of Special Scientific Interest (ASSI).

The Wildlife and Countryside Act 1981 (s. 28) charged the Nature Conservancy Council with a duty to select and designate SSSIs and to notify their existence to the relevant local planning authority, every owner and occupier of the land and the Secretary of State for the Environment. Objection could only be made when the scientific argument for the designation was unjustified.

The SSSI is the most important conservation designation of all because, as well as being a site designation in its own right, it is also used to protect other sites which are designated under national, European and international legislation including:

- National Nature Reserves
- Special Protection Areas (Wild Birds Directive)
- Candidate Special Protection Areas (Habitats Directive)
- Ramsar sites
- Biogenetic Reserves

Many SSSIs protect places of great natural beauty, but this is not always the case. At Nob End, in Bolton, there is a series of six locks (part of the Manchester, Bolton and Bury Canal), designed to raise barges 25 metres. Over a period of many years waste material from a chemical works was tipped on the site to a depth of 10 metres in places. The waste has weathered and oxidised over the years, producing a chalky layer that has been colonised by a unique biological community including nine orchid species. Nob End was granted SSSI status in 1988.

5.4.2   Definition, selection and notification of SSSIs

Sites of Special Scientific Interest are selected because of their scientific value, and not to enhance amenity or for recreation purposes. The system was created to establish a national network of representative samples of British habitats with the aim of maintaining biodiversity. There are over 6400 SSSIs and around 160 ASSIs currently designated.

The provisions relating to SSSIs in the WCA 1981 apply to England, Wales and Scotland. The amendments made by the Countryside and

---

Wildlife and Countryside Act s. 28

---

(1) Where the Nature Conservancy Council are of the opinion that any area of land is of special interest by reason of any of its flora, fauna, or geological or physiographical features, it shall be the duty of the Council to notify that fact –
 (a) to the local planning authority in whose area the land is situated;
 (b) to every owner and occupier of any of that land; and
 (c) to the Secretary of State.

Rights of Way Act 2000 (Schedule 9) apply to England and Wales as SSSIs are now a devolved matter. Schedule 9 replaced s. 28 of the Wildlife and Countryside Act 1981.

In addition the NCC must publish a notification in at least one local newspaper in the area in which the land is situated (s. 28(2)).

The notification shall specify:

- the nature of the special interest (s. 28(4)(a));
- any operations that the NCC believe are 'likely' to damage the special interest (s. 28(4)(b));
- the manner and time period within which representations or objections may be made to the NCC (s. 28(3));
- a statement of the NCC's views about the management of the land, including their views regarding the conservation and enhancement of the flora, fauna or other special features.

The time period for accepting objections to a notification is not less than three months (s. 28(3)) from the date of the giving of the notification. Within a period of nine months from the date of notification of the site to the Secretary of State, the NCC may:

- confirm the notification, with or without modifications (s. 28(5));
- withdraw the notification.

The purpose of this delay is to give the NCC an opportunity to enter into a management agreement with the owner or occupier of the land. If a notification is neither confirmed nor withdrawn by the end of the nine-month period it ceases to have effect (s. 28(6)(b)).

A notification under subsection 1(b) of land in England and Wales is a local land charge.

Section 28A makes provision for the conservation agencies to vary the matters specified in the notification at any time after its confirmation, apart from the area of land concerned. Further amendments make provision for the notification of additional land adjacent to an existing

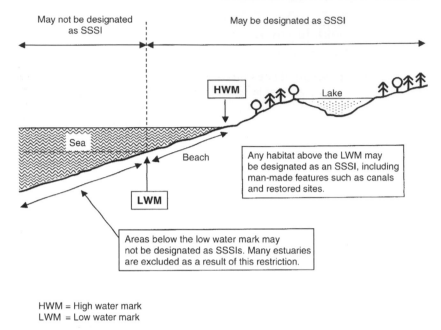

HWM = High water mark
LWM = Low water mark

Figure 5.2 Areas where SSSIs may and may not be designated.

SSSI (s. 28B) and the enlargement of a SSSI (s. 28C) where the NCC considers there to be a special interest. The notification process is similar to that for a new SSSI, but the additional notification may not be given until the original SSSI is confirmed by a notice issued under s. 28(5)(b). Provision is also made for the denotification of all or part of a site where the special interest has been lost (s. 28D). The persons whom the NCC must notify are:

■ the relevant local planning authority;
■ every owner and occupier of any of that land;
■ the Secretary of State;
■ the Environment Agency;
■ every relevant undertaker whose works, operations or activities may affect the land (s. 28D(2)).

The NCC must also publish a notification in at least one local newspaper in the area in which the land is situated (s. 28D(3)).
   Within a period of nine months the NCC must either:

■ withdraw the notification, or
■ confirm the notification (in relation to the original area or a smaller area)(s. 28D(5)).

If it does neither, the notification has no effect, so the SSSI remains.

The owner or occupier of a SSSI must not carry out (or cause or permit to be carried out) any of the potentially damaging operations listed in the s. 28 notification unless:

- one of them has, after service of the notification, given the NCC notice of a proposal to carry out the operation, and
- obtained written consent from the NCC, or
- the operation is carried out in accordance with the terms of an agreement with the NCC under s. 16 of the National Parks and Access to the Countryside Act 1949 or s. 15 of the Countryside Act 1968, or
- the operation is carried out in accordance with a management scheme under s. 28J or a management notice under s. 28K of the Wildlife and Countryside Act 1981 (s. 28E).

A consent given by the NCC may be subject to conditions and for a limited period (s. 28E(4)) but the NCC must give reasons. It must also give reasons if consent is refused. If consent is given it may later be modified or withdrawn. Provision is made for appeals against refusal of consent, withdrawal or modification of a consent (s. 28F). Section 28E does not apply where the owner or occupier is a public body (defined by s. 28G) while carrying out its functions as s. 28H will apply (see below).

Previously, under WCA 1981, if a period of four months had lapsed since the NCC was notified of the proposed operations and the NCC had not responded, the owner or occupier of the land was able to proceed with damaging operations without committing an offence, by default (s. 28(6)). This loophole has now been closed.

Failure by an owner or occupier to follow the procedures laid down in s. 28E is an offence under s. 28P(1). However it is a defence if:

- the operation is authorised by a planning permission;
- the operation needs planning permission and the permission of a public body (defined by s. 28G) and has both;
- the operation is carried out in an emergency.

5.4.3 Public bodies

Public bodies defined under s. 28G are:

- Ministers of the Crown and government departments;
- the National Assembly for Wales;
- local authorities;
- statutory undertakers;
- other public bodies.

These bodies have a duty to take reasonable steps, consistent with the proper exercise of their functions, to further the conservation and enhancement of the special features by virtue of which a SSSI has been designated (s. 28G(2)). This duty applies where the public body is exercising its statutory functions on a SSSI or on land outside the SSSI where those functions affect a SSSI.

A public body is required to notify the conservation agency when it proposes to carry out operations in the exercise of its function, which are likely to damage the features of special interest of a SSSI (s. 28H), including works outside a SSSI. The agency may assent to the operation (with or without conditions) or it may refuse its assent. If the agency does not send a notice within 28 days it shall be treated as having declined to assent. Where assent is refused or the conditions imposed by the conservation agency are unacceptable to the public body, the public body may proceed with the works provided that:

- it gives the agency not less than 28 days notice of the start of the operation;
- the notice states how it has taken into account advice given by the agency;
- the operation is carried out in such a way as to do as little damage as possible to the special features of the SSSI (taking into account any advice received from the conservation agency);
- the site is restored to its former condition, so far as is reasonably practicable, if damage is done (s. 28H(5)).

Where a public body has the power to grant permissions, authorisations or consents for others to carry out operations (on or outside a SSSI) which may damage the special features of a SSSI, it must give the conservation agency not less than 28 days' notice before granting permission (s. 28I). The authority must take into account the advice of the conservation agency before making a decision, including the need for any conditions. If the authority intends to grant permission against the advice of the agency it must notify the agency and allow 21 days before operations may commence. This gives the agency time to discuss mitigation measures with the applicant or to offer a management agreement.

5.4.4   Management schemes

The CRoW 2000 has introduced the concept of a management scheme to the provisions for protecting SSSIs. It is designed to provide a detailed statement of the measures required for the positive management of the SSSI. Originally the WCA 1981 did not require any positive measures to be taken to conserve SSSIs and consequently the status of many declined owing to neglect. The management scheme should provide more detail than the statement of views about the management of the site which is served with the notification of the SSSI under s. 28(4).

The conservation agency may formulate a management scheme for all or part of a SSSI under s. 28J(1). The scheme may be for:

- conserving the flora, fauna, or geological or physiographical features, or
- restoring them, or
- both.

The agency must consult with every owner and occupier of the land before serving them with a notice of a proposed scheme (s. 28J(3)), along with a copy of the proposed scheme (s. 28J(6)). Representations or objections to the scheme may be made to the agency during a period specified in the notice, which must not be less than three months (s. 28J(7)).

Within a period of nine months the agency may either:

- withdraw the notice, or
- confirm the management scheme (with or without modifications) (s. 28J(8)).

If, within the nine-month period, the conservation agency does neither of these, the notice ceases to have effect (s. 28J(9)(b)). The agency may cancel or propose a modification to a management scheme at any time (s. 28J(11)).

Where a conservation agency has formulated a management scheme which is not being implemented the agency may serve a management notice (s. 28K) requiring an owner or occupier to carry out work on the land or to do other things. The agency may make payments to any owner or occupier of land in relation to a management scheme (s. 28M(2)). However, if the work required by a management notice is not done, the agency may carry out the work and recover the costs. Persons who have been served with a management notice may appeal to the Secretary of State or the National Assembly for Wales, as appropriate (s. 29L).

5.4.5   Management agreements and compulsory purchase

The NCC was given a power, under the Countryside Act 1968 (s. 15), to enter into management agreements with owners and occupiers of SSSIs. This power was extended to enable agreements with owners and occupiers of adjoining land by the Environmental Protection Act 1990 (Schedule 9). Management agreements are contracts under which the NCC pays owners and occupiers to manage the land in the interests of nature conservation. In many cases this approach effectively compensated people for doing nothing, as opposed to requiring them to engage in positive management. In 1996, English Nature introduced its Wildlife Enhancement Scheme which encourages positive management of SSSIs and makes provision for standard payments for various conservation activities. The previous compensatory approach required each agreement to be negotiated on an individual basis.

Section 75(3) of the CRoW extends the agencies' powers so that management agreements may relate to other land which is not necessarily within or adjacent to a SSSI. This provision could be used, for example, to protect a water supply that is some distance from the SSSI.

Section 75(4) amends the Countryside Act 1968 by inserting a new section 15A giving the conservation agencies a power to compulsorily purchase the land referred to in s. 75(3) of the CRoW if they cannot reach an agreement for its conservation or if such an agreement has been broken. Having purchased the land the agencies can manage it themselves or dispose of it to someone else who will conserve the features of the SSSI.

The WCA 1981 also made provision for management agreements but courts were only given powers to make a restoration order (s. 31) in relation to damage done to a site which was subject to a nature conservation order (s. 29). Only 40 such orders have ever been made (Bell and McGillivray, 2000). It remains to be seen whether the new provisions contained in s. 28J will be actively used to restore SSSIs.

The conservation agencies are given a compulsory purchase power in relation to land notified as a SSSI (s. 28N). This new power may only be exercised where a management agreement cannot be concluded or where the terms of such an agreement have been breached. This power is in addition to that in s. 17 of the National Parks and Access to the Countryside Act 1949 which relates to nature reserves, and the power added to the Countryside Act by s. 75(3) of the CRoW.

5.4.6   Offences

Section 28P creates a number of offences in relation to SSSIs. It is an offence for an owner or occupier to carry out, cause or permit to be carried out any damaging operations.

Wildlife and Countryside Act 1981 s. 28P

(1) A person who, without reasonable excuse, contravenes section 28E(1)* is guilty of an offence and is liable on summary conviction to a fine not exceeding £20,000 or on conviction on indictment to a fine.

*It is an offence for an owner or occupier to cause or permit damaging operations on a SSSI.

A public body commits an offence if it carries out operations in contravention of s. 28H, without reasonable excuse. The penalty is the same as imposed under s. 28P.

It is a reasonable excuse (s. 28P(4)) for a person to carry out an operation (or to fail to comply with a requirement to send a notice about it) if:

■ the operation was authorised by planning permission;
■ the operation was necessary in an emergency (and details of the operation were notified to the NCC as soon as practicable after the commencement of the operation);
■ a permission or consent has been granted in accordance with s. 28I by a s. 28G authority (public body).

If an operation needs both planning permission and the permission of a s. 28G authority, both must be obtained in order to provide a reasonable excuse (s. 28P(5)).

Wildlife and Countryside Act 1981 s. 28P

(6) A person (other than a section 28G authority acting in the exercise of its functions) who without reasonable excuse –
   (a) intentionally or recklessly destroys or damages any of the flora, fauna, or geological or physiographical features by reason of which land is of special interest, or intentionally or recklessly disturbs any of those fauna, and
   (b) knew that what he destroyed, damaged or disturbed was within a site of special scientific interest,
   is guilty of an offence and is liable on summary conviction to a fine not exceeding £20,000 or on conviction on indictment to a fine.

Section 28P(6) makes a significant improvement to the protection of SSSIs by creating offences of intentionally or recklessly damaging a designated site or disturbing fauna on such a site. These offences may be committed by any person who knows the site is a SSSI (not just owners and occupiers), but do not, however, apply to public bodies. Subsection (7) provides for a similar defence of reasonable excuse as that provided by subsections 28P(4) and (5).

A person who without reasonable excuse fails to comply with a requirement of a management notice is guilty of an offence under s. 28P(8). Proceeding in England and Wales for an offence under s. 28P may only be taken by the conservation agency unless the Director of Public Prosecutions consents otherwise (s. 28P(10)).

If an owner of a SSSI disposes of any interest in the land or is aware of any change of occupation of the land, he is required to so advise the conservation agency within 28 days (s. 28Q). Failure to comply with this requirement is an offence.

Paragraph 3 of Schedule 9 of CRoW amends s. 31 of the Wildlife and Countryside Act 1981 and enables courts to order the restoration of a SSSI where a person has been convicted of damaging or destroying it under subsection 28P(1) or (6). The court may also order a public body to restore a SSSI where it has been convicted of an offence under subsection 28P(2) or (3). This applies whether the damaging operation took place on or off the land designated as a SSSI.

Section 28R gives the conservation agencies a power to make byelaws for the protection of a SSSI.

5.4.7   Entry onto land

Section 80 of the CRoW gives the conservation agencies powers of entry onto land (not limited to the actual or potential SSSI) for a range of purposes including:

- assessing whether land should be notified as a SSSI;
- formulating a scheme for the management of a SSSI;
- assessing the condition of the features of the site;
- ascertaining whether certain offences have been committed.

Entry to the land may be by boat or vehicle, and equipment or materials may be taken onto the land. The agency is liable to pay compensation for any damage caused by exercise of the power of entry. This power will be useful, for example, for monitoring the breeding success of rare birds.

5.4.8   Case law

Damaging operations

In relation to SSSIs, 'potentially damaging operations' were held to include the following wide range of activities in *Sweet* v. *Secretary of State and Nature Conservancy Council* [1989], a case brought before the changes made by CRoW 2000:

- cultivation (including ploughing, rotavation, harrowing and reseeding);
- mowing and other methods of cutting vegetation;
- application of manure, fertilisers and lime;
- burning;
- release of any wild, feral or domestic animal, reptile, amphibian, bird, fish or invertebrate;

- release of any plant or seed;
- storage of materials;
- use of materials;
- use of vehicles or craft likely to damage or disturb features of interest.

The definition also covers drainage, building operations and the application of pesticides, but does not cover neglect.

Who is an 'occupier'?

The issue of who is an 'occupier' was addressed by a number of cases under the original legislation. It is important because persons who are not owners or occupiers are not required to be notified of SSSIs by the NCC. Furthermore, they could not commit an offence under WCA 1981 s. 28 or enter into a management agreement.

The House of Lords decided, in *Southern Water Authority* v. *Nature Conservancy Council* [1992], that an occupier (for the purposes of s. 28 prior to amendment by CRoW 2000) must have some form of stable relationship with the land. In this case, a water authority did not commit a s. 28 offence when carrying out drainage operations while temporarily on a SSSI, even though it was knowingly carrying out potentially damaging operations.

Under s. 52(2C) of the WCA (as amended by CRoW 2000) the definition of 'occupier' includes commoners as defined by s. 22 of the

Figure 5.3 A retail and industrial park has been built on part of the Red Moss SSSI in Horwich near Bolton.

Commons Registration Act 1965, in relation to common land in England and Wales. Furthermore, the amendments made by CRoW include new offences which may be committed by public authorities and others which may be committed by anyone (see Section 5.4.5).

Political interference and judicial review

In cases where SSSI designations are overturned by the granting of planning permission and political interference is suspected an interested person or party, who could show *locus standi*, may seek judicial review. In April 1989 Poole Borough Council granted itself planning permission for a housing development on land forming Canford Heath (*R v. Poole Borough Council ex parte Beebee* [1991]). The land was included in a SSSI. Two representatives of the British Herpetological Society (BHS) and two representatives of the World Wide Fund for Nature (UK) applied for judicial review. Counsel for the respondents argued that since the NCC (the body set up by Parliament to protect the interest of SSSIs) had not applied for leave to move for judicial review the court should be slow to find that others had sufficient interest. Mr Justice Schiemann accepted that the BHS members had *locus standi* by virtue of their long-standing interest in the conservation of rare reptiles on the site but found that the WWF alone did not.

On 13 March 1991, Michael Heseltine, the then Environment Secretary, announced that he was proposing to revoke the deemed planning permission and called in two further applications for development on the edge of the heath. The development did not go ahead and shortly after these events additional safeguards for SSSIs were proposed in a consultation paper entitled *Planning Controls Over Sites of Special Scientific Interest* (1992). A recent judicial review of a SSSI designation citing the Human Rights Act 1998 was unsuccessful (see Chapter 10).

The role of the public in protecting SSSIs
Murdie (1993) suggests that the public have an important role to play in collecting evidence of damage to SSSIs, but stresses the importance the courts could attach to the method by which such evidence is obtained. He cautions against trespass and the use of unlawful methods and emphasises the need for accurate observation, and photographic or film evidence. Detailed information on any companies involved in damage would be particularly useful since s. 69 of the 1981 Act makes any 'director, manager, secretary or other similar officer of the body corporate' liable to prosecution as well as the corporation itself. However, it must be remembered that members of the public cannot bring actions in the courts in relation to s. 28P offences without the permission of the Director of Public Prosecutions.

5.4.9   Case study

Memorandum of Understanding between English Nature and Hanson Quarry Products

In February 2000 English Nature signed a Memorandum of Understanding with the international building materials business Hanson Quarry Products. Hanson operates 300 sites in the UK and is responsible for 85 SSSIs. Mineral extraction can benefit nature conservation by exposing important geological sequences and the recolonisation of disused quarries. The company is committed to achieving 'favorable conservation status' on all Hanson-controlled SSSIs through agreed site management plans. These sites include active quarries, disused quarries and land managed around extraction sites. Hanson has also agreed to make no new applications on land currently designated as a SSSI unless an overriding national need has been identified or it can demonstrate that the special interest can be sustained during extraction. The company also intends to incorporate national biodiversity targets and English Nature's natural area targets in all of its restoration and after-use schemes.

Over 700 SSSIs are related to mineral extraction because they are active quarries, disused quarries, or land managed by mineral companies. Poorly planned and inappropriate extraction can cause damage to nature conservation sites through loss of land, wildlife disturbance and hydrological impacts. Well-planned extraction, restoration and aftercare can prevent this damage.

Figure 5.4 The Ribble Estuary National Nature Reserve.

5.5 NATIONAL NATURE RESERVES

National Nature Reserves (NNRs) were established by the National Parks and Access to the Countryside Act 1949. They are areas of national, and in some cases, international importance, and are designated as SSSIs. These are currently over 370 NNRs designated in the UK.

The NCC has the power to designate and manage any area as a NNR. The NCC controls the site by:

- buying the land;
- leasing the land;
- entering into a nature reserve agreement under s. 16 of the 1949 Act.

If it is unable to conclude a satisfactory management agreement with the owner the NCC has the power to seek a compulsory purchase order (s. 17). In addition to the statutory restriction imposed by the designation of the site as a SSSI and the conditions imposed by any nature reserve agreement, the NCC may also make byelaws to protect the site (s. 20).

5.5.1 Case study

Arsonists and ancient heathland

Heathland is a rare and declining habitat in England and some 90 per cent of the heathland that existed in the eighteenth century has already disappeared. Heathland is managed by controlled burning, called swaling, which encourages regrowth. But blanket burning kills plant roots and seeds, causing so much damage that the habitat can take ten years to recover.

During March and April 2000 several dozen small fires damaged more than 1457 ha (3600 acres) of moor and heathland from north Devon to the southernmost tip of Cornwall. Such fires fragment the habitat and make it difficult for some species, particularly birds such as stonechats and linnets, to survive (Rowe, 2000)

One of the worst affected heathlands was Goss Moor in Cornwall, a SSSI and NNR, where firefighters took six hours to extinguish a blaze that destroyed 51 ha (125 acres). At the time of the fire the moor was a candidate Special Area of Conservation (cSAC). The heathlands alongside of the A30 were particularly badly damaged and as a result rare species such as the Dartford warbler (*Sylvia undata*) and the silver-studded blue butterfly (*Plebejus argus*) may become locally extinct. The police and English Nature believed that arsonists started the more serious fires. One theory suggests that the destruction of heathland has been orchestrated to prevent the area receiving further legal protection. If the European Commission designates this area as a SAC, plans to develop the A30, a notorious bottleneck during the tourist season, will be jeopardised and impoverished areas of Cornwall will be deprived of much needed investment.

5.6   FORESTS

5.6.1   Forest Nature Reserves

The management of the Forestry Commission's forests is the responsibility of Forest Enterprise (an executive agency of the Commission). Forest Enterprise has created 46 Forest Nature Reserves with the aim of conserving special forms of natural habitat, fauna and flora. The Forestry Commission estate includes a number of SSSIs and NNRs. The Forest Nature Reserves range in size from 50 ha to over 500 ha and include:

- Black Wood of Rannoch
- Cannop Valley Oaklands
- Forest of Dean
- Culbin Forest
- Glen Affric
- Wyre Forest

In Scotland, Forest Enterprise also manages 18 Caledonian Forest Reserves which are intended to protect and expand 16 000 ha of native oak and pine woodlands in the Scottish Highlands.

Figure 5.5 Forestry operations in Mabie Forest, Dumfries.

Forest Nature Reserves in Northern Ireland are designated and administered by the Forest Service, which is an agency of the Department of Agriculture and Rural Development for Northern Ireland. There are 36 Forest Nature Reserves in Northern Ireland, covering an area of around 1800 ha. In addition, there are 15 NNRs on property owned by the Service.

5.6.2   Community forests and the National Forest

The government has established community forests through a partnership between the Countryside Agency, the Forestry Commission and local authorities. Twelve community forests have been developed in England:

- Forest of Avon
- Tees Forest
- Forest of Mercia
- Great North Forest
- Great Western
- Marston Vale
- Greenwood
- Mersey Forest
- Red Rose Forest
- South Yorkshire Forest
- Thames Chase
- Watling Chase

In addition, the National Forest has been established in the Midlands. In 1990 a 200 square mile area was earmarked as the site for a forest of 30 million trees.

The aim of the project is to promote the creation, regeneration and multipurpose use of wooded landscapes around urban areas. The establishment of community forests is facilitated through planning policy guidance. Community forests are non-statutory designations and development plans should ensure that any development proposal within the forests respects the woodland setting. These areas include private property, but access agreements have been made between the Countryside Agency and some landowners.

5.7   LOCAL WILDLIFE SITES

5.7.1   Local Nature Reserves

Local Nature Reserves (LNRs) are places of wildlife or geological interest that are of special local interest. They are set aside as places for wildlife and people, and are usually greater than 2 ha in size.

There are over 600 LNRs in England covering a total area of over 29 000 ha. They include coastal headlands, ancient woodlands, flower meadows, and areas which have returned to nature, including derelict railway lines, abandoned landfill sites and industrial areas.

National Parks and Access to the Countryside Act 1949 s. 15

... 'nature reserve' means land managed for the purpose –
(a) of providing, under suitable conditions and controls, special opportunities for the study of, and research into, matters relating to the fauna and flora of Great Britain and the physical conditions in which they live, and for the study of geological and physiographical features of special interest in the area, or
(b) of preserving flora, fauna or geological or physiographical features of special interest in the area, or for both those purposes.

LNRs are designated under s. 21 of the National Parks and Access to the Countryside Act 1949, by local authorities, or where the local authority has delegated the power, by parish or town councils (Local Government Act 1972, s. 101). In order for a local authority to declare a LNR it must be within the area controlled by the local authority, and the authority must have a legal interest in the land. The local authority may own the land, lease it or have a nature reserve agreement with the owner.

National Parks and Access to the Countryside Act 1949 s. 21

(1) The council of a county or county borough or in Scotland a planning authority shall have power to provide, or secure the provision of, nature reserves on any land in their area (not being held by, or managed in accordance with an agreement entered into with, the Nature Conservancy Council) as to which it appears to the council expedient that it should be managed as a nature reserve.

Figure 5.6 Wilton Quarries – a Local Nature Reserve in Bolton.

When a local authority intends to make a LNR declaration, it should:

■ draw up a formal declaration, including a map showing the reserve boundary;
■ obtain agreement from the relevant local authority committees;
■ publish the declaration in a local newspaper and make copies of the declaration and a map available to the public free of charge;
■ formally notify the statutory nature conservation agency of the LNR declaration.

In Northern Ireland the equivalent designation is Local Authority Nature Reserve (LANR).

5.7.2   Local wildlife sites and the Wildlife Trusts

The Wildlife Trusts is a network of 46 Wildlife Trusts and 52 Urban Wildlife Groups. It manages 2300 nature reserves covering some 60 000 hectares spread across the UK. The smallest is Hethel Old Thorn, the oldest hawthorn in the country, which was planted in Norfolk in the thirteenth century. The largest is Benmore Coigach, a 6000 hectare mountain wilderness in Scotland.

Kent Wildlife Trust has identified about 500 Sites of Nature Conservation Interest (SNCI) in the county, which it believes to be of county importance. Most such sites are recognised by local authorities and are protected under their Local Plans.

The precise title used to identify these non-statutory sites varies with the local authority concerned, for example:

- County Biological Heritage Site (Lancashire Borough of Chorley)
- Site of Biological Importance (City of Salford)
- Site of Biological Interest (Metropolitan Borough of Knowsley)
- Site of Community Wildlife Interest (Metropolitan Borough of St Helens)
- Site of Local Biological Interest (Metropolitan Borough of Sefton)
- Site of Nature Conservation Value (City of Liverpool)

Case law

R v. Sefton Metropolitan Borough Council, ex parte British Association of Shooting & Conservation Ltd (2000)
Section 111 of the Local Government Act 1972 gives a power to any local authority to manage land in its ownership. Sefton MBC was entitled under s. 120(1)(b) of the Act to refuse to renew shooting rights over that land on the grounds that such refusal was for the benefit, improvement or development of its area.

5.7.3   Country parks

The Countryside Act 1968 confers powers on local authorities to provide 'country parks' for public enjoyment of the countryside. Local authorities may acquire land compulsorily or by agreement to create a country park, or they may set up such a park on land owned by others by agreement with them.

Country parks may contain areas which are designated as SSSIs. They are managed by local authorities, and the Environment and Heritage Service, in Northern Ireland. Many country parks are located in the countryside, but some are to be found in urban areas. Moses Gate Country Park in Bolton, Greater Manchester, provides a variety of habitats for wildlife in an urban setting, including Crompton Lodges, the River Croal, forest, ponds, a canal and grassland. Over 100 species of birds have been recorded at the park including cormorant, kingfisher, goosander, sparrowhawk, cuckoo and tawny owl.

Pennington Flash Country Park, Leigh, contains a large area of artificially created water adjacent to an area of coal mine waste. On 8 March 1994 a black-faced bunting (*Emberiza spodocephala*) was discovered at the park. It stayed for five weeks and attracted some 6000 birdwatchers, including parties from Finland and Belgium. This was the first British record of the species, which normally breeds in Siberia.

5.7.4   Privately owned nature reserves

Many conservation organisations such as the RSPB, National Trust, Wildfowl and Wetlands Trust and the Wildlife Trusts own and manage nature reserves. Some large companies also have land that they have set aside as conservation areas. Although some of these areas are designated (e.g. as SSSIs, NNRs or Ramsar sites), many have no statutory protection. To a large extent they rely upon property law for their protection.

The Hope Carr Nature Reserve was created by North West Water Ltd (now United Utilities) on the edge of Leigh. The 15 ha (37 acre) site was originally part of Hope Carr Farm and was bought by the local sewage board in 1898. Fields which were once used to grow silage and others which were used for the disposal of sewage sludge have now been reclaimed. A variety of habitats, dominated by a network of wetland, and including lakes and ponds now support over 100 species of birds.

5.8   PROTECTION OF IMPORTANT GEOLOGICAL SITES

Some important geological sites are protected as SSSIs (see Section 5.4). Locally important sites may be protected within Local Nature Reserves (see Section 5.5) or as Regionally Important Geological Sites. Limestone pavements are particularly threatened and may be given special protection under Limestone Pavement Orders.

Figure 5.7 The Marshside Nature Reserve, a RSPB reserve at Southport.

5.8.1   Limestone pavements

Geology

Limestone is a rock with a high calcium carbonate content (at least 50 per cent). There are many types of limestone, all of which have similar characteristics owing to their similar chemical composition. Most of them are partly or wholly of organic origin and contain the hard parts of organisms, such as coral skeletons and mollusc shells.

As rainwater passes through the atmosphere it absorbs carbon dioxide, forming a weak solution of carbonic acid. When rain falls in limestone areas water seeps into fissures in the rock and slowly dissolves it, forming underground streams and caves. On the flat tops or terraces of bare limestone hills rainwater forms fissures (grikes) between ridges of bare rock (clints). Soil is blown into the fissures enabling plants to take root. The narrowness of the fissures provides excellent shelter for the plants in these characteristically windy places. The resultant patchwork of bare rock surfaces separated by vegetation is called a limestone pavement. These features occur mainly in the Yorkshire Dales (the Craven Pennines) and south Westmorland.

The classic example of a limestone pavement is the surface above Malham Cove in Yorkshire. Much of the limestone scenery in the nearby Ingleborough area, including limestone pavements, is protected within a National Nature Reserve (Figure 5.8).

Figure 5.8 Limestone pavement, at Ingleborough in the Yorkshire Dales, exhibiting the characteristic clint and grike structure.

Limestone pavements and the law

Limestone pavements are a relatively rare geological feature and they have been extensively damaged by people gathering stone for garden rockeries. In Britain, they may be protected using the Wildlife and Countryside Act 1981, s. 34 by Limestone Pavement Orders.

For the purposes of the Wildlife and Countryside Act, limestone pavement has a legal definition (s. 34(6)).

Wildlife and Countryside Act 1981 s. 34

---

(6) ... 'limestone pavement' means an area of limestone which lies wholly or partly exposed on the surface of the ground and has been fissured by natural erosion.

The NCC or the Countryside Agency has a duty to notify the local planning authority of any limestone pavements of special interest (s. 34(1)). This should then be taken into account in any planning application.

Wildlife and Countryside Act 1981 s. 34

---

(1) Where the Nature Conservancy Council or the Commission* are of the opinion that any land in the countryside which comprises a limestone pavement is of special interest by reason of its flora, fauna or geological or physiographical features, it shall be the duty of the Council or the Commission* to notify that fact to the local planning authority ...

*Now the Countryside Agency.

The Secretary of State or the planning authority may make a Limestone Pavement Order where the character of any area notified under subsection (1) would be likely to be adversely affected by the removal of limestone or any disturbance whatsoever. Such an order would prohibit such removal or disturbance.

Any person convicted of damaging a limestone pavement which is protected by a limestone pavement order shall be liable to a fine unless the damage is authorised by a planning permission.

Wildlife and Countryside Act 1981 s. 34

(2) Where it appears to the Secretary of State or the relevant authority* that the character or appearance of any land [limestone pavement] notified under subsection (1) would be likely to be adversely affected by the removal of the limestone or by its disturbance in any way whatever, the Secretary of State or that authority may make an order [… a 'limestone pavement order'] designating the land and prohibiting removal or disturbance of limestone on or in it; …

*Planning authority, as defined by s. 34(6), or National Park Authority (Environment Act 1995 s. 69(1)).

Wildlife and Countryside Act 1981 s. 34

(4) If any person without reasonable excuse removes or disturbs limestone on or in any land designated by a limestone pavement order he shall be liable –
   (a) on summary conviction, to a fine not exceeding £20,000,
   (b) on conviction on indictment, to a fine.
(5) It is a reasonable excuse in any event for a person to remove or disturb limestone or cause or permit its removal or disturbance, if the removal or disturbance was authorised by a planning permission …

Any limestone pavement protected by a Limestone Pavement Order is also designated as a priority habitat under the Conservation (Natural Habitats, etc.) Regulations 1994.

5.8.2   Regionally Important Geological and Geomorphological Sites (RIGS)

RIGS are the most important geological and geomorphological sites outside those sites that have statutory protection such as SSSIs. They are selected on a local or regional basis using the following criteria:

■ the value of the site for educational purposes;
■ the value of the site for professional and amateur study by Earth scientists;

- the historical value of the site in terms of important advances in Earth science knowledge, events or human exploitation;
- the aesthetic value of the site in the landscape, particularly in promoting public awareness and appreciation of Earth sciences.

RIGS are equivalent in status to local wildlife sites and other non-statutory wildlife designations. They can be listed in development plans drawn up by local authorities. The planning system can be used to protect RIGS if they are recommended to the local planning authority by a RIGS group. The conservation and management of these sites usually depends upon agreements with landowners.

5.8.3   Protection of caves

Caves are an important part of our natural and archaeological heritage. They are an important scientific resource, providing evidence of human cultural development, landscape development and climate change. Caves may be protected within the statutory designations used for other important habitats, such as NNRs and SSSIs. They may also be protected as Scheduled Ancient Monuments (SAMs).

Caves are a primary feature of karst or limestone landscapes. One of the most important features of these landscapes are shakeholes or depressions, many of which have been lost through land reclamation and tipping. Other threats to caves come from quarrying and mining, water abstraction and inappropriate land management practices.

The National Cave, Karst and Mine Register

The National Caving Association is establishing a database of information on all caves, abandoned mines and karst features in Great Britain. The register will contain information on conservation, recreation and research. The conservation objectives of the register are to:

- enable immediate response to planning applications or other threat;
- provide a depository for Cave Conservation Plans;
- provide a means of monitoring the conservation state of caves;
- assess and monitor potentially damaging operations;
- assess potential conflicts between conservation and access;
- provide data for sympathetic management of sites;
- develop and monitor Local Agenda 21 action plans;
- study the biodiversity of cave and mine systems.

The database includes a wide range of information including:

- Maps and surveys
- Access data
- Conservation monitoring

Figure 5.9 Typical limestone (karst) scenery, Ingleborough in the Yorkshire Dales.

- Biology
- Geomorphology and geology
- Hydrology
- Palaeontology

5.9   GEOLOGICAL STRUCTURES AND LANDSCAPE FEATURES AS ANCIENT MONUMENTS

Some important geological sites are protected as Scheduled Ancient Monuments (see Section 5.13.1), e.g. caves in the Peak District National Park:

- Haboro Cave, Brassington, Derbyshire
- Eldebush Cave, Wetton, Staffordshire

5.9.1   Case study

A conservation paradox

An interesting paradox has arisen recently in Leicestershire. English Heritage decided to preserve an ancient rabbit warren as a national monument (Anon., 1998a). The warren was probably built in the 1280s in what was Leicester Forest, which was owned by the earls of

Leicester. Such warrens were purpose-built with breeding chambers designed to make catching the animals easy, and they supplied meat and skins for medieval people. Nesting places were made of stone slabs or cut into the subsoil.

The warren is an important part of our heritage and it is hoped that it will be used as an educational resource. However, under the law, the handful of rabbits that still inhabit the ancient mound enjoy little protection. They can be legally shot as pests. The prospect of the warren being declared an ancient monument was reported on the front page of *The Times* (Anon., 1998b).

5.10   PROTECTION OF FRESHWATER ENVIRONMENTS

Control of freshwater pollution is the responsibility of the Environment Agency (in England and Wales), the Scottish Environment Protection Agency, and the Rivers Agency in Northern Ireland.

5.10.1   The ecological importance of freshwater habitats

Freshwater habitats include rivers, lakes, reservoirs and ponds. Ponds are important refuges for amphibians (frogs, newts and toads), fish, invertebrates and a wide variety of plants. Ponds may also be an important source of food and water for birds and mammals. Rivers are linear ecosystems and are important migration and dispersal routes for fish and other aquatic organisms including otters.

Figure 5.10 Recreational activity threatens some freshwater habitats.

Freshwater habitats and the law

Some freshwater habitats are protected as SSSIs, for example some rivers and canals, or as NNRs or Ramsar sites, SPAs or under the other designations discussed elsewhere. Freshwater species are particularly susceptible to water pollution, especially when it occurs in an enclosed body of water such as a pond or lake where dilution and dispersal are more difficult than in running waters.

Much of the law relating to the protection of freshwater habitats from pollution is based on the concept of 'controlled waters'. 'Controlled waters' are defined by s. 104 of the Water Resources Act 1991 (which applies in England and Wales) and include almost all coastal and inland waters. There are four categories of controlled waters:

- inland waters (rivers, surface and underground streams, canals, lakes and reservoirs, including those that are temporarily dry);
- groundwaters (water in wells, boreholes and in underground strata);
- coastal waters (the sea up to the line of the highest tide and tidal waters up to the freshwater limit);
- territorial waters (the sea within three miles (5.56 km) of the baseline from which the territorial limit is measured).

Land-locked waters that do not drain into any other controlled waters are excluded from the definition. Lakes and ponds are not normally to be treated as controlled waters unless they drain into other waters that are controlled. However, the Secretary of State may include or exclude any waters by order. This has been done in the Controlled Waters (Lakes and Ponds) Order 1989 (SI 1989/1149).

Section 85(1) of the Water Resources Act (WRA) creates a general water pollution offence, but the Act does not define 'poisonous, noxious or polluting matter'.

Water Resources Act 1991 s. 85

---

(1) A person contravenes this section if he causes or knowingly permits any poisonous, noxious or polluting matter or any solid waste matter to enter any controlled waters.

Under s. 90 of WRA 1991 it is an offence for a person to cause or permit a substantial amount of vegetation to be cut or uprooted in any freshwaters, or to be cut or uprooted so near to any such waters that it falls into them, and he fails to take all reasonable steps to remove the vegetation from those waters. However, this section does not apply to any activity undertaken with the consent of the Environment Agency.

Defences available include: if the entry of polluting material occurred in an emergency in order to avoid danger to life or health, or if a discharge is made in accordance with a consent issued by the Environment Agency. Such consents are required for sewage and industrial discharges to rivers.

Under the Salmon and Freshwater Fisheries Act 1975 it is unlawful to poison or injure fish by polluting water.

Salmon and Freshwater Fisheries Act 1975 s. 4

(1) ... any person who causes or knowingly permits to flow, or puts or knowingly permits to be put, into any waters containing fish or into any tributaries of waters containing fish, any liquid or solid matter to such an extent as to cause the waters to be poisonous or injurious to fish or the spawning grounds, spawn or food of fish, shall be guilty of an offence.

A number of EC Directives are aimed at the control of freshwater pollution and the protection of aquatic habitats. The Directive on the quality of freshwaters needing protection or improvement in order to support fish life (78/659/EEC) is discussed in Chapter 7.

Case study

In September 2001 the Tesco supermarket chain was fined over £30 500 for an offence under s. 85(1) of the Water Resources Act 1991, after allowing its shopping trolleys to pollute a river. This was the first prosecution of its kind brought by the Environment Agency. During a cleaning operation in April 1999, the Environment Agency removed 51 trolleys from the River Chelmer in Chelmsford, Essex, 33 of which belonged to Tesco. The store pleaded guilty at Witham Magistrates' Court, but the case was referred to Chelmsford Crown Court for sentencing.

5.11   PROTECTION OF MARINE WILDLIFE AND HABITATS

It has been estimated that half of the UK's species live in the sea and coastal waters. Marine organisms are threatened by pollution, over-exploitation and habitat loss. There are conflicts of interest between marine wildlife and industrial activity such as electricity generation from offshore wind farms and exploration for oil and gas. In addition, coastal

PANEL 5.2   CANALS AND WILDLIFE

The following are examples of some of the ways in which British Waterways has helped to protect wildlife and habitats:

- Installation of bat bricks in canal bridge walls
- Reintroduction of water voles to the Kennet and Avon Canal after restoration work, following a two-year stay at Bristol Zoo
- Establishment of tree management plans for waterways
- 10 000 chub introduced into the canal between Wombourne (near Wolverhampton) and Milford (near Stafford)
- Installation of hundreds of bat boxes and bird boxes
- Provision of crevices in walls for use by crayfish
- Recording of wildlife on the Grand Union Canal North by staff and customers using a system of waterway wildlife sighting cards
- Breaching caused by water voles on the Kennet and Avon Canal used to allow the flooding of adjacent fields and the creation of a lake (instead of protecting the canal bank by piling)
- Translocation of white clawed crayfish from Whetstone Lane Lock to another location on the Grand Union Canal, 100 metres away, during restoration work
- Two 30° escape ramps attached to the wall of a spillweir on the Trent and Mersey Canal, above the Anderton Boat Lift, to prevent ducklings from being washed over the weir
- Staff training in archaeology, heritage, bat conservation and countryside skills including, hedge laying, tree management, bank protection and site-specific environmental management
- Translocation of the pheasant's eye (*Adonis annua*) from alongside the Oxford Canal towpath which was threatened by bank works
- Seeds of the floating water plantain (*Luronium natans*) collected from Huddersfield Broad Canal and deposited in the Millennium Seed Bank at the Royal Botanic Gardens, Kew
- Soft bank protection used as an erosion control measure on the Grand Union Canal Waterway which will assist in the establishment of fringe vegetation for wildlife
- Construction of a duck raft as a resting place for mallard on the Crinan Canal
- Construction of two log-pile otter holts and 530 metres of vole-friendly bank protection on the Birmingham and Fazeley Canal
- Piling a 300 metre embankment on the Grantham Canal two metres back from the water's edge to protect grass-wrack pondweed and other fringe vegetation
- Publication of a conservation plan for the Kennet and Avon Canal to guide restoration work and ensure a sustainable future for the canal

Source: Various issues of *BW Monthly*, The British Waterways Staff Newsletter.

erosion and global warming with the associated sea level rise, are threats to the coastline in general and to some protected sites in particular.

5.11.1   Marine Nature Reserves

Marine Nature Reserves (MNRs) are the marine equivalent of National Nature Reserves and may be designated under the Wildlife and Countryside Act 1981 and, in Northern Ireland, the Wildlife (Northern Ireland) Order 1985.

Wildlife and Countryside Act 1981 s. 36

(1) Where, in the case of any land covered (continuously or intermittently) by tidal waters or parts of the sea which are landward of the baselines from which the breadth of the territorial sea adjacent to Great Britain is measured or are seaward of those baselines up to a distance of three nautical miles, it appears to the Secretary of State expedient, on an application made by the Nature Conservancy Council, that the land and waters covering it should be managed by the Council for the purpose of –
 (a) conserving marine flora or fauna or geological or physiographical features of special interest in the area; or
 (b) providing, under suitable conditions and control, special opportunities for the study of, and research into, matters relating to marine flora and fauna and the physical conditions in which they live, or for the study of geological or physiographical features of special interest in the area,
he may by order designate the area comprising that land and those waters as a marine nature reserve; and the Council shall manage any area so designated for either or both of those purposes.

MNRs are protected by byelaws made by the relevant local authority (s. 36) and by the NCC (s. 37). Byelaws made by the NCC may prohibit or restrict:

■ entry into, or movement within, the reserves of persons or vessels;
■ the killing, taking, destruction, molestation or disturbance of animals or plants;
■ damage or disturbance of the sea bed or any object in the reserve;
■ depositing of rubbish.

Byelaws may also provide for the issue of entry permits to the reserve and permits to do anything which would otherwise be unlawful.

Only three MNRs have been designated so far:

- Lundy, Bristol Channel
- Skomer, Dyfed
- Strangford Lough, Northern Ireland

Heritage Coasts are discussed as non-statutory landscape designations in Section 5.3.3. The Marine Wildlife Conservation Bill is currently being considered by Parliament (see Chapter 10). Some marine species are also protected under other legislation such as the Wildlife and Countryside Act 1981 and the Conservation of Seals Act 1970 and various European and international laws (see Chapter 7).

5.12   FARMING, WILDLIFE AND ACCESS TO THE COUNTRYSIDE

Agricultural practices in the UK have caused significant damage to wildlife populations and habitats. The intensification of farming has led to the removal of hedgerows to facilitate the use of large machinery which itself has compacted the soil, reducing drainage and increasing runoff. The growing of single crop species over large areas has resulted in a loss of biodiversity and a general homogenisation of the countryside.

Wildlife thrives on heterogeneity: a small woodland here, a pond there, a hedgerow, a ditch, a wild flower meadow, a mountain stream. Farming cannot expect to survive in isolation from the rest of the environment and there are signs that both EC and UK agricultural policies plan to change the nature of our countryside by using economic measures to encourage the restoration of a variety of habitats.

In the aftermath of the foot-and-mouth outbreak in February 2001 some of the farmers who lost their livestock, either from the disease itself or because of the contiguous culls implemented to control the disease, declared that they would not return to livestock farming. This was partly because of concerns that MAFF (now DEFRA) would not be any better prepared to deal with any future outbreak, but also because the economic climate at the time made it difficult for livestock farmers to run profitable businesses.

There are already some indications that agricultural subsidies may be used in future to support biodiversity rather than overproduction. In November 2000 the government published a White Paper entitled *Our Countryside: The Future – A Fair Deal for Rural England*. This recognised that there has been pressure for unwelcome poor quality development in the countryside which has encroached on some valued landscapes, and that wildlife habitats and wildlife diversity have declined. The White Paper announced that the newly created England Rural Development

---

PANEL 5.3   FARMING AND WILDLIFE

Some of the adverse effects of agricultural activities on wildlife and habitats are:

- Loss of hedgerows and associated plant and animal species
- Pollution of rivers, ponds and ditches by runoff containing herbicides, pesticides and nitrogen compounds from farm waste
- Contamination of soil with excess manure
- Heavy machinery causes compaction and smearing of soil particles, destroying crumb structure, reducing air content, and making root penetration difficult
- Loss of plant diversity, especially 'weed' species
- Loss of insect diversity, especially butterflies
- Loss of bird diversity, especially birds of prey
- Damage to soil structure caused by ploughing
- Plant monocultures (single species) grown over large areas, thereby simplifying ecosystems
- Loss of wetlands through land drainage

Programme (ERDP) would invest £1.6 billion in the countryside by 2006. Money for agri-environment schemes would double and the ERDP would promote environmentally sensitive practices. It also pledged £152 million for a new Rural Enterprise Scheme to help farmers diversify, and additional funds for woodland grants (Anon., 2000).

5.12.1   Agri-environment designations

Environmentally Sensitive Areas (ESAs)

Environmentally Sensitive Areas are areas of high landscape and conservation value. They were first introduced under Article 19 of EC Regulation 797/85 which authorised member states to introduce special national schemes in environmentally sensitive areas. This was superseded by Article 21 of EC Regulation 2328/91 and amendments.

The requirement to designate ESAs was implemented in the UK by the Agriculture Act 1986 and the first ESAs were established in 1987. Farmers may enter into voluntary agreements for 10 years to manage the land according to specific prescriptions in return for annual and capital incentive payments. Capital grants may also be available. ESAs later became part of the measures implemented under the Agri-environment Regulation 2078/92. They include the Lake District and Dartmoor.

Water protection zones

The Secretary of State is given a power to establish these zones by s. 93 Water Resources Act 1991. Their purpose is to prevent or control the pollution of controlled waters, but they may not refer to the entry of nitrate into these waters as a result of its use on land for agricultural purposes. The Secretary of State may designate such areas in England by order and prohibit or restrict activities specified in the order. An order may confer power on the Environment Agency to determine which activities are controlled and under what circumstances.

Nitrate sensitive areas

The relevant minister is given a power to create these areas by s. 94 WRA 1991. The purpose of these areas is to prevent or control the entry of nitrate used for agricultural purposes into any controlled waters. The minister may by order control activities within the designated area and he may enter into a management agreement with the owner and pay compensation.

5.12.2   Farmyard manure and wildlife

There is increasing concern about the damage caused to wildlife by ammonia released from manure. This material damages habitats and plant life, and it has been suggested that farmers may be forced to dig manure into the soil or pay for expensive chemical treatments to manure heaps (dubbed a 'bullshit tax' in Whitehall). The government intends to publish a consultation document on options to deal with the threat to the environment posed by manure in 2002 (Elliot, 2001).

Material with a high biological oxygen demand (BOD) such as food wastes, animal slurry, and even milk, remove oxygen from water resulting in fish kills. On 29 November 2001 two million litres of pig slurry from Mill of Carden Farm near Oyne poured into the Gadie Burn in Aberdeenshire. The Burn is a tributary of the River Urie which flows into the River Don. The main A96 Aberdeen to Inverurie road was closed for several hours as a result of the incident. Officers of the Scottish Environment Protection Agency (SEPA) used earthwork barriers and hay bales to minimise the amount of slurry entering the water course. The River Don District Salmon Fishery Board was advised by SEPA to take appropriate precautionary measures.

5.12.3   Destruction of wildlife for disease control

For the purposes of disease control in the countryside it may be necessary to cull certain wildlife species.

Figure 5.11 During the 2001 foot-and-mouth disease outbreak many footpaths in Britain were closed to the public.

The Animal Health Act 1981 provides powers to the Minister of State for Agriculture, Fisheries and Food (now DEFRA) to reduce the risk or control the spread of disease. The legislation specifically lists foot-and-mouth disease, swine fever, cattle plague, pleuro-pneumonia and poultry diseases. Orders may be made:

- for rabies control;
- for the creation of eradication areas;
- to restrict animal movements;
- to restrict animal imports and exports;
- to require the dipping of sheep to control sheep scab.

Criminal offences include:

- failure to keep an infected animal separate;
- failure to report an infected animal to the police;
- obstructing the police.

Where a disease has become established the Minister of State is empowered to make an order requiring the destruction of wild animals

Figure 5.12 Foot-and-mouth disease may be spread by some wildlife, including deer. During the 2001 outbreak deer parks either closed to the public or confined their animals.

where this is necessary to eliminate or reduce the spread of disease. The minister must consult the appropriate conservation agency before making such an order.

Case study

In April 2001, during an outbreak of foot-and-mouth disease a farmer in Netherton, Perthshire, was granted a licence to shoot pink-footed geese (*Anser brachyrhynchus*) as a precaution to prevent the spread of the disease to one of Scotland's largest herds of Aberdeen Angus cattle. The geese were migrating to the Arctic from the Solway Firth, which was badly affected by the disease at the time (Anon., 2001).

5.12.4 Restriction of access to the countryside for the purposes of conservation

The Countryside and Rights of Way Act 2000 will give members of the public access to large areas of the countryside in England and Wales from which they have been previously excluded. However, this new found 'right to roam' brings with it increased threats to protected species and habitats.

PANEL 5.4   THE EFFECTS OF THE 2001 FOOT-AND-MOUTH DISEASE OUTBREAK ON ECOSYSTEMS

A large-scale outbreak of foot-and-mouth disease (FMD) occurred in the UK in 2001. As a result much of the countryside was closed to the public and restrictions were placed on livestock movements. In affected areas large numbers of livestock were culled. This resulted in:

- Reduced grazing pressure in areas where livestock were culled and areas to which livestock could not legally be moved
- Increased grazing pressure in areas where livestock were confined
- Reduced human disturbance to areas from which members of the public were excluded

Lack of grazing affected a wide range of habitats, including:

- Lowland grassland
- Lowland heathland
- Coastal marshland
- Sand dunes
- Upland meadows and pastures
- Upland calcareous grassland and limestone pavement
- Upland heath
- Blanket bog
- Montane habitats

The impacts of the FMD outbreak include:

- Decline of some vulnerable species, e.g. high brown fritillary butterfly (*Argynnis adippe*) on the Malvern Hills
- Increase in some vulnerable species, e.g. marsh saxifrage (*Saxifraga hirculus*) in Moor House and Upper Teesdale National Nature Reserve
- Heavily trampled ground
- Over-enriched ground resulting from excess dung and the remains of supplementary feed
- Increased sediment load in freshwater habitats and increased organic waste runoff
- Possible pollution risks from pyres, burial sites, and chemicals used in the disposal of carcasses, e.g. possible risk to lichens (from air pollution) and freshwater habitats (from leachates from pyre sites)
- Increase in deer populations due to cessation of culling, causing damage to woodlands and some grazing-intolerant plants like oxlip (*Primula elatior*)
- Long-term loss of wildlife habitats in lowland landscapes dependent upon grazing
- Reduced grazing of upland habitats, which have been overgrazed by sheep, may provide an opportunity to reverse the damage done to the benefit of upland bird species such as merlin (*Falco columbarius*) and red grouse (*Lagopus lagopus*)

Source: Robertson, Crowle and Hinton (2001).

Figure 5.13 The restrictions on livestock movements imposed during the 2001 foot-and-mouth outbreak prevented the grazing of many upland areas, which consequently began to revert to their natural vegetation. This photograph illustrates the effect of a fence on vegetation in the West Pennine Moors where sheep are only able to graze the vegetation on the left side.

Under s. 26, CRoW allows the relevant authority by direction to exclude or restrict access to any land to which the Act gives access indefinitely or for a fixed period for conservation purposes. The relevant authorities (s. 21(5)) are:

■ The Countryside Agency (for land in England)
■ The Countryside Council for Wales (for land in Wales)
■ The National Park Authority (for land within any National Park)
■ The Forestry Commissioners (for certain woodlands)

The purposes for which rights of access may be excluded or restricted are (s. 26(3)):

■ conserving flora, fauna or geological or physiographical features of the land in question;
■ preserving
  – any scheduled monument (defined by s. 1(11) of the Ancient Monuments and Archaeological Areas Act 1979);
  – any other structure, work, site, garden or area which is of historic, architectural, traditional, artistic or archaeological interest.

Section 26(4) of the Act imposes an obligation on the relevant body to have regard to the advice of the relevant advisory body in considering whether to give a direction under s. 26. It is also required to consult any local access forum that may exist if the direction will exclude or restrict access indefinitely or for a period exceeding six months (s. 27(1)).

At the end of December 2001, as the Countryside Agency began to produce draft maps of areas to which the public will be given access, farmers were showing concern that their farming activities would be restricted by the new law. One farmer who specialises in re-seeding heather on former grassland, expressed concern that heather could be included as 'open country' and therefore opened up to the public. He intended to stop re-seeding until the mapping exercise was completed in three years' time. Other farmers may be tempted to improve their land specifically so that it will not be included in the provisional maps. The provisional maps completed so far appear in some cases to be based on existing inaccurate habitat maps and data which were not compiled for this purpose (Clover, 2001). Once the provisional maps are published, landowners will be able to appeal to the Secretary of State where they feel mistakes of land classification have been made.

5.12.5   Genetically modified organisms and the threat to the countryside

There is concern among some ecologists that pollen from genetically modified crops may contaminate organic crops and that the presence of these crops may pose an ecological threat to the wider countryside. There have been a number of cases of protestors taking direct action to destroy GM crops. Such actions clearly involve trespass, but the courts have recently been asked to consider whether or not such action amounts to aggravated trespass.

Protestors were acquitted of aggravated trespass after they had damaged genetically modified crops on farms in Dorset and Cambridgeshire. Mrs Justice Rafferty, in the Queen's Bench Division, held that the offence of aggravated trespass (s. 68, Public Order Act 1994) entailed the intended interruption of lawful activity being carried out by people present on the land. The presence of people was necessary before an offence was committed. The disruption of farming by crop destruction was, in itself, insufficient (*Tilly* v. *DPP, DPP* v. *Tilly and Others* (2001)).

5.13   HERITAGE CONSERVATION

The protection of heritage sites is often inextricably linked to the protection of habitats, especially during development operations. As with landscape and habitat protection, heritage sites are protected by a number of statutory and non-statutory designations.

Heritage designations cover a wide range of features which are of archaeological, historic or cultural importance. They occur in rural and urban settings and include:

- individual objects, e.g. shipwrecks or standing stones;
- landscape features, e.g. ditches and dykes;
- buildings, e.g. manor houses, castles, mills, and bridges;
- areas, e.g. battlefield sites and historic towns.

All identified sites and monuments of archaeological interest are listed in the Sites and Monument Record (SMR). This is maintained by local authorities and includes sites which enjoy statutory protection, and sites that do not.

5.13.1   Statutory heritage designations

Scheduled Ancient Monuments (SAMs)

Scheduled Ancient Monuments are designated under the Ancient Monuments and Archaeological Areas Act 1979. There are over 600 000 known archaeological sites in England alone but only a very small proportion are scheduled. Lists of SAMs are held on a county basis on the Sites and Monument Record.

Once a site is scheduled the Secretary of State must give consent before any works may be carried out, including repair work. Consent can only be granted for detailed proposals.

Areas of Archaeological Importance (AAIs)

Areas of Archaeological Importance are designated under the Ancient Monuments and Archaeological Areas Act 1979. The designation covers the historic centres of:

- Canterbury
- Chester
- Exeter
- Hereford
- York

Within AAIs potential developers must give notice of proposals to disturb ground, tip on it, or flood it. Development may then be delayed while the site is investigated and possibly excavated.

Conservation areas

Local planning authorities have a duty to designate as conservation areas any 'areas of special architectural or historic interest, the character or appearance of which it is desirable to preserve or enhance', under the Planning (Listed Buildings and Conservation Areas) Act 1990. This designation introduces a general control over the demolition of unlisted (*sic*) buildings. It also gives special protection to trees in designated areas (see Chapter 6).

Listed buildings

The Planning (Listed Buildings and Conservation Areas) Act 1990 imposes a duty on the Secretary of State to compile a list of buildings of special architectural or historic interest. Once listed, planning permission is required for any activity which would affect its character as a building of special interest. Some listed buildings also have SAM status, in which case the ancient monument legislation takes precedence.

Protected wrecks

The Protection of Wrecks Act 1970 protects wrecks of historic interest below the high water mark. Such wrecks are designated by the Secretary of State, and their positions are marked on admiralty charts.

5.13.2   Non-statutory heritage designations

English Heritage and Cadw have adopted three non-statutory designations which help to identify and conserve other aspects of the historic environment:

- Historic Battlefield Sites
- Historic Landscapes
- Historic Parks and Gardens

The heritage bodies keep registers of the sites in each of these categories, and local planning authorities should consider their location when devising planning policies and considering planning applications.

REFERENCES

Anon. (1998a) Rabbit warren gets protection. Mammal Briefs. *Mammal News*, No. 116, Winter 1998/99. Mammal Society, London.

Anon. (1998b) Ancient monument status for rabbit warren. *The Times*, 20 April 1998, p. 1.

Anon. (1999) Beachy Head cliff crashes into the sea. *The Times*, 12 January 1999, p. 1.

Anon. (2000) *Our Countryside: The Future – A Fair Deal for Rural England*. Cm 4909, The Stationery Office Ltd, London.

Anon. (2001) Geese 'cull' plan. *The Times*, 21 April 2001, p. 14.

Bell, S. and McGillivray, D. (2000) *Environmental Law. The Law and Policy Relating to the Protection of the Environment* (5th edn). Blackstone Press Ltd, London.

Clover, C. (2001) Farmers hit by right to roam blunder. *The Daily Telegraph*, 31 December 2001, p. 1.

Elliott, V. (2001) Heap of tax on manure. *The Times*, 25 October 2001, p. 19.

Hill, S.L. and Fleming, B.F. (1999) *Urban Meresy Basin Natural Area Profile. A Nature Conservation Profile*. English Nature, North West Team, Wigan, Lancashire.

Huxley, J.S. (1947) *Conservation of Nature in England and Wales*. Cm 7122, HMSO, London.

Murdie, A. (1993) *Environmental Law and Citizen Action*. Earthscan Publications Ltd, London.

Noel, S. (1993) *The Guinness Guide to Nature in Danger*. Guinness Publishing Ltd, Enfield, Middlesex.

Roberston, H.J., Crowle, A. and Hinton, G. (eds) (2001) *Interim Assessment of the Effects of the Foot and Mouth Disease Outbreak on England's Biodiversity*. English Nature Research Reports, No. 430. English Nature, Peterborough.

Rowe, M. (2000) Who on earth would destroy all this? *The Independent on Sunday*, 23 April 2000, p. 17.

6

The protection of trees and hedgerows

Trees and hedgerows are important features of the landscape. Some trees and some hedges are protected by law, and even those that are not may contain protected animal species. Although hedgerows may contain tree species, hedges and trees are treated separately by the law.

6.1 THE PROTECTION OF TREES

6.1.1 Introduction

England has less tree cover than any other country in Europe, apart from Ireland. Most European countries typically have 20 to 25 per cent of their land under trees, while the figure for England is just 7.5 per cent. The Countryside Agency hopes that this figure will have doubled to 15 per cent by 2050. In 1990 the Agency (at that time the Countryside Commission) earmarked an area of 200 square miles in the heart of England as the site for a new National Forest. To this end a government-appointed company is currently in the process of planting 30 million trees. In addition the Agency, in partnership with the Forestry Commission, is developing 12 community forests on the outskirts of major towns and cities in England.

Trees are important from both an amenity and an ecological point of view. Trees and woodlands have an amenity value as part of the land-scape. They provide a three-dimensional structure to the land and may be a useful part of the overall design of an urban landscape, a park or a new road network. Trees are also an integral part of many terrestrial ecosystems.

In considering the importance of trees we need to draw a clear distinction between their amenity value and their ecological value. From a purely amenity point of view the species to which a particular tree belongs may be irrelevant. A single individual tree or an area of woodland may be worth protecting simply because it is considered to be a valuable component of the landscape. However, from an ecological standpoint some species and some communities of trees are more important than others. In general terms, native trees are more important than introduced species in increasing the diversity of other species in the ecosystem.

Trees affect the local microclimate by reducing wind speed and by intercepting sunlight thereby shading the soil surface. They also affect the global climate by playing an important part in the carbon dioxide balance of the atmosphere. Increased concentrations of carbon dioxide in the atmosphere are in part responsible for global warming.

Trees and the Kyoto Protocol

In 1997 the parties to the UN Framework Convention on Climate Change agreed a protocol to reduce greenhouse gas emissions at Kyoto, Japan. The Kyoto Protocol sets emission targets and requires inventories to be kept which will allow the calculation of carbon emissions and carbon sequestrations (see Section 7.2.8)

The Kyoto Protocol recognises the importance of trees in sequestering carbon from the atmosphere and thereby acting as a carbon sink, allowing countries to credit their greenhouse gas inventory with the emissions captured when they create or replant forests. When a country destroys a forest, the resultant carbon emissions are added to its inventory. The role of trees in sequestering carbon in the UK has been discussed by Cannell (1999).

6.1.2   Natural history

A tree is a tall woody plant with a single, usually stout, stem. Trees may be divided into conifers (softwoods) that have narrow leaves (needles) and broad-leaved trees (hardwoods). However, this simple classification is not completely accurate as some broad-leaved trees, e.g. willows, have narrow leaves, while some hardwood is soft and some softwood is hard.

The root system of a tree may be as extensive as the aerial parts. As in other vascular plants, the roots obtain both water and nutrients from the soil. If the root system is damaged or its growth is restricted, the growth of the aerial parts may be affected and, in extreme cases, the tree may die. However, damage to the roots may have no obvious immediate effect and death may occur some considerable time later. If a tree has to be moved this should be done when the plant is semi-dormant and not during the summer when it is losing water.

PANEL 6.1

Pedunculate oak   *Quercus robur*

Identification/size
Trees up to 40 metres in height. Leaves on short stalks, with acorn cups on long stalks. Twigs grey-brown, with light brown winter buds, spirally set, with a cluster near the tip. Flowers of both sexes appear on the same tree in May. Male catkins are pale green. Female flowers are green with long stalks.

Habitat
Occurs on a wide range of soils, especially deep, rich soils.

Distribution
Occurs throughout most of the British Isles. Rarely occurs at altitudes over 300 metres.

Ecology
Often abundant and forms dense woodlands. Commonest British tree and often dominant in broad-leaved woodland. May live for several hundred years. Approximately 300 species of insects and a similar number of lichens are found associated with the pedunculate oak.

Water and nutrients move up and down a tree in tubular vessels that make up a tissue known as the vascular bundle. This tissue is located just beneath the bark and if it is broken all the way around the circumference, the tree will die. This process is known as 'ringing' and may occur when animals such as deer, rabbits or squirrels damage the bark, but it can also occur because of human activity.

When it rains trees act as temporary stores of water, preventing the rain from falling directly onto the soil, thereby preventing flooding of water courses. Tree roots bind the soil together, preventing soil erosion. Roots also remove water from the soil and return it to the atmosphere through the leaves by the process of transpiration. The loss of trees from catchment areas combined with the covering of the soil surface by roads and buildings results in the flooding of many urban areas and the overloading of waste water treatment works by runoff that would otherwise enter the groundwater.

Trees provide important habitats for other species. Native tree species are particularly important in this respect as they have evolved alongside other native species, providing them with food and shelter. For example, some 284 species of insects and 324 species of lichens have been recorded in association with the oak, a native species. In contrast, only 15 insect species and 183 lichen species have been recorded in association with the introduced sycamore (Springthorpe and Myhill, 1985).

Almost all of the woodland in Britain is secondary woodland. At some time in the past it has been cut or managed and what now remains is secondary growth that has replaced the original ecosystem. There are very few remaining areas of primary or ancient woodland. These areas are not only characterised by the tree species they contain but also by their distinctive communities of smaller flowering plants, woodland slugs, snails and other animal species. In particular, these woodlands tend to contain a large number of species that are poor colonisers and therefore tend to be rare in the relatively unstable envi-

Figure 6.1 Even dead trees provide important habitats for wildlife and should be retained where they are not in danger of falling down.

ronments created by man. Such primary woodlands are considered to have a very high ecological value.

Our legislative system provides protection for both native and introduced tree species through a variety of mechanisms. It also protects some trees because of their amenity value and may require the planting of trees in some circumstances.

6.1.3   Trees and the law

There is no specific legislation protecting particular tree species in the UK. However, trees are wild plants and as such covered by the Wildlife and Countryside Act 1981. Under s. 13(1)(b) of WCA 1981 it is an offence for an unauthorised person to uproot any wild plant (not included in Schedule 8). The definition of a wild plant under s. 27(1) of the Act would include any tree which ordinarily grows in Great Britain in a wild state. An authorised person as defined by s. 27(1) includes any owner or occupier or any person authorised by the owner or occupier, any person authorised in writing by the local authority, the Environment Agency, a water undertaker or a sewerage undertaker.

Tree Preservation Orders

County councils and local planning authorities may protect trees by making a Tree Preservation Order (TPO) under the Town and Country Planning (Trees) Regulations 1999 (SI 1999/1892). A TPO may be used to protect:

- a single tree;
- a group of trees; or
- an entire woodland.

It may also include trees that form part of a hedge or hedgerow, but does not include hedges, bushes or shrubs. Trees which have been felled may leave a stump which is still alive and therefore still capable of being covered by a TPO. However, uprooted trees would not be covered since they cannot recover and will die.

A Tree Preservation Order protects trees from being:

- cut;
- lopped;
- uprooted;
- wilfully damaged; or
- wilfully destroyed.

The form to be taken by a TPO is laid down in the schedule to the regulations. Where the order relates to a group of trees the number in the group must be specified (reg. 2(1)(b)). The order must include a map prepared to a scale sufficient to give a clear indication of the position of the trees, groups of trees or woodlands to which the order relates (regs 2(2) and (3).

It is an offence to:

■ cut down, top, lop, uproot, wilfully damage or wilfully destroy; or

■ cause or permit the cutting down, topping, lopping, uprooting, wilful damage or wilful destruction of,

any individual tree or tree within a group of trees or woodland specified in a TPO. No offence is committed if the local planning authority has consented to any prohibited operations. Permission is not necessary for work to a tree that is the subject of a TPO if it is:

■ dead or dying;
■ dangerous;
■ creating a nuisance;
■ cultivated for the production of fruit, in the course of business or trade;
■ interfering with the work of a statutory undertaker on operational land (e.g. the Post Office, the Civil Aviation Authority, a water or sewerage undertaker, a public gas transporter);
■ preventing the implementation of development for which detailed planning permission exists;
■ the subject of an obligation imposed by another Act;
■ the subject of a Forestry Commission felling licence;
■ carried out by or at the request of the Environment Agency to implement a permitted development.

For a detailed description of exemptions see Article 5 of the schedule to the Town and Country Planning (Trees) Regulations 1999 and the Town and Country Planning Act 1990. Where permission is given for a tree to be cut down a local authority is likely to require a replacement tree to be planted. A TPO will attach to any replacement tree.

Tree felling licences

Trees can only be felled after obtaining a licence from the Forest Authority, which is part of the Forestry Commission. This requirement does not apply to trees that are:

■ in gardens, orchards, churchyards or public open spaces;
■ fruit trees;
■ dead or diseased;
■ small trees which are:
   – 8 cm or less in diameter at a height of 1.3 m from the ground,
   – thinnings with a diameter of 10 cm or less,

– coppice or underwood with a diameter of 15 cm or less;

or to the lopping, topping, pruning or pollarding of trees.

In most cases where a felling licence is granted the Commission will require the planting of replacement trees. Unlicensed felling is punishable by a maximum fine of £2000. Under the Forestry Act 1986 the Commission may require the planting and maintenance of a suitable tree following a conviction for unlawful felling.

Where an application for a felling licence includes a tree covered by a TPO the Forestry Commission will consult the local planning authority before deciding whether to grant the licence. A felling licence is the equivalent of a TPO consent.

Case law

Damage to property caused by trees

Protecting trees can be expensive if they damage property. In October 2001 the House of Lords considered a case for damages resulting from the growth of tree roots that necessitated remedial work to a group of mansion blocks in London (*Delaware Mansions Ltd and Another* v. *Westminster City Council* (2001)). Damage to three of the blocks was caused by the roots of a tree on an adjoining highway, before the owner bought the blocks in 1990. The highway authority refused to remove the tree because there had been no new damage since that date.

The risk of damage to the land and the foundations was foreseeable by Westminster City Council, due to their proximity to the tree. If it had been removed the total cost of repair to the building would have been £14 000. However, during a period of drought in 1989 the ground beneath the blocks became desiccated as a result of incursion of the roots, requiring underpinning and piling costing £570 734.

The House of Lords found that where there was a continuing nuisance of which the defendant knew or ought to have known, reasonable remedial expenditure could be recovered by the owner who had to incur it. Westminster Council was given ample notice and a reasonable opportunity of abatement and was therefore liable.

Injunctions

Injunctions may be imposed to prevent contravention of a TPO. A Kent farmer who was a persistent offender, was imprisoned for failing to comply with the terms of such an injunction (Bell and McGillivray, 2000).

Defining a tree

A stump of a tree which is still alive (after the tree has been felled) may still be covered by a TPO. A TPO may cover anything ordinarily called a tree and may also cover a coppiced woodland (*Bullock* v. *Secretary of State for the Environment* (1980)).

Trees and the planning system

Where detailed planning permission exists protected trees that are in the way of a development which is about to start may be cut back or cut down. Tree work cannot be carried out where only outline planning permission has been granted.

Trees growing on development sites can be protected by TPOs or by conditions attached to the planning permission, or both. Section 197 of the Town and Country Planning Act 1990 places a duty upon local planning authorities to make adequate provision for trees when planning permission is granted. Apart from protecting existing trees the duty allows local authorities to require that new trees be planted which may be covered by a TPO. The order takes effect once the trees are planted.

Town and Country Planning Act 1990 s. 197

It shall be the duty of the local planning authority –
(a) to ensure, wherever it is appropriate, that in granting planning permission for any development adequate provision is made, by the imposition of conditions, for the preservation or planting of trees; and
(b) to make such orders under s. 198* as appear to the authority to be necessary in connection with the grant of such permission, whether for giving effect to such conditions or otherwise.

*Tree Preservation Orders.

Section 89 of the National Parks and Access to the Countryside Act 1949 gives a power to local authorities to plant trees on land in their area for the purpose of preserving or enhancing the area's natural beauty and to undertake remedial work on derelict land. The powers include compulsory purchase of land for the purpose of fulfilling the local authorities' responsibilities with respect to tree planting and derelict land. Section 114 of the Act makes it clear that for these purposes the planting of trees includes the planting of bushes, planting or sowing of flowers and the sowing of grass and the laying of turf.

If this interpretation of 'trees' is used to interpret s. 197 of the Town and Country Planning Act 1990 then it would appear that local authorities have a power to require developers to reconstruct entire ecosystems as a condition of planning permission. In any event, a local authority may attach any condition it pleases to a particular permission so the legal interpretation of the term 'tree planting' is largely irrelevant.

Some developers take a responsible approach to tree preservation and may even use the existence of the trees as a feature of their development. Within the grounds of the former Victorian psychiatric hospital in Winwick, Cheshire, stand lines of protected trees which now enhance the site of a new residential development. The developers have erected signs around the perimeter of the site drawing attention to the protected trees (Figure 6.2). In contrast, Gilbert (1991) quotes one site where the approved plans showed 28 trees but at the end of the development only seven remained.

In spite of the responsibilities of local authorities to encourage tree planting, there is little evidence in many of the new out-of-town developments that developers are being required to incorporate significant numbers of trees into their designs (Figure 6.3).

Figure 6.2 Protected trees have enhanced the appearance of this new housing development in Winwick, near Warrington. The site was formerly occupied by a Victorian hospital and the trees were planted in avenues along the hospital's roads and in a row within the perimeter wall.

Figure 6.3 Bolton Wanderers' football ground, the Reebok Stadium, in Horwich. Although planning authorities have a duty to make provision for trees most new car parks are featureless landscapes designed only to maximise their capacity to accommodate cars.

Trees in conservation areas

Trees in designated conservation areas are subject to a statutory restriction which prohibits the cutting down, lopping, topping, uprooting, wilful damage or wilful destruction of the tree, under the Town and Country Planning Act 1990. This is effectively a statutory TPO, but is more limited than most individual TPOs and the Secretary of State may exempt certain works under s. 212. In the case of these statutory TPOs prohibited acts may proceed six weeks after notification has been given to the local planning authority, unless the authority has by then imposed a TPO. Unless prior authorisation is given, it is an offence to carry out prohibited acts without notifying the planning authority and waiting six weeks.

Trees, highway works and building sites

Trees are highly susceptible to damage during construction and highway maintenance work. Most tree roots are near the surface and are easily damaged by heavy construction vehicles. Tree trunks can be severely damaged by quite gentle blows that may damage the vascular tissue. This damage occurs beneath the bark and may not be apparent until long after the contractor has left the site. Gilbert (1991) has suggested that a single collision with a dumper truck can cause a huge scar; two such collisions could ring bark a tree, preventing the flow of water and nutrients, and ultimately causing its death.

The former DETR has produced guidelines for highway contractors and others whose activities are likely to disturb or damage trees. This suggests that a protection zone should be set up around each tree to protect its roots before construction or maintenance work begins (Figure 6.4). The size of this 'precautionary area' is determined by the following equation:

Radius of precautionary area = 4 × tree girth at chest height

The precautionary area should not be used for storing chemicals or other materials, machinery or vehicles. Excavation within this area should be done by hand or by using trenchless techniques. Mechanical

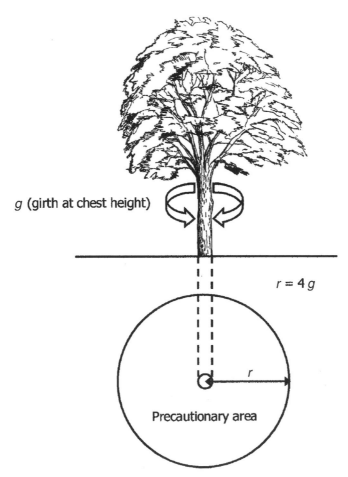

Figure 6.4 Method of calculating the 'precautionary area' around the base of a tree.

compaction of the roadbase and base course should not take place within the precautionary area.

Tree roots can be protected using tree pits, guards, grills and root barriers. Care should be taken not to damage the bark by leaning paving slabs against it or chaining machinery to it. Where work is to be left open it is important to protect exposed tree roots. In winter they should be wrapped in dry sacking overnight; in summer they should be covered at all times with damp sacking. Care should be taken not to expose trees to toxic chemicals, especially when backfilling material around the roots, and when using herbicides near trees.

The roadside tree database may contain reports on specific trees. This information is available from tree officers working for local authorities.

Trees and highways

The Highways Act 1980 imposes a duty upon the Highway Authority (HA) to maintain the highway (s. 41). Section 154 gives the HA powers to serve a notice on the owners of trees which constitute a danger to users of the highway. This would include:

- dangerous trees that could fall on the road;
- trees and hedges that block a driver's view;
- trees and hedges that interfere with the light level from a street lamp.

The notice gives the tree owner 14 days to carry out the necessary works. If the works are not carried out, the HA may do the works themselves and reclaim the costs from the tree owner.

The HA may restrict tree planting and order the removal of trees and other vegetation on private land where they represent a danger on a road bend (s. 79). Licences to plant and maintain vegetation in the highway may be granted to a member of the public by the HA (s. 142).

Registers

Tree preservation orders are a local land charge and as such appear in the Local Land Charges Register maintained by local authorities. The authorities also keep a register of applications and decisions relating to work carried out on protected trees and hold registers of notices concerning trees in conservation areas (under section 211 of the Town and Country Planning Act 1990). The Forestry Commission maintains a public register of all trees that have been felled or planted.

The Tree Register is a registered charity that maintains a database of over 125 000 notable trees in Britain and Ireland, including details of rare, unusual and historically significant trees. It contains historical records of trees from reference works going back over 200 years. Information held by the Register could be useful in highlighting the local or national importance of a particular tree, and to give support for protection and retention during development. The Register can also help to select suitable trees for use in urban landscaping schemes.

6.1.4   Mitigation measures

An application to a local authority for a licence to damage or remove a tree protected by a TPO is more likely to succeed if the proposed damage or loss would be mitigated by plans to plant replacement trees. If the trees that would be lost or damaged have no particular conservation value the local authority may consider that, on balance, the amenity value of the replacement trees more than compensates for the loss. From a conservation point of view, replacement trees should preferably be native species.

6.1.5   Case study

Tree translocations

Some large companies have planted significant numbers of new trees in recent years. During 1998 Tarmac Quarry Products planted 60 000 trees and 400 000 willow whips. In total the company has planted over half-a-million trees across the UK. Other companies have recognised the value of existing trees and gone to extraordinary lengths to preserve them.

The value of mature trees is increasingly being appreciated by developers. Civic Trees is a company that specialises in the supply, planting and transplanting of semi-mature trees. The company was founded in 1963 and operates its own fleet of Tree Spades and specially adapted vehicles. The Tree Spades are imported from the United States and are capable of moving trees with a stem circumference of up to 60 cm. The company's founder, Chris Newman, has developed the Newman Frame which is used to support rootballs for trees with a stem girth of up to 3000 mm.

In the autumn of 2000, Civic Trees moved a 35-tonne elm tree that was protected by a TPO to make room for a supermarket extension in Thame, Oxfordshire. The rootball of the tree was dug out using a mini-excavator and spades before being encased in a Newman Frame to bind the soil to the roots. The tree was then lifted onto a low loader with a 120-tonne crane and moved to a new site 80 feet away. After replanting, the elm was supported with guy wires. The whole operation cost £10 000 – half the £20 000 fine that could have been imposed on summary conviction for felling it illegally (Anon., 2000).

Some housing developers are also going to extraordinary lengths to preserve trees. Six large trees aged around 100 years, including limes and magnolias, were recently uprooted at a site off the King's Road in Chelsea. Three 20 metre high limes were replanted immediately and the remainder were placed in temporary holes until the project was finished.

In 1999 Civic Trees completed a £100 000 contract to plant 150 trees on an estate near Blackburn, Lancashire. The owner wanted the trees, mostly oaks and limes, laid out in the formation of the opposing armies of a Civil War battle that had been fought nearby.

6.2   THE PROTECTION OF HEDGEROWS

6.2.1   Introduction

Hedgerows are an entirely artificial feature of the British landscape. However, some have been established for several hundred years and are important ecological refuges for both wild plants and animals.

Hedgerows were originally planted in order to divide the land into individual small fields. However, the increased mechanisation of farming has resulted in the creation of larger and larger farms and the consequent removal of hundreds of miles of hedges. It has been estimated that between 1945 and 1970 about 8000 km of hedgerows were lost on average each year in the UK. Between 1945 and 1990, some 25 per cent of Britain's hedgerows were destroyed. In 1996 the government estimated that losses from neglect and grubbing out were causing the disappearance of 16 000 km of hedgerows per year.

6.2.2   Natural history

It has been estimated that more than 600 plant, 1500 insect, 65 bird and 20 mammal species use hedgerows in Britain. The older the hedgerow, the larger the number of plant species. Ecologists are able to determine the approximate age of a hedgerow by examining the species present using the formula:

Age of hedge (in centuries) = Number of woody plant species
in a 30 m length of hedge

Hedgerows function as dispersal corridors for plants and a wide variety of animals, from ground beetles to small mammals, and even badgers and foxes. They also provide food and shelter for birds, insects and other species. The grasses and herbs at the bottom of the hedge are particularly important because of the high animal and plant species richness found there. Elm trees occur more frequently in hedges than anywhere else and provide food for about 100 insect species.

The cutting of hedgerows should be carefully timed as it may damage their usefulness as a wildlife habitat. Some farmers cut their hedges when birds are nesting, with obvious results. A useful practice is to trim only part of a hedge in any year, leaving berries on the branches as food for birds.

6.2.3   Hedgerows and the law

Section 97 of the Environment Act 1995 gives appropriate ministers a power to make regulations to protect important hedgerows in England and Wales. In June 1997, the Hedgerows Regulations (SI 1997/1160) came into force.

These regulations apply to hedgerows growing in, or adjacent to:

■ any common land (as defined in the Commons Registration Act 1965);
■ protected land (SSSI or nature reserve established by a local authority);
■ land used for agriculture (including horticulture);
■ land used for forestry;
■ land used for the breeding or keeping of horses, ponies or donkeys.

They do not apply to hedgerows which are within the curtilage of, or mark the boundary of the curtilage of, a dwelling-house (reg. 3(3)).

The regulations only apply if the hedgerow has:

■ a continuous length of at least 20 metres; or
■ a continuous length of less than 20 metres and at each end meets another hedgerow (by intersection or junction) (reg. 3(1)).

For the purposes of the regulations a hedgerow ends at the point where it meets another hedgerow (reg. 3(4)). For the purposes of measurement, any gaps not exceeding 20 metres, or gaps which result from a breach of the regulations, are treated as part of the hedgerow (reg. 3(5)) (Figures 6.5 and 6.6).

The regulations give local planning authorities the power to prevent the removal of 'important' hedgerows by requiring owners of hedgerows to submit for approval a 'hedgerow removal notice'. Where a utility operator requires the removal of a hedgerow from land it does not own it must submit a removal notice.

The notice (Schedule 4) requires the applicant to provide:

■ A map showing the hedgerow to which the notice relates (normally to a scale of 1:2500) indicating those sections which were planted less than 30 years ago.

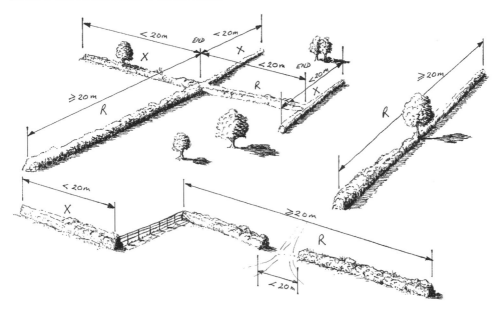

Figure 6.5 Application of the Hedgerows Regulations 1997 to some hypothetical sketched hedgerows. The regulations apply to hedgerows labelled 'R' but not to those marked 'X'.

Figure 6.6 Any gaps in a hedgerow that are less than 20 m long are treated as part of the hedgerow for the purposes of the Hedgerow Regulations 1997.

- The reason for the proposed removal.
- Evidence of the date of planting.

The planning authority is required to consult the relevant parish council in England or community council in Wales on proposals to remove any hedgerows within their boundaries.

The authority must normally respond to a removal notice within 42 days of receiving it. If the authority has not issued a retention notice within this time the hedgerow may be removed.

The planning authority may only issue a 'hedgerow retention notice' in respect of 'important' hedgerows as defined by regulation 4. The retention notice prohibits the removal of the hedgerow and must specify each criterion (of those listed in Schedule 1) which applies to the hedgerow to which it relates. The authority may decide not to issue a notice if it is satisfied that there are circumstances which justify the hedgerow's removal (reg. 5(5)). The owner of a hedgerow who has been issued a hedgerow retention notice may appeal to the Secretary of State, who may hold a local inquiry.

What is an 'important' hedgerow?

An 'important' hedgerow (reg. 4), for the purposes of s. 97 of the Environment Act 1995 and the Hedgerows Regulations 1997, is one which:

1. has existed for at least 30 years; and
2. satisfies at least one of the following criteria (listed in Part II of Schedule 1 of the regulations):
   (a) Archaeology and history
      (i)   The hedgerow marks the boundary of a historic parish or township (existing before 1850).
      (ii) The hedgerow incorporates or is associated with an archaeological feature.
      (iii) The hedgerow marks the boundary of a pre-1600 AD estate or manor, or is visibly related to any building or other feature of such an estate or manor.
      (iv) The hedgerow is an integral part of a field system predating the Enclosure Acts or is visibly related to any building or other feature associated with such a system.
   (b) Wildlife and landscape
      (i)   The hedgerow contains species listed or categorised as follows:
            – Wildlife and the Countryside Act 1981, Part I (protection at all times) Schedule 1 (birds protected by special penalties), Schedule 5 (animals which are protected) or Schedule 8 (plants which are protected).

- Categorised as a declining breeder in 'Red Data Birds in Britain'.
- Categorised as 'endangered', 'extinct', 'rare' or 'vulnerable' in the Red Data Books for vascular plants, insects, stoneworts and invertebrates other than insects, or

(ii) The listed species have been officially recorded in the hedgerow within five years before, with respect to animals and birds, and within ten years before, with respect to plants, the making of the regulations (in March 1997).

(iii) The hedgerow includes a specified number of woody species (listed in Schedule 3), or specific named tree species, or a combination of woody species and particular features such as banks, walls, ditches etc. The number of woody species is reduced for some of the northern counties of England.

(iv) The hedgerow is adjacent to a bridleway or footpath, a road used as a public path, or a byway open to all traffic and includes at least four woody species and at least two of the features listed in paragraph 7(4)(a) to (g), e.g. banks, walls, ditches etc.

The criteria for identifying 'important' hedgerows are extremely complex and reference should be made to the regulations themselves for a detailed definition.

Permitted work

The removal of such an 'important' hedgerow may be permitted (reg. 6) if it is required:

1. for making a new opening in substitution for an existing opening which gives access to land, provided that the existing opening is filled by planting a hedge within eight months;
2. to obtain temporary access to land in an emergency;
3. to obtain access to land where another means is not available or costly;
4. for the purposes of national defence;
5. for carrying out certain developments for which planning permission has been granted;
6. for the purposes of flood defence or land drainage;
7. to prevent the spread of or to eradicate plant diseases and tree pests;
8. to enable the Secretary of State to carry out his or her functions in respect of any highway for which he or she is the highway authority;
9. to prevent interference with the electricity supply;
10. for the proper management of the hedgerow.

Offences

It is an offence to:

1. intentionally or recklessly remove, or cause or permit another person to remove, a hedgerow in contravention of the regulations (reg. 7(1));
2. fail to plant a hedge to fill the existing opening in a hedgerow within eight months of the creation of a new opening (reg. 7(2)).

A local planning authority may apply to the court for an injunction to restrain an actual or apprehended offence under these regulations.

The regulations give any person authorised in writing by the local planning authority or the Secretary of State a right to enter land to carry out various functions in relation to the protection of hedgerows, with and without a warrant (regs 12 and 13) and to take samples from hedgerows and soil (reg. 14).

The local planning authority may issue a notice requiring an owner (or utility operator) to plant another hedgerow where it appears that a hedgerow has been illegally removed (reg. 8). Replacement hedgerows shall be treated as 'important' hedgerows for the purposes of the regulations for the period of 30 years beginning with the date of completion of planting.

It should be noted that, although TPOs do not apply to hedges, trees covered by a TPO may be part of a hedge which itself is not an 'important' hedge and therefore not protected by the Hedgerow Regulations.

Records

The local planning authority is required under regulation 10 to compile and keep for public inspection free of charge a record of:

- Hedgerow removal notices received
- Hedgerow retention notices issued
- Notices giving permission to remove hedgerows (under reg. 5(1)(b)(i))
- Determinations of appeals heard by the Secretary of State

The Enclosure Acts

The Hedgerows Regulations are clearly intended to protect only the most ecologically important hedges, thereby devaluing all others. Colin Seymour has found an interesting solution to the problem of protecting other 'ordinary' hedges.

During the fourteenth century, enclosure of common land and its appropriation as private property began in Britain, initially for sheep rearing and later for corn production. The first Enclosure Act was passed in 1603 and subsequent Enclosure Acts eventually applied to around a quarter of all land in England. Much enclosed land was delimited by hedges.

At Hull County Court in January 1997, Seymour was successful in preventing the removal of a hedge by Flamborough Parish Council, his eighty-first success out of 81 such cases contested. The Council was ordered to preserve a hawthorn hedge for ever because it was still bound by the Flamborough Enclosure Act 1765 (*Seymour* v. *Flamborough Parish Council* (1997)). The judge emphasised that the court could only grant a remedy where there was a public interest, such as those hedges bordering a public place, or those on rights of way. It is believed that there are more than 4000 other Enclosure Acts still in force which could be used to protect some 40 000 miles of hedges in Britain (Anon., 1997).

6.2.4   Planting new hedges

The former DETR published an information leaflet entitled 'The right hedge for you' offering advice on laying hedges around domestic boundaries (Anon., 1999). It lists the numerous benefits of hedges, including acting as:

- wind breaks;
- filters (absorbing dust and noise);
- security (if plants with thorns are used);
- shelter and food for wildlife;
- inexpensive and long-lived barriers (compared with most fencing).

Unfortunately, of the 17 plant species and varieties suggested for use in new hedges by the DETR only six are native. While it is probably true that any hedge is better than no hedge at all, the ecological value of native species is far greater than that of non-native species.

The native species suggested are:

- Beech   *Fagus sylvatica*
- Hawthorn   *Crataegus monogyna*
- Hazel   *Corylus avellana*
- Holly   *Ilex aquifolium*
- Hornbeam   *Carpinus betulus*
- Yew   *Taxus baccata*

The form of hedges is not uniform throughout the UK. Different regions have their own style of hedge, for example the Devon and Somerset Hedge, and the Lancashire Hedge. Tarmac Quarry Products has planted more than 40 km of hedgerow and sponsors the National Hedgelaying Championships.

A number of organisations can supply information on native plant species and their availability. The Natural History Museum maintains a database of species native to various regions in the UK accessible via the internet and based on postcode locations. The charity Flora Locale maintains a list of suppliers of native seed.

REFERENCES

Anon. (1997) Ancient laws for today's hedges. *Natural World*, The Wildlife Trusts, Summer/Spring 1997, p. 12.

Anon. (1999) The right hedge for you. DETR, The Stationery Office Ltd, London.

Anon. (2000) Just moving the oak tree to the left, darling. *The Times Weekend Supplement*, 9 December 2000, p. 15.

Bell, S. and McGillivray, D. (2000) *Environmental Law. The Law and Policy relating to the Protection of the Environment*, 5th edn. Blackstone Press Ltd, London.

Cannell, M.G.R. (1999) Growing trees to sequester carbon in the UK: answers to some common questions. *Forestry*, 72(1): 237–247.

Gilbert, O.L. (1991) *The Ecology of Urban Habitats*. Chapman & Hall, London.

Springthorpe, G.D. and Myhill, N.G. (eds) (1985) *Wildlife Rangers Handbook*. Forestry Commission.

7

European and international wildlife law

The UK is bound by a variety of European and international laws whose purpose is to protect wildlife and wild places. Some of these laws have nature conservation as their primary objective, while others are principally concerned with the protection of some other element of the environment and a conservation gain may be achieved as a consequence.

7.1   EUROPEAN LAW

A number of EC laws are designed primarily to protect wildlife and wild places. Others are principally intended to protect other areas of the environment, such as water, but have a secondary effect on nature conservation. This section considers the Wild Birds Directive and the Habitats Directive and their effects upon domestic legislation. The Environmental Impact Assessment Directive is considered in Chapter 9.

7.1.1   Wild Birds Directive

Council Directive 79/409/EEC on the conservation of wild birds (the Wild Birds Directive) requires member states to take various conservation measures to protect birds. It relates to the conservation of all species of birds which occur naturally in the wild within the European territory of the member states (except Greenland), and covers their:

- Protection
- Management
- Control
- Exploitation

The Directive applies to:

- Birds
- Eggs
- Nests
- Habitats

Wild Birds Directive 1979 Art. 2

Member States shall take prerequisite measures to maintain the population of the species referred to in Article 1 at a level which corresponds in particular to ecological, scientific and cultural requirements, while taking account of economic and recreational requirements, or to adapt the population of these species to that level.

Member states are required to take measures to preserve or re-establish suitable habitats for all birds using a variety of methods.

Wild Birds Directive 1979 Art. 3

2. The preservation, maintenance and re-establishment of biotopes and habitats shall include primarily the following measures:
    (a) creation of protected areas;
    (b) upkeep and management in accordance with the ecological needs of habitats inside and outside the protected zones;
    (c) re-establishment of destroyed biotopes;
    (d) creation of biotopes.

Certain listed species (Annex I) must be the subject of special conservation measures, under Article 4, including the establishment of Special Protection Areas (SPAs) (Table 7.1). Similar measures must be taken for regularly occurring migratory species not listed in Annex I (Art. 4(2)), and particular attention must be paid to wetlands, especially those of international importance.

Member states are required to notify the Commission when they designate SPAs and provide all relevant information so that it may take appropriate measures to ensure that they form a coherent whole which meets the protection requirements of the protected species (Art. 4(3)). They must also protect SPAs from pollution or deterioration (Art. 4(4)).

Table 7.1 Special Protection Areas in the UK (April 2002)

Country	Number of sites	
	Classified SPAs	*Proposed SPAs*
England	76	7
Wales	14	4
Scotland	131	7
Northern Ireland	10	2
England – Scotland	1	0
England – Wales	2	2
United Kingdom	234	22

Source: *www.jncc.gov.uk/idt*

Wild Birds Directive 1979 Art. 4

1. The species listed in Annex I shall be the subject of special conservation measures concerning their habitat in order to ensure their survival and reproduction in their area of distribution.

     In this connection, account shall be taken of:
   (a) species in danger of extinction;
   (b) species vulnerable to specific changes in their habitat;
   (c) species considered rare because of small populations or restricted local distribution;
   (d) other species requiring particular attention for reasons of the specific nature of their habitat.
   … Member States shall classify in particular the most suitable territories in number and size as special protection areas for the conservation of these species, taking into account their protection requirements in the geographical sea and land area where this Directive applies.

Wild Birds Directive 1979 Art. 4

4. In respect of the protection areas … Member States shall take appropriate steps to avoid pollution or deterioration of habitats or any disturbances affecting the birds. … Outside these protection areas, Member States shall also strive to avoid pollution or deterioration of habitats.

Article 5 of the Directive requires the establishment of a general system of protection for all species of birds referred to in Article 1, prohibiting the killing or capture of birds, the damage or taking of eggs and nests, disturbance and the keeping of protected species. Article 6 prohibits the sale and transport of some species, whether alive or dead, including any parts or derivatives. Article 8 prohibits large-scale capture and killing.

Species which may be hunted under national legislation are listed in Annex II (Art. 7). Derogations from the provisions of Articles 5, 6, 7 and 8 are allowed for reasons of air safety, to protect public health and safety, to protect crops, livestock, fisheries, forests and water. Derogations are also allowed for other purposes, including conservation (Art. 9). Article 11 requires that any introduction of a bird species to an area where it does not naturally occur in the wild does not prejudice the local flora and fauna, and requires member states to consult the Commission.

Article 14 allows any member state to introduce stricter measures than provided for under the Directive. The Wildlife and Countryside Act 1981 and the Wildlife (Northern Ireland) Order 1985 implement many of the provisions of the Directive in the UK.

The conservation agencies are responsible for the designation of SPAs in the UK. No site is classified under this Directive unless it has first been notified as a SSSI. The SPA designation recognises the international importance of the site and protects it as a European site under the Conservation (Natural Habitats etc.) Regulations 1994. Development proposals that would be detrimental to the nature conservation interest of a SPA will only be permitted in exceptional circumstances.

EC case law

Case C-355/90: Commission v. Spain [1993] *(Marismas de Santoña)*

The Spanish government was in breach of Article 4 by failing to designate the Marismas de Santoña (Santoña Marshes), an important wetland area, as a SPA. This case established that a member state is under a duty to designate as a SPA any area that fulfils the ornithological criteria in the Directive.

Case C-44/95: R v. Secretary of State for the Environment, ex parte Royal Society for the Protection of Birds [1997] *(Lappel Bank)*

The government was challenged by the RSPB after failing to designate as a SPA an area within the Medway Estuary and Marshes known as Lappel Bank, arguing that economic considerations were relevant. The ECJ followed the decision in Santoña Marshes and held that the duty to designate was unaffected by economic considerations.

Case C-3/96: Commission v. *Netherlands* [1999]
The ECJ held that the Netherlands had breached its obligation under the Directive by failing to designate *sufficient* SPAs. A study had determined that 70 sites should have been designated but only 23 sites had been designated.

Case C-57/89: Commission v. *Germany* [1991] *(Leybucht Dykes)*
In the Leybucht Dykes case the ECJ held that a reduction in the area of a SPA was only justified on very limited grounds (for example in the interest of public health or public safety), and that works could not be permitted for recreational or economic reasons. This decision created a strong presumption against development in a SPA. However, the effect of this ruling was mitigated by an amendment to Article 4(4) of the Wild Birds Directive by Article 6(4) of the Habitats Directive (see page 263).

Case study – South Pennines Special Protection Area

Some 21 000 hectares of heather moorland and peat uplands in the South Pennines have been given Special Protection Area status because of their importance as a breeding area for upland birds. These include 12 species that are listed as being of international importance and in decline, while another four species are important migrants needing help and protection. The Standing Conference of South Pennine Authorities (SCOSPA) has produced an 'Integrated Management Strategy and Conservation Action Programme' for the area with the help of funding from the EU *Life* programme. SCOSPA comprises county and district authorities in the Pennines, two water companies and one voluntary organisation. It works closely with national statutory bodies and specialist environmental organisations. In order to protect the site SCOSPA has worked with landowners, farmers, commoners and individuals who hold shooting rights on the moors.

7.1.2   Habitats Directive

The main aim of Council Directive 92/43/EEC of 21 May 1992 on the conservation of natural habitats and wild fauna and flora (the Habitats Directive) is to promote the maintenance of biodiversity (Art. 2), while taking into account economic, social, cultural and regional requirements. It identifies certain priority species and habitats in need of special protection, and recognises that many of the threats to our natural heritage are of a transboundary nature.

The Directive requires the establishment of a coherent European ecological network of special areas of conservation (SACs) entitled 'Natura 2000' (Art. 3) (Table 7.2). This network will include:

Habitats Directive 1992 Art. 2

1. The aim of this Directive shall be to contribute towards ensuring bio-diversity through the conservation of natural habitats and of wild fauna and flora in the European territory of the Member States to which the Treaty applies.

2. Measures taken pursuant to this Directive shall be designed to maintain or restore, at a favourable conservation status, natural habitats and species of wild fauna and flora of Community interest.

3. Measures taken pursuant to this Directive shall take account of economic, social and cultural requirements and regional and local characteristics.

Table 7.2 Candidate Special Areas of Conservation in the UK (December 2001)

Country	Number of sites	
	Candidate SACs (cSACs)	Possible SACs (pSACs)
England	213	5
Wales	85	0
Scotland	219	5
Northern Ireland	43	0
England – Scotland	3	0
England – Wales	4	3
United Kingdom	567	13

Source: *www.jncc.gov.uk/idt*

- sites hosting the natural habitat types listed in Annex I;
- habitats of the species listed in Annex II;
- the Special Protection Areas (SPAs) established by the Wild Birds Directive.

Annexes I to III list:

Annex I   Natural habitat types of Community interest whose conservation requires the designation of Special Areas of Conservation (see Appendix 7).

Annex II   Animal and plant species of Community interest whose conservation requires the designation of Special Areas of Conservation.

Annex III   Criteria for selecting sites eligible for identification as sites of Community importance and designation as Special Areas of Conservation.

Habitats Directive 1992 Art. 4

1. On the basis of the criteria set out in Annex III (stage 1) and relevant scientific information, each Member State shall propose a list of sites indicating which natural habitats in Annex I and which species in Annex II that are native to its territory the sites host. ... The list shall be transmitted to the Commission, within three years of the notification of this Directive, together with information on each site ...

2. On the basis of the criteria set out ... the Commission shall establish, in agreement with each Member State, a draft list of sites of Community importance drawn from the Member States' lists identifying those which host one or more priority natural habitat types or priority species ...

   The list of sites selected as sites of Community importance ... shall be adopted by the Commission ...

4. Once a site of Community importance has been adopted ... the Member State concerned shall designate that site as a special area of conservation as soon as possible and within six years at most ...

Habitats Directive 1992 Art. 5

1. In exceptional cases where the Commission finds that a national list ... fails to mention a site hosting a priority natural habitat type or priority species which, on the basis of relevant and reliable scientific information, it considers to be essential for the maintenance of that priority natural habitat type or for the survival of that priority species, a bilateral consultation procedure shall be initiated between the Member State and the Commission for the purposes of comparing the scientific data used by each.

2. If, on expiry of a consultation period not exceeding six months, the dispute remains unresolved, the Commission shall forward to the Council a proposal relating to the selection of the site as a site of Community importance.

Where the Commission believes that a member state has not included on its list of proposed SACs a habitat that meets the criteria for selection the Commission may add it to the list (Art. 5) (see Figure 7.1).

Once a referral has been made to the Council a decision must be made within three months. During the consultation period the site is subject to Article 6(2) and must be protected from habitat deterioration and disturbance of the species for which the area is being considered for designation. Once a SAC is designated member states must establish conservation measures, management plans and appropriate statutory, administrative or contractual measures corresponding to the ecological requirements of the protected habitat or species (Art. 6(1)).

Habitats Directive 1992 Art. 6

2. Member States shall take appropriate steps to avoid, in the special areas of conservation, the deterioration of natural habitats and the habitats of species as well as disturbance of the species for which the areas have been designated ...

PANEL 7.1

Wildcat   *Felis sylvestris*

Identification/size
Similar to large tabby cat. Bushy, blunt-ended tail, with dark tip and several dark rings. Dark stripes on body. Head, body and tail 75–105 cm.

Habitat
Forests, scrub and plantations.

Distribution
In the British Isles, confined to Scottish Highlands.

Ecology and behaviour
Secretive and largely nocturnal. Territorial and a good climber. Feeds on small mammals, birds, amphibians and fish. Mating season February–March, litter size 2–6. Kittens born in a den in a hollow tree or abandoned badger sett or fox burrow. Interbreeds extensively with feral and domestic cats.

The Directive requires an 'appropriate assessment' of any plan or project which might significantly affect a SAC. The competent national authority may only agree to the plan or project after establishing that it will not adversely affect the integrity of the site (Art. 6(3)).

Habitats Directive 1992 Art. 6

3. Any plan or project not directly connected with or necessary to the management of the site but likely to have a significant effect thereon, ... shall be subject to appropriate assessment of its implications for the site in view of the site's conservation objectives. ... The competent national authorities shall agree to the plan or project only after having ascertained that it will not adversely affect the integrity of the site concerned and, if appropriate, after having obtained the opinion of the general public.

Notwithstanding Article 6(3), where there is a negative assessment of the implications for the site and where no alternative solution is available, a project or plan may be carried out for 'imperative reasons of overriding public interest, including those of a social or economic nature', provided that compensatory measures are taken (Art. 6(4)).

Obligations to protect sites arising under Articles 6(2), (3) and (4) of the Habitats Directive replace any obligations arising under the first sentence of Article 4(4) of the Wild Birds Directive in respect of areas classified as SPAs (Art. 7).

Habitats Directive 1992 Art. 6

4. ... Where the site concerned hosts a priority habitat type and/or a priority species, the only considerations which may be raised are those relating to human health or public safety, to beneficial consequences of primary importance for the environment or, further to an opinion from the Commission, to other imperative reasons of overriding interest.

Article 10 requires member states to endeavour to encourage the management of landscape features which are important for wild animals and plants in their land-use planning and development policies. Such features are those which are essential for migration, dispersal and genetic exchange. These include linear features, such as rivers

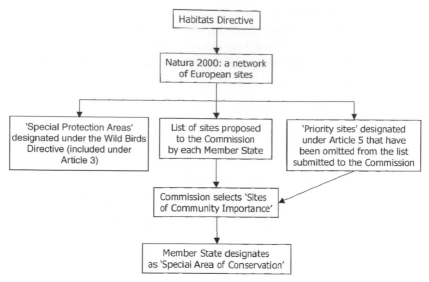

Figure 7.1 The designation of Special Areas of Conservation (Habitats Directive).

and their banks, and field boundaries such as hedgerows, along with features that function as 'stepping stones' such as ponds and small woods.

The Directive requires surveillance of the conservation status of the natural habitats and species referred to in Article 2, with particular regard to priority habitat types and priority species (Art. 11).

Articles 12, 13 and 14 make provision for the strict protection of species (Annex IV), the management of the exploitation of species (Annex V) and the prohibition of certain methods of taking, killing and transporting some animal species (Annex VI):

Annex IV   Animal and plant species of Community interest in need of strict protection (see Appendix 4).

Annex V   Animal and plant species of Community interest whose taking in the wild and exploitation may be subject to management measures (see Appendix 4).

Annex VI   Prohibited methods and means of capture and killing and modes of transport.

Article 19 makes provision for derogations similar to those allowed under the Wild Birds Directive, provided that the derogation is not detrimental to the maintenance of the populations of the species concerned at a favourable conservation status in their natural range.

The re-introduction of species listed in Annex IV is encouraged by Article 22 (a), where this may contribute to the re-establishment of these species at a favourable conservation status. The deliberate introduction of non-native species into the wild must be regulated (Art. 22(b)) and may be prohibited if necessary.

The Habitats Directive is implemented for Great Britain by the Conservation (Natural Habitats etc.) Regulations 1994. Some provisions are also implemented by the Wildife and Countryside Act 1981 and the Wildlife (Northern Ireland) Order 1985. All candidate SACs are protected as SSSIs.

Marine Special Areas of Conservation (Marine SACs)

The government is required to identify areas for protection as Special Areas of Conservation under the Habitats Directive, including marine areas. The UK Marine SACs Project has established management schemes for 12 candidate Marine SACs (Table 7.3). The special features of these sites include:

- Sandbanks which are slightly covered by sea water at all times
- Estuaries
- Mud and sand flats not covered by sea water at low tide

Table 7.3 The locations of Marine SACs and their special features

Location	Sandbanks	Estuaries	Mud and sand flats	Inlets and bays	Lagoons	Reefs	Submerged sea caves	Grey seal	Common seal	Bottlenose dolphin
Berwickshire and North Northumberland Coast			●			●	●	●		
Chesil and The Fleet					●					
Morecambe Bay			●	●						
Solway Firth	●									
Plymouth Sound	●	●		●						
The Wash and North Norfolk Coast	●		●	●					●	
Loch Maddy				●	●					
Papa Stour						●	●			
Sound of Arisaig	●									
Strangford Lough				●						
Pen Llŷn a'r Sarnau		●				●				
Cardigan Bay										●

Source: English Nature: *www.english-nature.org.uk/uk-marine/* (accessed 4.12.01).

- Large shallow inlets and bays
- Lagoons
- Reefs
- Submerged or partly submerged sea caves
- Grey seal
- Common seal
- Bottlenose dolphin

7.1.3 The Conservation (Natural Habitats etc.) Regulations 1994

The Conservation (Natural Habitats etc.) Regulations 1994 (SI 1994/2716) implement the Habitats Directive for Great Britain. Regulation 7 requires the Secretary of State to propose a list of sites to the Commission as required by the Directive. Once a site of Community importance has been adopted in Great Britain in accordance with the Directive, the Secretary of State is required to designate that site as a Special Area of Conservation as soon as possible and within six years at most (reg. 8(1)). The Secretary of State must establish priorities for the designation of sites in the light of their importance and the threats to which they are exposed (reg. 8(2)).

Conservation (Natural Habitats etc.) Regulations 1994 (SI 1994/No. 2716) reg. 8

---

(2) The Secretary of State shall establish priorities for the designation of sites in the light of –
   (a) the importance of the sites for the maintenance or restoration at a favourable conservation status of –
      (i) a natural habitat type in Annex I to the Habitats Directive, or
      (ii) a species in Annex II to the Directive,
     and for the coherence of Natura 2000; and
   (b) the threats of degradation or destruction to which those sites are exposed.

Regulation 9 makes provision for the Commission to add sites in addition to those proposed by the Secretary of State in pursuance of paragraph 2 of Article 5 of the Directive.

Under the Regulations, a 'European site' is defined as:

- a Special Area of Conservation;
- a site of Community importance which has been placed on the list referred to in the third sub-paragraph of Article 4(2) of the Directive

(i.e. the list sent to the Commission, which may contain additional sites added by the Commission);

■ a site hosting a priority habitat type or priority species in respect of which consultation has been initiated under Article 5(1) of the Directive, during the consultation period or pending a decision of the Council under Article 5(3), regulation 10(1)(c); or

■ Special Protection Area designated under the Wild Birds Directive.

The Conservation (Natural Habitats etc.) (Amendment) (England) Regulations 2000 (SI 2000/192) added a category to the definition of a European site in England, namely any site included in the list that has been transmitted to the Commission by the Secretary under regulation 7 until:

■ the draft list of sites is established under Article 4(2) of the Directive (draft list of sites of Community importance), where in any case the site is not included in that list, or

■ the list of sites selected as sites of Community importance (under Article 4(2)) is adopted by the Commission.

Sites which are European sites by virtue only of regulation 10(1)(c) are not within regulations 20(1) and (2), 24 and 48, which relate to the approval of certain plans and projects, and therefore receive a reduced level of protection from potentially damaging operations. The term 'European marine site' applies to a European site which consists of, or so far as it consists of, marine areas (reg. 2(1)).

Under regulation 11, registers of European sites must be compiled and maintained in Britain. These registers must be available for public inspection free of charge. In England and Wales an entry in the register is a local land charge (reg. 14). Any sites added to or removed from the register must be notified to the appropriate nature conservation body (reg. 12) which must then notify:

■ every owner or occupier of land within the site (including land covered by water);

■ every local planning authority in whose area any part of the site is located;

■ other persons or bodies as directed by the Secretary of State.

The appropriate nature conservation body may enter into a management agreement with every owner, lessee and occupier of land which forms part of a European site or land adjacent to such a site, for the management, conservation, restoration or protection of the site as a whole or any part of it (reg. 16(1)). Compensation payments may be made for any restrictions imposed.

Under regulation 18, any notification in force in relation to a European site under s. 28 of the Wildlife and Countryside Act 1981 (areas of special scientific interest) has effect for the purposes of the Conservation (Natural Habitats etc.) Regulations. Restrictions on the carrying out of damaging operations within a European site are imposed by regulation 19.

Regulation 22 gives the Secretary of State a power to make a special nature conservation order in respect of any land within a European site after consultation with the appropriate nature conservation body. Such an order will specify operations which appear to the Secretary of State likely to destroy or damage the special features for which the site has been designated. In England and Wales a special nature conservaton order is a local land charge. Restrictions on the carrying out of operations specified in the order by any person (not just the owner or occupier) are given in regulation 23 and provisions for compensation for any loss of value of any interest in the land resulting from the making of an order are made by regulation 25. A person convicted of an offence under regulation 23 may be made the subject of an order requiring him to restore the land to its former condition. Failure to comply with such an order is an offence under regulation 26(5). If any operations specified by the order are not carried out within the period specified in the order the appropriate nature conservation body may enter the land, carry out the operations and recover the costs from the person who is the subject of the order (reg. 26(6)).

Under regulation 28 the appropriate nature conservation body may make byelaws for the protection of a European site under s. 20 of the National Parks and Access to the Countryside Act 1949. Such byelaws may:

- prohibit or restrict entry to, or movement within the site of, persons, vehicles, boats or animals;
- protect the animals, plants or soil;
- prohibit the depositing of rubbish and leaving of litter;
- restrict or prohibit the lighting of fires or any activities likely to cause fire.

The appropriate conservation body may acquire land within a European site by compulsory purchase where:

- it is unable to conclude a management agreement on reasonable terms, or
- where a management agreement has been broken which prevents or impairs the satisfactory management of the site (reg. 32).

Regulation 33 makes provision for the appropriate conservation body to indicate the position of a European marine site with markers. In addition it requires the conservation body to advise other relevant authorities of:

- the conservation objectives of a site, and
- any operations which may cause deterioration of natural habitats or the habitats of species, or disturbance of species, for which the site has been designated.

Provision for the making of management schemes by the relevant authorities and the making of byelaws by the nature conservation bodies to protect European marine sites are made by regulations 34 and 36 respectively. For the purposes of these regulations, in relation to marine sites, the relevant authorities may include:

- a nature conservation body;
- a relevant council;
- the Environment Agency;
- a harbour authority;
- a lighthouse authority;
- a navigation authority;
- a local fisheries committee etc. (reg. 5).

Regulation 37 requires planning authorities to include in their development policies a consideration of the conservation value of linear and continuous landscape features (such as rivers) and features which act as 'stepping stones' for wildlife (such as ponds and small woods).

Part III of the regulations contains a range of offences aimed at the protection of wild animals and wild plants of European protected species. These are similar to those created by the Wildlife and Countryside Act 1981. Regulation 44 makes provision for the licensing of activities affecting European protected species which are otherwise prohibited by the regulations. European protected species of animals are those species listed in Annex IV(a) to the Habitats Directive and are listed in Schedule 2 of the regulations. European protected species of plants are listed in Annex IV(b) to the Directive and Schedule 4 of the regulations (see Appendix 4). Species of animals listed in Schedule 3 of the regulations may not be taken or killed in certain ways. However, the prohibited methods may be used if a licence has been granted under regulation 44.

Part IV of the regulations is concerned with the adaptation of planning and other controls for the protection of European sites. Regulation 48 requires the assessment of the implications for a European site of any proposed plan or project.

A person applying for such a consent must provide any information required by the competent authority for the purposes of the assessment (reg. 48(2)). The competent authority is also required to consult the appropriate nature conservation body (reg. 48(3)) and may also consult the general public (reg. 48(4)). The competent authority should agree to the plan or project only if it will not adversely affect the European site (reg. 48(5)).

**Conservation (Natural Habitats etc.) Regulations 1994
(SI 1994/2716) reg. 48**

---

(1) A competent authority, before deciding to undertake or give
any consent, permission or other authorisation for, a plan or
project which –
  (a) is likely to have a significant effect on a European site in
      Great Britain (either alone or in combination with other
      plans or projects), and
  (b) is not directly connected with or necessary to the
      management of the site, shall make an appropriate
      assessment of the implications for the site in view of that
      site's conservation objectives.

**Conservation (Natural Habitats etc.) Regulations 1994
(SI 1994/2716) reg. 48**

---

(5) In the light of the conclusions of the assessment, and subject
to regulation 49, the authority shall agree to the plan or
project only after having ascertained that it will not adversely
affect the integrity of the European site.

Regulation 49 gives the competent authority the power to agree
to the plan or project notwithstanding a negative assessment if no
alternative solution exists and if it must be carried out for 'imperative
reasons of overriding public interest'. These reasons must relate to:

■ human health;
■ public safety;
■ beneficial consequences of primary importance to the environment;
  or
■ other reasons which in the opinion of the European Commission are
  imperative reasons of overriding public interest.

The types of planning permission to which regulations 48 and 49 apply
and other issues relating to planning, authorisations and consents are
dealt with by regulations 54 to 85.

Part V of the regulations makes a number of supplementary
provisions relating to, among other things, management agreements,
potentially damaging operations, payments and compensation,

powers of entry of nature conservation bodies, enforcement and the procedure for making byelaws.

Case law

R v. Secretary of State for Transport, ex parte Berkshire, Buckinghamshire and Oxfordshire Naturalists Trust [1997]
This case concerned an application for judicial review of the construction of the Newbury Bypass. The Naturalists Trust argued that the decision to proceed with the bypass frustrated any future decision to submit the site as a candidate SAC. The site was important because of the presence of terrestrial pulmonate snails. The site was not a SSSI and for this reason could not be a candidate SAC. (The UK government takes the view that all terrestrial candidate SACs must first be SSSIs). Furthermore, English Nature and the Highways Agency had made considerable efforts to translocate the snails. The application was rejected but the judge expressed his regret that the protection of the natural environment keeps coming second.

R v. Secretary of State for Scotland and Others, ex parte World Wildlife Fund UK Ltd and Royal Society for the Protection of Birds (1998)
The court dismissed an application for judicial review challenging the manner of delineating the boundaries for a special protection area and a special area of conservation in the Cairngorms. The boundaries were of conservation significance as they related to a planning application for a proposed funicular railway for skiers. Part of the reason for excluding the area from the candidate site was that it was already developed. It was held that there was room for discretion in the drawing of boundaries of a SAC provided the discretion was exercised on ornithological grounds only. This decision suggests that the courts are unlikely to interfere with the drawing of boundaries around sites unless they are completely irrational.

Case C-371/98: R v. Secretary of State for the Environment, Transport and the Regions, ex parte First Corporate Shipping Limited (World Wide Fund for Nature UK and Avon Wildlife Trust, interveners) (2000)
Economic considerations should not be taken into account by member states when submitting candidate SAC sites to the Commission. When selecting and deciding the boundaries of sites to be proposed to the Commission for designation as SACs under the Habitats Directive, member states must have regard to conservation matters only, and not economic or cultural ones. This was the ruling of the European Court of Justice (ECJ) on a reference by the Queen's Bench Divisional Court

for a preliminary ruling on a question of interpretation of the Habitats Directive.

The statutory port authority of the port of Bristol, the applicant, owned considerable land in the neighbourhood and had invested £220 million in the development of its facilities and employed between 3000 and 5000 persons. The applicant sought judicial review of the Secretary of State's stated intention to propose the Severn Estuary to the Commission as a site eligible for designation as a SAC under Article 4(1) of the Habitats Directive, submitting that he was obliged to take account of economic, social and cultural requirements under Article 2(3).

The purpose of the Directive is to establish a network of protected sites of Community importance (Natura 2000). The ECJ ruled that, since, when a member state drew up its national list of proposed SACs, it was not in a position to have detailed knowledge of the situation of habitats in the other states, it could not of its own accord, because of economic, social or cultural requirements, or regional or local characteristics, delete sites which were relevant to the conservation objective at national level without jeopardising the realisation of that objective at Community level.

R v. Secretary of State for Trade and Industry, ex parte Greenpeace
(No. 2) [2000]
The High Court ruled that the provisions of the Habitats Directive extend to the continental shelf. Greenpeace challenged the awarding of oil exploration licences because of the potential damage to marine life including some distant water species listed in the Directive. The Court also considered the place of the Directive within other measures to which the UK is a party which clearly extend beyond territorial waters, such as the UN Convention on Biological Diversity.

7.1.4   The Water Directives

A number of Directives seek to improve the quality of freshwater and marine waters and as such have a significant impact upon the environment of aquatic organisms.

Quality of Freshwater to Support Fish Life Directive

The Directive on the quality of freshwaters needing protection or improvement in order to support fish life (78/659/EEC) aims to establish water quality requirements for freshwater fish. It regulates sampling frequency, measuring methods and measuring requirements, along with conditions in which the quality requirements are to be achieved. The waters referred to in the Directive are either suitable for salmonids (salmon and trout) or cyprinids (coarse fish).

Quality of Freshwater to Support Fish Life Directive Art. 1

3. The aim of this Directive is to protect or improve the quality of those running or standing fresh waters which support or which, if pollution were reduced or eliminated, would become capable of supporting fish belonging to:
    (a) indigenous species offering a natural diversity, or
    (b) species the presence of which is judged desirable for water management purposes by the competent authorities of the Member States.

Urban Waste Water Treatment Directive

The Directive 91/271/EEC on urban waste water treatment sets minimum requirements for waste water treatment, thereby protecting both freshwater and marine ecosystems from pollution from sewage. It also banned the dumping of sewage sludge in the sea from the end of 1998.

Changes to the law on the disposal of sewage and sewage sludge have had beneficial effects on the marine environment. The Urban Waste Water Treatment Directive has required the construction of waste water treatment plants in places such as Liverpool and Blackpool, where sewage was previously discharged untreated into the sea.

This Directive has been implemented by the Urban Waste Water Treatment (England and Wales) Regulations 1994 (SI 1994/2841) and the Urban Waste Water Treatment (Scotland) Regulations 1994 (SI 1994/2842).

Urban Waste Water Treatment Directive Art. 4

1. Member States shall ensure that urban waste water entering collecting systems shall before discharge be subject to secondary treatment or an equivalent treatment ...

Bathing Waters Directive

Directive 76/160/EEC concerning the quality of bathing waters lays down various parameters, mostly bacteriological, with which all 'traditional' bathing waters must comply. Until recently, one of the most polluted coastlines in the UK was the Fylde Coast. In 1993 the ECJ

> ## Bathing Waters Directive Art. 3
>
> 1. Member States shall set, for all bathing areas or for each individual bathing area, the values applicable to bathing water for the parameters given in the Annex.

Figure 7.2 The sea at Weston-super-Mare has consistently complied with EC Bathing Waters Directive standards. The small island to the right of the pier is Steep Holm, a SSSI.

found the UK to be in breach of EC standards on Blackpool and Southport beaches (*Commission* v. *United Kingdom* [1993]). In November 2001 the Environment Agency announced that the sea at Blackpool had met EC bathing water standards for the first time in 22 years (Anon., 2001a). The Directive was implemented in England and Wales by the Bathing Waters (Classification) Regulations 1991 (SI 1991/1597).

The Shellfish Waters Directive

Directive 79/923/EEC on the quality required of shellfish waters has been implemented by the Surface Waters (Shellfish) (Classification) Regulations 1997 (SI 1997/1332) and the Surface Waters (Shellfish) Directions 1997. The regulations prescribe a system for classifying controlled waters (coastal or brackish) which need protection or improvement in order to support shellfish. They also incorporate the reference methods of measurement and sampling frequency laid down by the Directive.

7.1.5   Trade in endangered species regulation in the EU

Trade in endangered species within the EU is regulated by the Regulation on the implementation in the Community of the Convention on International Trade in Endangered Species of Wild Fauna and Flora (338/97/EC). This treaty is discussed below in Section 7.2.9.

7.1.6   Environmental impact assessment in the EU

The requirement for the assessment of the environmental effects of major development projects in the EU is laid down in Directive 85/337/EEC on the assessment of the effects of certain public and private projects on the environment. This Directive and its implementation in UK law is discussed in Chapter 9.

7.2   INTERNATIONAL PROTECTION OF WILDLIFE AND HABITATS

Some of the most important nature conservation sites in the UK are designated as protected areas under international law. They may also benefit from European and UK designations and will generally be SSSIs.

The Secretaries of State normally call in and determine any planning applications which significantly affect conservation sites of national or international importance.

7.2.1   Biosphere Reserves

Biosphere Reserves (see Panel 7.2) are protected areas of terrestrial and coastal ecosystems which represent important examples of the world's biomes. Collectively the sites form a world network, and they are particularly useful in the monitoring of long-term ecological changes in the biosphere as a whole.

Biosphere Reserves were devised by UNESCO as part of its Man and the Biosphere (MAB) Programme (Project No. 8). Criteria and guidelines for selection of sites were produced by a UNESCO task force in 1974. Sites are nominated by national governments and must meet a set of minimum criteria and conditions before being admitted to the world network. All Biosphere Reserves receive statutory protection as National Nature Reserves.

PANEL 7.2   BIOSPHERE RESERVES

Each Biosphere Reserve is intended to fulfil three basic functions:

- Conservation of landscapes, ecosystems, species and genetic variation
- Sustainable development
- Support for research, monitoring, education and information exchange

There are currently 393 Biosphere Reserves in 94 countries, including 13 in the UK:

- Beinn Eighe
- Braunton Burrows
- Caerlaverock
- Cairnsmore of Fleet
- Claish Moss
- Dyfi
- Isle of Rhum
- Loch Druidibeg
- Moor House – Upper Teesdale
- North Norfolk Coast
- Silver Flowe – Merrick Kells
- St Kilda (also a World Heritage Site)
- Taynish

Examples of sites listed in the UK *Man and the Biosphere* directory:

Braunton Burrows

Location	51 deg. 06'N, 04 deg. 12'W
Area (hectares)	596
Elevation range (m)	0–30
Biome	Dune system
Biogeographical province	British Islands
Year of designation	1976

Principal monitoring and research themes

- Biological inventory
- Ecosystem restoration (dune systems)
- Coastal geomorphology
- Effects of natural and human influences on biota

Caerlaverock

Location	54 deg. 58'N, 04 deg. 17'W
Area (hectares)	5010
Elevation range (m)	200–710
Biome	Mud flat and grassland
Biogeographical province	British Islands
Year of designation	1976

Principal monitoring and research themes

- Ecosystem dynamics of tidal marshes
- Wildlife management research (barnacle geese)
- Faunal populations (emphasis on birds and herpetofauna)

7.2.2   Biogenetic Reserves

In 1973, the European Ministerial Conference on the Environment recommended the establishment of a European network of reserves to conserve representative examples of European flora, fauna and natural areas. In 1976, the European Network of Biogenetic Reserves was established by the Council of Europe.

The selection of Biogenetic Reserves is based on two criteria (Council of Europe Resolution (76)17):

■ their nature conservation value: they must contain specimens of flora or fauna that are typical, unique, rare or endangered;
■ the effectiveness of their protected status: this must be sufficient to ensure the long-term conservation or management of a site.

The Council of Europe undertakes specialist studies for each biotope, identifying the most characteristic sites in each case. The aim is to have all of these sites included in the network in the long term.

The network of reserves develops in two different ways:

■ countries can propose reserves by applying to the Secretariat and Committee of Experts of the Council of Europe; or
■ the Council of Europe may identify sites (whether already protected or not) which it would like to include in the network.

To date, the network consists of 340 reserves, covering almost 4 million hectares. In the UK there are currently 18 Biogenetic Reserves, covering an area of some 8000 hectares. All of these reserves receive statutory protection as SSSIs, and most are NNRs.

7.2.3   Ramsar Convention

The Convention on Wetlands of International Importance Especially as Waterfowl Habitat (the Ramsar Convention) was adopted in 1971 and ratified by the UK in 1976.

The aim of the Convention is to protect wetlands, their flora and fauna, and to promote their wise use. Parties are obliged to designate wetlands in their territory for inclusion in the List of Wetlands of International Importance, based on eight criteria (Panel 7.3). They are also obliged to include wetland conservation considerations in their national land-use planning.

Article 2(1) of the Convention requires the contracting parties to designate suitable wetlands for inclusion in a 'List of Wetlands of International Importance' which is maintained by the International Union for the Conservation of Nature and Natural Resources (IUCN) (Art. 8).

Very broad definitions of both wetlands and wildfowl are used in the Convention (Art. 1).

Ramsar Convention 1971 Art. 1

1. ... wetlands are areas of marsh, fen, peatland, or water,
    whether natural or artificial, permanent or temporary, with
    water that is static or flowing, fresh, brackish or salt, including
    areas of marine water the depth of which at low tide does not
    exceed six metres.
2. ... waterfowl are birds ecologically dependent on wetlands.

Ramsar Convention 1971 Art. 2

2. Wetlands should be selected for the List on account of their
    international significance in terms of ecology, botany, zoology,
    limnology or hydrology. In the first instance wetlands of
    international importance to waterfowl at any season should
    be included.

Ramsar sites are designated as SSSIs in Britain and the equivalent
ASSIs in Northern Ireland. There are 142 Ramsar sites in the UK including estuaries, meres, bogs, reservoirs, lochs, mountains, plateaux,
forests, moors, heathlands, fens, flats, mosses, marshes, bays, islands,
harbours and stretches of coastline.

On 2 February 2000 (World Wetlands Day), the then DETR minister
Chris Mullin marked the 29th anniversary of the signing of the
Convention in Ramsar, Iran, by designating the Northumbria Coast
and Breydon Water Extension in Norfolk as Ramsar sites. Since then
other sites have been designated at Slieve Beagh, Black Bog, Gary Bog
and Fairy Water Bogs (Northern Ireland), Lewis Peatlands, Inner Clyde
Estuary, Firth of Tay and Eden Estuary (Scotland), Lee Valley, Thames
Estuary and Marshes, South West London Waterbodies (England) and
the South East Coast of Jersey, Channel Islands.

Under the Wildlife and Countryside Act 1981 s. 37A (as amended by
the Countryside and Rights of Way Act 2000 s. 77) the Secretary of State
is required to notify the designation of wetlands as Ramsar
sites to English Nature and/or the Countryside Council for Wales,
depending upon their location. These agencies are, upon receipt of
such notification, required to notify:

- the relevant local planning authorities;
- every owner and occupier of any of the wetland;
- the Environment Agency;

PANEL 7.3   CRITERIA FOR IDENTIFYING WETLANDS OF
INTERNATIONAL IMPORTANCE

A wetland should be considered internationally important if it ...

Group A – Sites containing representative, rare or unique wetland types

Criterion 1   ... contains a representative, rare, or unique example of a natural or near-natural wetland type found within the appropriate biogeographical region.

Group B – Sites of international importance for conserving biological diversity

Criterion 2   ... supports vulnerable, endangered, or critically endangered species or threatened ecological communities.

Criterion 3   ... supports populations of a plant and/or animal species important for maintaining the biological diversity of a particular biogeographical region.

Criterion 4   ... supports plant and/or animal species at a critical stage in their life cycles, or provides refuge during adverse conditions.

Criterion 5   ... regularly supports 20 000 or more waterbirds.

Criterion 6   ... regularly supports 1 per cent of the individuals in a population of one species or subspecies or waterbird.

Criterion 7   ... supports a significant proportion of indigenous fish subspecies, species or families, life history stages, species interactions and/or populations that are representative of wetland benefits and/or values and thereby contributes to global biodiversity.

Criterion 8   ... is an important source of food for fish, spawning ground, nursery and/or migration path on which fish stocks, either within the wetland or elsewhere, depend.

- every relevant undertaker (s. 4(1) of the Water Industry Act 1991) and every internal drainage board (s. 61C(1) of the Land Drainage Act 1991) whose works, operations or activities may affect the wetland (s. 37A(2)).

Ramsar sites in the UK form part of an international network of such sites, including:

- Danube Delta, Romania
- Lake Naivasha, Kenya
- Okavango Delta System, Botswana
- Larnaca Salt Lake, Cyprus

The Convention's short-term goal is to have at least 2000 sites on the Ramsar List by 2005, double the current number.

PANEL 7.4   RAMSAR SITES

There are 142 designated sites distributed throughout the UK as follows:

- England   66
- Wales   7
- Scotland   50
- Northern Ireland   15
- England–Wales   3
- England–Scotland   1

Some of the largest sites in the UK (i.e. those over 10 000 ha) are:

- Caithness and Sutherland Peatlands   143 539 ha
- The Wash   62 212 ha
- Lewis Peatlands   58 984 ha
- Lough Neagh and Lough Beg   50 166 ha
- Morecambe Bay   37 405 ha
- Upper Solway Flats and Marshes   30 706 ha
- New Forest   28 003 ha
- Severn Estuary   24 701 ha
- Strangford Loch   15 581 ha
- Humber Flats, Marshes and Coast   15 203 ha
- Ribble and Alt Estuaries   13 423 ha
- The Dee Estuary   13 085 ha
- Foulness   10 933 ha

The smallest site in the UK is Llyn Idwal, a glacial lake in Snowdonia which covers just 14 ha.

Profiles of selected Ramsar sites

Name	Rostherne Mere
Location	England
Area	80 ha
Designations	Nature Reserve, SSSI
Site no.	221
Habitat	Part of a series of open water and peatland areas, including peat bog and marsh, set in a glaciated landscape. Vegetation consists of fringing reedbeds, woodland and farmland.
Species	Wintering waterbirds include nationally important numbers of various duck species.
Human activities	Agriculture and bird hunting.

Name	Rutland Water
Location	England
Area	1 360 ha
Designations	Special Protection Area, SSSI
Site no.	533
Habitat	Large, artificial freshwater reservoir, fringed by wetland habitats and semi-natural mature woodland.
Species	Regionally important for breeding and passage birds. Wintering waterbirds regularly exceed 20 000, including internationally important numbers of ducks and nationally important numbers of several species of ducks, geese and swans.
Human activities	Recreation

Name	Solent and Southampton Water
Location	England
Area	5 415 ha
Designations	National Nature Reserve, SSSI
Site no.	965
Habitat	Estuaries, coastal habitats including tidal flats, saline lagoons, shingle beaches, reefs, saltmarsh, reedbeds, grazing marsh and damp woodland. Long periods of slack water at high and low tide.
Species	Internationally important numbers of wintering waterfowl (over 50 000), breeding gull and tern populations, rare invertebrates and plants.
Human activities	Recreation, tourism, fishing, marine aquaculture and hunting.

Name	Chesil Beach and The Fleet
Location	England
Area	748 ha
Designations	Special Protection Area, SSSI, AONB, Heritage Coast
Site no.	300
Habitat	Shingle storm beach of international geomorphological importance. Shallow lagoon supporting saltmarsh and reedbeds.
Species	Internationally important for wintering ducks, geese and swans. Nationally important for breeding birds.
Human activities	Tourism, recreation and near a major port.

Name	Claish Moss
Location	Scotland
Area	568 ha
Designations	UNESCO Biosphere Reserve, Nature Reserve, SSSI
Site no.	218
Habitat	One of Britain's best examples of a patterned raised mire system. Water is of high quality. Divided by streams into distinct units.
Species	Rare plants, 14 moss species, interesting invertebrate fauna and nationally important assemblage of nine dragonfly species.
Human activities	Annual deer cull to prevent overpopulation.

Source: *The Annotated Ramsar List of Wetlands of International Importance, Ramsar Country Profiles: United Kingdom, www.ramsar.org/profiles_uk.htm.*

7.2.4   World Heritage Convention

World Heritage Sites are of international importance and are designated under the Convention Concerning the Protection of the World Cultural and Natural Heritage (World Heritage Convention). The Convention was adopted by the United Nations Educational, Scientific and Cultural Organisation (UNESCO) in 1972 and ratified by the UK government in 1984.

The Convention distinguishes between 'cultural' and 'natural' heritage. Article 1 defines 'cultural heritage' as including monuments, buildings, architectural structures, works of monumental sculpture and painting, archaeological sites, inscriptions and cave dwellings. 'Natural heritage' is defined by Art. 2.

World Heritage Convention 1972 Art. 2

For the purposes of this Convention, the following shall be considered as 'natural heritage':

Natural features consisting of physical and biological formations or groups of formations, which are of outstanding universal value from the aesthetic or scientific point of view;

Geological and physiographical formations and precisely delineated areas which constitute the habitat of threatened species of animals and plants of outstanding universal value from the point of view of science or conservation;

Natural sites or precisely delineated areas of outstanding universal value from the point of view of science, conservation or natural beauty.

Most sites are managed and conserved by a single organisation. Hadrian's Wall is an exception, with 41 organisations responsible for its protection, including English Nature and a number of local authorities. It stretches 73 miles, between Wallsend on the River Tyne, and Bowness on the Solway Firth, and marked the northernmost frontier of the Roman Empire.

The Convention requires parties to protect the cultural and natural heritage through the establishment of appropriate agencies, planning policies, research and legislation (Art. 5).

Article 11 requires every state party to the Convention to submit to the World Heritage Committee an inventory of property within its territory to be included in the 'World Heritage List'.

PANEL 7.5   WORLD HERITAGE SITES

There are currently 690 World Heritage Sites, divided into two categories: natural and cultural. They include many internationally famous sites such as:

- Galapagos Islands (Ecuador)
- Serengeti National Park and Ngorongoro Conservation Area (Tanzania)
- Grand Canyon National Park (USA)
- Great Barrier Reef (Australia)
- Bialowieza National Park (Poland)

The sites in the UK are:

- Giant's Causeway and Causeway Coast
- Durham Castle and Cathedral
- Ironbridge Gorge
- Studley Royal Park, including the Ruins of Fountains Abbey
- Stonehenge, Avebury and associated sites
- Castles and Town Walls of King Edward in Gwynedd
- St Kilda (also a Biosphere Reserve)
- Blenheim Palace
- City of Bath
- Hadrian's Wall
- Westminster Palace, Westminster Abbey and St Margaret's Church
- Henderson Island
- Tower of London
- Canterbury Cathedral, St Augustine's Abbey and St Martin's Church
- Old and New Towns of Edinburgh
- Gough Island Wildlife Reserve
- Maritime Greenwich
- Heart of Neolithic Orkney
- Blaenavon Industrial Landscape
- Devon and East Dorset coast (designated 2001)
- New Lanark Industrial Landscape, South Lanarkshire (designated 2001)
- Saltaire Industrial Landscape, West Yorkshire (designated 2001)
- Derwent Valley Mills Industrial Landscape, Derbyshire (designated 2001)

Hadrian's Wall

World Heritage Convention 1972 Art. 5

To ensure that effective and active measures are taken for the protection, conservation and presentation of the cultural and natural heritage ... each state party ... shall endeavour ...:

a. to adopt a general policy which aims to give the cultural and natural heritage a function in the life of the community and to integrate the protection of that heritage into comprehensive planning programmes;
b. to set up ... services for the protection, conservation and presentation of the cultural and natural heritage ...;
c. to develop scientific and technical studies and research and to work out such operating methods as will make the state capable of countering the dangers that threaten its cultural or natural heritage;
d. to take the appropriate legal, scientific, technical, administrative and financial measures necessary for the identification, protection, conservation, presentation and rehabilitation of this heritage.

7.2.5 Bonn Convention

The Convention on the Conservation of Migratory Species of Wild Animals (Bonn Convention) was adopted in 1979 and came into force in the UK on 1 October 1985.

The principal objective of the Bonn Convention is to protect migratory species. In order to achieve this the Convention provides strict protection for listed species in danger of extinction and aims to persuade 'range states' to conclude agreements for the conservation of other species that have an unfavourable conservation status.

The Convention lists endangered migratory species in Appendix I (Art. III(1)) and migratory species which have an unfavourable

---

Bonn Convention 1979 Art. I

1. For the purpose of this Convention:
   a. 'Migratory species' means the entire population or any geographically separate part of the population of any species or lower taxon of wild animals, a significant proportion of whose members cyclically and predictably cross one or more national jurisdictional boundaries.

conservation status and which either require protection by international agreements or would benefit from international cooperation in Appendix II (Art. IV(1)).

Contracting parties that are range states of Appendix I species must restore habitats important to these species, remove barriers to their migrations and control factors responsible for their decline.

Appendix I species may only be taken by parties that are range states for scientific or conservation purposes, to accommodate the needs of traditional subsistence users, and under extraordinary circumstances (Art. III(5)).

The UK has signed and ratified three regional agreements under the convention:

Bonn Convention 1979 Art. II

3. ..., the Parties:
    b. shall endeavour to provide immediate protection for migratory species included in Appendix I; and
    c. shall endeavour to conclude agreements covering the conservation and management of migratory species included in Appendix II.

Bonn Convention 1979 Art. III

1. Appendix I shall list migratory species which are endangered.

4. Parties that are Range States of a migratory species listed in Appendix I shall endeavour:
    a. to conserve and, where feasible and appropriate, restore those habitats of the species which are of importance in removing the species from danger of extinction;
    b. to prevent, remove, compensate for or minimise, as appropriate, the adverse effects of activities or obstacles that seriously impede or prevent the migration of the species; and
    c. to the extent feasible and appropriate, to prevent, reduce or control factors that are endangering or are likely to further endanger the species, including strictly controlling the introduction of, or controlling or eliminating, already introduced exotic species.

- The Agreement on the Conservation of Small Cetaceans of the Baltic and North Seas (ASCOBANS) 1991
- The Agreement on the Conservation of Bats in Europe (EUROBATS) 1991
- The Agreement on the Conservation of African–Eurasian Migratory Waterbirds (AEWA) 1995

The UK is working towards becoming a signatory to the Agreement on the Conservation of Cetaceans in the Black Sea, Mediterranean Sea and Contiguous Atlantic Area (ACCOBAMS) 1996, on behalf of Gibraltar.

7.2.6   Berne Convention

The Convention on the Conservation of European Wildlife and Natural Habitats (Berne Convention) was adopted in 1979 and ratified by the UK in 1982. The aim of the Convention is to conserve flora, fauna and their natural habitats in Europe. The Convention places particular emphasis on the protection of endangered and vulnerable species (endemic and migratory) and those species and habitats whose conservation requires the cooperation of several European states.

Parties to the Convention undertake to maintain populations of the species to which the agreement relates, or to take steps to increase the populations of these species where necessary. Account may be taken of cultural, social and recreational requirements. The requirements of particular subspecies, varieties or forms (Art. 2) must also be considered. Parties must ensure that local and national planning policies take account of wildlife. Chapter I of the Convention (Articles 1–3) contains general provisions.

Berne Convention 1979 Art. 3

1. Each Contracting Party shall take steps to promote national policies for the conservation of wild flora, wild fauna and natural habitats, with particular attention to endangered and vulnerable species, especially endemic ones, and endangered habitats, in accordance with the provisions of this Convention.

2. Each Contracting Party undertakes, in its planning and development policies and in its measures against pollution, to have regard to the conservation of wild flora and fauna.

The Convention lists protected species in three appendices:

- Appendix I   Protected plants
- Appendix II   Protected animals
- Appendix III   Protected animals not included in Appendix II

Chapter II of the Convention is concerned with the protection of habitats (Art. 4).

Berne Convention 1979 Art. 4

1. Each Contracting Party shall take appropriate and necessary legislative and administrative measures to ensure the conservation of the habitats of the wild flora and fauna species, especially those specified in Appendices I and II, and the conservation of endangered natural habitats.

2. The Contracting Parties in their planning and development policies shall have regard to the conservation requirements of the areas protected under the preceding paragraph, so as to avoid or minimise as far as possible any deterioration of such areas.

Parties must give special attention to the protection of areas which are important to the migratory species listed in Appendices I and II (Art. 4(3)) and coordinate their efforts to protect the natural habitats referred to in Article 4 when they are situated in frontier areas.

Chapter III of the Convention makes provision for the protection of animal and plant species (Arts 5–9).

Berne Convention 1979 Art. 5

Each Contracting Party shall take appropriate and necessary legislative and administrative measures to ensure the special protection of the wild flora species specified in Appendix I. Deliberate picking, collecting, cutting or uprooting of such plants shall be prohibited. Each Contracting Party shall, as appropriate, prohibit the possession and sale of these species.

Article 7 requires contracting parties to protect Appendix III species and to regulate their exploitation by the use of closed seasons, temporary or local prohibitions of exploitation and the control of their sale.

Under Article 9, derogations from Articles 4, 5, 6 and 7 are allowed:

- for the protection of flora and fauna;
- to prevent serious damage to crops, livestock, forests, fisheries, water and other property;
- in the interests of public health and safety, air safety or other overriding public interests;
- for the purposes of research, education, repopulation or re-introduction;

Berne Convention 1979 Art. 6

Each Contracting Party shall take appropriate and necessary legislative and administrative measures to ensure the special protection of the wild fauna specified in Appendix II. The following will in particular be prohibited for these species:
a. all forms of deliberate capture and keeping and deliberate killing;
b. the deliberate damage to or destruction of breeding or resting sites;
c. the deliberate disturbance of wild fauna, particularly during the period of breeding, rearing and hibernation, ...
d. the deliberate destruction or taking of eggs from the wild or keeping these eggs even if empty;
e. the possession of and internal trade of these animals, alive or dead, including stuffed animals, and any recognisable part or derivative thereof, ...

■ to permit the keeping or exploitation of certain wild animals and plants in small numbers (on a limited basis and under strictly supervised conditions).

The Convention encourages the re-introduction of native species (Art. 11(2)(a)), requires the strict control of the introduction of non-native species (Art. 11(2)(b)) and makes special provisions for migratory species (Art. 10).

7.2.7   International Whaling Convention

The International Convention for the Regulation of Whaling established the International Whaling Commission (Art. III) with the aim of conserving whale stocks. The Convention includes a Schedule which contains details of the conservation and utilisation measures agreed by the contracting parties, and forms an integral part of the Convention (Art. I).

International Whaling Convention 1946 Art. V

1. The Commission may amend from time to time the provisions of the Schedule by adopting regulations with respect to the conservation and utilisation of whale resources, ...

Article V(1) allows the fixing of:

(a) protected and unprotected species;
(b) open and closed seasons;
(c) open and closed waters, including the designation of sanctuary areas;
(d) size limits for each species;
(e) time, methods, and intensity of whaling (including the maximum catch of whales to be taken in any one season);
(f) types and specifications of gear and apparatus and appliances which may be used;
(g) methods of measurement; and
(h) catch returns and other statistical and biological records.

All cetacean species in UK waters are completely protected under Schedule 5 of the Wildlife and Countryside Act 1981, the Wildlife (Northern Ireland) Order 1985 and Schedule 2 of the Conservation (Natural Habitats etc.) Regulations 1994 (implementing the Habitats Directive (Annex IV)). Trade in cetaceans and cetacean products are regulated under CITES, and by the Regulation on the implementation in the Community of the Convention on International Trade in Endangered Species of Wild Fauna and Flora (338/97/EC) and the Regulation on common rules for the import of whales or other cetacean products (348/81/EEC).

Whales, dolphins and porpoises are also included in the Berne and Bonn Conventions.

7.2.8   Kyoto Protocol

The Kyoto Protocol to the UN Framework Convention on Climate Change 1997 sets targets for the reduction of various gases which are believed to be responsible for global climate change. Carbon dioxide is one of the most important of these gases, and the agreement allows contracting parties to take carbon sinks into account when calculating

---

Kyoto Protocol 1997 Art. 3

3. The net changes in greenhouse gas emissions by sources and removals by sinks resulting from direct human-induced land-use change and forestry activities, limited to afforestation, reforestation and deforestation since 1990, measured as verifiable changes in carbon stocks in each commitment period, shall be used to meet the commitments under this Article of each Party included in Annex I ...

the reduction required, in certain circumstances. Forests absorb large amounts of CO_2 from the atmosphere as trees photosynthesise, and the process of afforestation may significantly affect a state's commitments to CO_2 reduction. It has been suggested that managed forests in Russia could justify a 25 per cent reduction in its overall target (Bell and McGillivray, 2000).

Local authorities have considerable powers to protect and encourage the planting of trees (see Chapter 6).

7.2.9   CITES Convention

A wide variety of endangered species of animals and plants, and their products, is illegally imported into the UK every year. HM Customs and Excise seized over a million individual items between 1996 and 2002 (Lawson, 2002). The worldwide illegal wildlife trade is estimated to be worth £3 billion each year (Anon., 2000).

Trade occurs in various species of mammals, birds, reptiles, amphibians, insects and plants. It involves live and stuffed organisms, and also their products and derivatives, such as skins, ornaments, wool, medicines and foodstuffs. In the UK, both native and exotic species are involved in the wildlife trade. Exotic parrots and finches are traded in large numbers and demands from falconry are increasing. The UK imports over one million live reptiles and amphibians each year, and there is an enormous trade in tropical fish. There has been a recent increase in the use of traditional and herbal medicines which often contain animal and plant derivatives.

TRAFFIC (Trade Records Analysis of Flora and Fauna in Commerce) is the joint trade monitoring programme of the World Wide Fund for Nature (WWF) and the World Conservation Union (IUCN) which works in cooperation with the CITES Secretariat. It was formed in 1976 and is run by full-time staff with an international network of informants. According to TRAFFIC (Anon., 1999), CITES-listed animal and plants illegally imported into UK in 1996 included:

■ Live birds	16 000
■ Live mammals	2 800
■ Live reptiles and amphibians	25 000
■ Live plants	3 000
■ Reptile skins	90 000
■ Pieces of coral	42 200

Illegal trade in UK native species has particularly affected:

- ■ Birds of prey (especially the peregrine falcon (*Falco peregrinus*))
- ■ Butterflies
- ■ Rare orchids
- ■ Wild bulbs (e.g. snowdrops (*Galanthus nivalis*))

Wildlife trade and the law

International trade in wildlife is controlled by the Convention on International Trade in Endangered Species of Wild Fauna and Flora (CITES) which was signed in Washington DC in March 1973 and entered into force in July 1975. CITES prohibits international commercial trade in the rarest species and requires licences from the country of origin for exports of some other rare species. The Convention regulates trade in whole animals and plants, living or dead, and recognisable parts and derivatives. Around 30 000 species (including about 25 000 plants) are covered by the Convention, listed in three appendices:

- Appendix I
  Includes all species threatened with extinction which are or may be affected by trade.
- Appendix II
  Includes all species which may become threatened with extinction if trade is not strictly regulated (and other species which must be subject to strict regulation in order to achieve this objective).
- Appendix III
  Includes other species which any party strictly protects within its own jurisdiction and which requires the cooperation of other parties in the control of trade.

Trade in endangered species is regulated by the requirement for import and export licences. The strictest restrictions apply to Appendix I species (Art. III).

CITES 1973 Art. III

3. The import of any specimen of a species included in Appendix I shall require the prior grant and presentation of an import permit and either an export permit or re-export certificate. An import permit shall only be granted when the following conditions have been met:
    a. a scientific authority of the state of import has advised that the import will be for purposes which are not detrimental to the survival of the species involved;
    b. a scientific authority of the state of import is satisfied that the proposed recipient of a living specimen is suitably equipped to house and care for it; and
    c. a management authority of the state of import is satisfied that the specimen is not to be used primarily for commercial purposes.

The term 'specimen' is defined in Article I, and means an animal or plant, whether alive or dead, and includes recognisable parts and derivatives.

The EU laws which regulate wildlife trade are as follows:

- Regulation on common rules for the import of whales or other cetacean products (348/81/EEC).
- Directive concerning the importation into member states of skins of certain seal pups and products derived therefrom (83/129/EEC).
- Regulation on the protection of species of wild fauna and flora by regulating trade therein (338/97/EC).
- Regulation prohibiting the use of leghold traps in the Community and the introduction into the Community of pelts and manufactured goods of certain wild animal species originating in countries which catch them by means of leghold traps or trapping methods which do not meet international humane trapping standards (3254/91/EEC).

Exotic species are frequently being smuggled into the UK via other European Union countries. Permits to import CITES-listed species into the UK are not required if these species are being imported from another EU country. So, if they are able to enter EU countries with weaker border controls, they may be transported unimpeded into the UK.

The UK has recently invested more resources in combating wildlife crime with the establishment of:

- Customs CITES liaison officers
- CITES Enforcement Teams at Heathrow Airport and the Port of Dover
- Police Wildlife Liaison Officer Network
- Partnership for Action Against Wildlife Crime
- National Wildlife Crime Intelligence Unit (see Section 3.1.7)

Case studies

Penalties for wildlife trading have often been too low to have a deterrent effect. However, there are occasional high profile cases that result in significant custodial sentences.

Man imprisoned for selling rare macaws
On 14 April 2000 a 61-year-old man from Northallerton, North Yorkshire, was jailed for two-and-a-half years on four counts of smuggling endangered birds into the UK. Customs officers seized nine birds from his premises in 1998, including three Lear's macaws. The rare macaws were held at a secret location until after the trial and were then returned to the Brazilian government for release back into the wild.

London shopkeepers convicted of selling bushmeat

In June 2001 two London shopkeepers were convicted of smuggling bushmeat into Britain and sentenced to four months' imprisonment. They were arrested after selling a monkey to a journalist who was posing as a customer (Anon., 2001b). Mobolaji Osakuade and Rose Kinnane were both charged with 12 counts of importing bushmeat, contrary to s. 170(2) of the Customs and Excise Management Act 1979, and selling specimens of species listed on Appendix II of CITES, contrary to regulation 8(2) of the Control of Trade in Endangered Species Enforcement Regulations 1985 (SI 1985/1155) (Anon., 2002a). Included in the specimens for sale were whole lion carcases at £5000 each (Anon., 2002b).

Operation Retort: animal dealer jailed

An animal dealer was jailed for six-and-a-half years by Isleworth Crown Court on 18 January 2002 after 23 endangered birds of prey were discovered in suitcases at Heathrow Airport (Horsnell, 2002). The birds had been on a flight from Thailand and included eagles, owls and kites, with a black-market value of £35 000. When RSPCA inspectors raided the defendant's home they recovered birds, mammals and reptiles from 14 endangered species. In total more than 60 animals, dead and alive, were seized, from 29 species including several primates. The trial followed 'Operation Retort', the biggest ever joint operation against wildlife crime mounted between the police and HM Customs.

Zoo owner convicted of displaying CITES Appendix I species without a licence

The owner of Southport Zoo, Douglas Petrie, pleaded guilty to the commercial display of species listed on Appendix I of CITES without a licence in July 2001. He was fined £5000 and the 37 specimens listed in the charges were confiscated. They included cotton top tamarins, ocelots, owls, scarlet macaws, and tortoises (Anon., 2002a). These animals were discovered by a routine inspection and as a result the police and RSPCA were issued with a warrant to search the zoo. However, the conviction was largely based on a technicality as Petrie had applied for the appropriate licence some days before the first inspection, but had unlawfully displayed the animals while waiting for the application to be processed (Gould, 2001).

7.2.10 Biodiversity Convention

The United Nations Conference on Environment and Development was held in June 1992, in Rio de Janeiro, Brazil and has come to be known as the 'Earth Summit'. At this summit the Convention on

Biological Diversity 1992 was signed by 155 states and the EU, along with a programme of action for governments which is called 'Agenda 21'. The Convention came into force on 29 December 1993, and requires parties to take action to conserve biodiversity. This has been implemented in the UK by the 'Biodiversity Action Plan'. Agenda 21 is currently being implemented by local government.

PANEL 7.6   STRUCTURE OF THE CONVENTION ON BIOLOGICAL DIVERSITY 1992

	Preamble
Article 1	Objectives
Article 2	Use of Terms
Article 3	Principle
Article 4	Jurisdictional Scope
Article 5	Cooperation
Article 6	General Measures for Conservation and Sustainable Use
Article 7	Identification and Monitoring
Article 8	*In-situ* Conservation
Article 9	*Ex-situ* Conservation
Article 10	Sustainable Use of Components of Biological Diversity
Article 11	Incentive Measures
Article 12	Research and Training
Article 13	Public Education and Awareness
Article 14	Impact Assessment and Minimising Adverse Impacts
Article 15	Access to Genetic Resources
Article 16	Access to and Transfer of Technology
Article 17	Exchange of Information
Article 18	Technical and Scientific Cooperation
Article 19	Handling of Biotechnology and Distribution of its Benefits
Article 20	Financial Resources
Article 21	Financial Mechanism
Article 22	Relationship with Other International Conventions
Article 23	Conference of the Parties
Article 24	Secretariat
Article 25	Subsidiary Body on Scientific, Technical and Technological Advice
Articles 26–42	These articles contain technical provisions relating to the operation of the Convention

---

The definition of biological diversity in the Convention is very wide-ranging so that the Convention aims to conserve variability within and between species and ecosystems (Art. 2).

Convention on Biological Diversity 1992 Art. 1

The objectives of this Convention ... are the conservation of biological diversity, the sustainable use of its components and the fair and equitable sharing of the benefits arising out of the utilisation of genetic resources, including by appropriate access to genetic resources and by the appropriate transfer of relevant technologies, taking into account all rights over those resources and to technologies, and by appropriate funding.

Convention on Biological Diversity 1992 Art. 2

'Biological diversity' means the variability among living organisms from all sources including, *inter alia*, terrestrial, marine and other aquatic eco-systems and the ecological complexes of which they are part; this includes diversity within species, between species and of eco-systems.

Article 3 of the Convention reaffirms that states have the sovereign right to exploit their own resources under international law, but also places upon states the responsibility of ensuring that activities within their jurisdiction or control do not cause damage to the environment of other states or areas beyond the limits of national jurisdiction. This provision would, for example, make the UK government responsible for the activities of UK companies operating abroad.

Convention on Biological Diversity 1992 Art. 6

Each Contracting Party shall, in accordance with its particular conditions and capabilities:

(a) Develop national strategies, plans or programmes for the conservation and sustainable use of biological diversity ...

(b) Integrate, as far as possible and as appropriate, the conservation and sustainable use of biological diversity into relevant sectoral or cross-sectoral plans, programmes and policies.

Convention on Biological Diversity 1992, Annex I –
Identification and Monitoring

1. Eco-systems and habitats: containing high diversity, large numbers of endemic or threatened species, or wilderness; required by migratory species; of social, economic, cultural or scientific importance; or, which are representative, unique or associated with key evolutionary or other biological processes;

2. Species and communities which are: threatened; wild relatives of domesticated or cultivated species; of medicinal, agricultural or other economic value; of social, scientific or cultural importance; of importance for research into the conservation and sustainable use of biological diversity, such as indicator species; and

3. Described genomes and genes of social, scientific or economic importance.

The Convention requires the contracting parties to restore species and habitats to a favourable conservation status and requires governments to have regard to the conservation of biological diversity in the carrying out of their functions.

Article 7 of the Convention requires states to identify components of biological diversity important for its conservation and sustainable use (having regard to the list in Annex I), monitor these components paying particular attention to those requiring urgent conservation measures, and to collect appropriate data. This article further requires states to identify and monitor activities and processes which have or are likely to have adverse effects on the conservation or sustainable use of biological diversity.

The *in-situ* conservation of biological diversity is required by Article 8, including:

- establishment of protected areas;
- management of biological resources important for conservation within and outside protected areas;
- promotion of sustainable development in areas adjacent to protected areas;
- regulation, management and control of the risks associated with the release of living modified organisms;
- prevention of the spread of alien species;
- development of legislation to protect threatened species and populations.

Convention on Biological Diversity 1992 Art. 8

Each Contracting Party shall, as far as possible and as appropriate:

(f) Rehabilitate and restore degraded eco-systems and promote the recovery of threatened species, *inter alia*, through the development and implementation of management plans or other management strategies;

(h) Prevent the introduction of, control or eradicate those alien species which threaten eco-systems, habitats or species.

PANEL 7.7

Otter   *Lutra lutra*

Identification/size
Up to 140 cm long (including tail), with male larger and heavier than female; shiny brown fur, with yellowish brown (or whitish) around the chin and neck; flat head with rounded, small ears; stiff whiskers; long tail, thick at base; feet clawed and webbed.

Habitat
Freshwater ecosystems (usually with thick bank vegetation) and remote rocky seashores.

Distribution
Occurs across much of Britain and Ireland in small numbers, but population is increasing.

Ecology and behaviour
Eats mainly fish, but will also take small mammals, and birds, amphibians and crustacea; lives in a holt in a river bank; partly aquatic and mainly nocturnal; ungainly on land.

Article 8(f) is being implemented in the UK by Biodiversity Action Plans. These were originally established as a matter of policy but have since been given statutory force by the Countryside and Rights of Way Act 2000.

UK Biodiversity Action Plans and Agenda 21

Biodiversity Action Plans

Some of the international obligations imposed by the Biodiversity Convention have now been given effect by the Countryside and Rights of Way Act 2000. Section 74(1) imposes a duty on government ministers, government departments and the National Assembly for Wales to have regard to the purpose of conserving biological diversity in accordance with the Biodiversity Convention, in the carrying out of their functions. The Secretary of State (in England) and the National Assembly for Wales (together described as the 'listing authorities') are required to publish lists of important species and habitats (s. 74(2)) after consultation with English Nature or the Countryside Council for Wales, respectively.

Countryside and Rights of Way Act 2000 s. 74

---

(3) ... it is the duty of the listing authority to take, or promote the taking by others, of such steps as appear to the authority to be reasonably practicable to further the conservation of the living organisms and types of habitat included in any list published by the authority ...

Section 74(7) defines conservation as including 'the restoration or enhancement of a population or habitat'. This effectively imposes a legal duty on the government to restore threatened species and habitats that has never before existed in English law. In practical terms this duty applies to those species and habitats included in the UK Biodiversity Action Plan (Anon., 1995) which was drawn up to comply with Articles 6 and 8 of the Biodiversity Convention.

Species were considered for inclusion in the action plans if they qualified for at least one of the following categories:

- threatened endemic and other globally threatened species;
- species where the UK has more than 25 per cent of the world or appropriate biogeographical population;
- numbers or range have declined by more than 25 per cent in the last 25 years;

- species found in fewer than 15 ten km squares in the UK;
- listed in EU Wild Birds or Habitats Directives, the Berne, Bonn or CITES Conventions, or under the Wildlife and Countryside Act 1981 and the Nature Conservation and Amenity Lands (Northern Ireland) Order 1985 (SR 1985/170 (NI1)).

This approach produced an initial list of 1250 species from which 400 were drawn. This second list contained species that were either globally threatened or rapidly declining in the UK. A shortlist of 116 species was then drawn up for early action (Anon., 1995).

Some examples of species for which Species Action Plans have been produced under the UK Biodiversity Action Plan are:

■ Southern wood ant	*Formica rufa*
■ Heath tiger beetle	*Cicindela sylvatica*
■ Tadpole shrimp	*Triops cancriformis*
■ Natterjack toad	*Bufo calamita*
■ Burbot	*Lota lota*
■ Sand lizard	*Lacerta agilis*
■ Bittern	*Botaurus stellaris*
■ Brown hare	*Lepus europaeus*
■ Nail fungus	*Poronia punctata*
■ Northern prongwort	*Herbertus borealis*
■ Fen orchid	*Liparis loeselii*

Habitats were treated as key habitats if they qualified under one or more of the following criteria:

- habitats for which the UK has international obligations;
- habitats at risk (as a result of a high rate of decline in the last 20 years) or rare;
- areas which are functionally critical, particularly marine areas;
- areas important for key species.

Examples of habitats for which Habitat Action Plans have been produced under the UK Biodiversity Action Plan are:

- Acid grasslands
- Fens
- Chalk rivers
- Limestone pavements
- Mudflats
- Reedbeds
- Saline lagoons
- Upland hay meadows
- Upland heathland
- Wet woodland

PANEL 7.8   SPECIES ACTION PLAN

Species Action Plan   Red squirrel (*Sciurus vulgaris*)

Current status
- Replaced by introduced grey squirrel (*Sciurus carolinensis*) throughout most of England and Wales
- Widespread in Scotland and Northern Ireland
- Current population estimated at 100 000
- Listed on Appendix III of the Berne Convention, protected by Schedules 5 and 6 of the Wildlife and Countryside Act 1981 and Schedules 5 and 6 of the Wildlife (Northern Ireland) Order 1985

Current factors causing loss or decline
- Spread of grey squirrels
- Habitat fragmentation
- Disease

Current action
- JNCC is drafting a UK strategy for Red Squirrel Conservation
- A Species Recovery Programme is run by English Nature, the 'Red Alert' campaign is intended to raise public awareness. A Squirrel Forum has been established
- Forestry Commission is researching a hopper designed to selectively poison grey squirrels
- Forest habitat manipulation studies are in progress, three forests identified where red squirrel conservation management is a priority
- Experimental translocations have taken place, planning for a full-scale translocation is in progress

Action plan objectives and targets
- Maintain and enhance current populations through good management
- Re-establish red squirrel populations, where appropriate

Proposed action with lead agencies
- Policy and legislation
  - Achieve agreement on UK Red Squirrel Strategy
  - Develop regional guidelines for management of squirrel populations
  - Review geographical restrictions on use of warfarin
  - Consider needs of red squirrels when reviewing or preparing Indicative Forestry Strategies
- Site safeguard and management
  - Prepare and implement site management plans for all sites with viable populations
  - Attempt to create or maintain 2000 ha of conifer reserves in Wales for red squirrels
- Species management and protection
  - Develop strategies, within the national framework, to guide work
  - Assess experimental translocation projects for wider use
  - Attempt to prevent expansion of grey squirrel range to key areas occupied by reds

■ Advisory
  – Advise land managers on appropriate management
  – Develop guidance on forestry design to benefit red squirrels
■ Future research and monitoring
  – Continue research on feeding ecology, control, translocation, habitat manipulation, phylogenetics etc.
  – Establish a survey method and Squirrel Monitoring Scheme to determine population levels
  – Pass survey information to the JNCC so that it may be incorporated in a national database
■ Communications and publicity
  – Clear information explaining the relationship between red and grey squirrels should be made available to the public and landholders

Lead partner
■ Joint Nature Conservation Committee

Source: UK Biodiversity Action Plan (*www.ukbap.org.uk*)

The UK Biodiversity Action Plan currently lists 391 Species Action Plans (SAPs) and 45 Habitat Action Plans (HAPs). The UK Biodiversity Group recently reported that, although 54 per cent of SAPs and HAPs showed progress towards their targets, 43 priority species and one priority habitat were reported to be in decline (Anon., 2001c).

It is clearly important that limited resources are focused on species and habitats that are especially vulnerable. However, such an approach runs the risk of ignoring less threatened species and habitats until their status deteriorates. It is also important to recognise that very little scientific data exists on the status of many species, particularly marine species and invertebrates, so it is inevitable that only well-known species are targeted for assistance. The action plan for the otter (*Lutra lutra*) was brought forward because of its 'high popular appeal' (Anon., 1995).

Bell and McGillivray (2000) have criticised what they refer to as a conservation 'stamp collecting' approach in the UK, focusing on protected sites and isolating conservation from the wider economic pressures. However, it is to be hoped that the movement away from conservation 'policies' and towards conservation 'duties' will encourage the government to integrate biodiversity considerations into sectoral policies such as fisheries and agriculture.

Agenda 21
As a result of international concern for the state of the environment, on 22 December 1989 the United Nations General Assembly called for a global meeting to devise strategies to halt and reverse the effects of environmental degradation. In particular, there was a recognition of the need to promote sustainable and environmentally sound development in all countries.

PANEL 7.9   HABITAT ACTION PLAN

Habitat Action Plan   Coastal sand dunes

Current status
- Physical and biological status
  - Sand dunes develop where there is a sufficiently large beach plain that dries out between high tides
  - Sand dunes form in relatively exposed locations
  - Sand dune vegetation forms a number of zones, fixed dunes and dune heath are priority habitats under the EC Habitats Directive.
  - Fixed dune communities are maintained by grazing
  - Sand dune communities vary geographically
  - Dune grassland and slack support a wide variety of flowering plants, including orchid species, and are rich in butterflies, moths, bees, wasps and other invertebrates
  - Major dune systems are widely distributed within the UK

Links with other action plans
- The machair habitat action plan (in western Scotland), and the lowland heathland action plan are related to this plan
- Nine BAP priority species have significant populations on sand dunes (four insects and five plants)

Current factors affecting the habitat
- Erosion and progradation
- Falling water tables
- Grazing
- Recreation
- Sea defence and stabilisation
- Beach management
- Forestry
- Military use
- Ownership
- Other human influences

Current action
- Legal status
  - Much of the resource is designated as SSSI or ASSI
  - Twenty-one sites in UK have been selected as candidate SACs under the EC Habitats Directive
- Management, research and guidance
  - UK government publications including Planning Policy Guidance
  - Coastal fora and other partnerships established
  - EU *Life* programme – integrated coastal zone management
  - Conservation management, especially on NNRs, SSSIs and ASSIs
  - International conferences on dune management promoted by the networks of the European Union for Coastal Conservation and Eurosite; European Golf Association Ecology Unit has promoted management of dune golf links
  - Sand Dune Survey of Great Britain was initiated in 1987

Action plan objectives and targets
- Protect the existing sand dune resource
- Encourage new dunes to accrete to offset expected net losses
- Seek opportunities for restoration of dune habitat lost to forestry, agriculture or other human losses
- Encourage natural development of dune systems and control natural succession where necessary
- Maintain dune grassland, heath and lichen communities on most dune systems; create Atlantic dune woodland on up to five sites

Proposed action with lead agencies
- Policy and legislation
  - Develop and promote planning policies and procedures which aim to prevent further loss of dune systems
  - Develop and promote agri-environment schemes to encourage restoration and sustainable management
  - Develop and promote incentives to encourage landward movement of dunes
  - Develop and promote coastal management policies so as to conserve dune systems and include in Shoreline Management Plans
- Site safeguard and management
  - Notify any remaining areas of dune habitat which meet national criteria as SSSIs and ASSIs by 2004, and ensure proper management of designated sites
  - Use management agreements to encourage sustainable grazing of dune ecosystems
  - Encourage sympathetic management of golf courses on sand dunes
  - Promote and encourage the restoration of open dune vegetation on afforested dune systems
  - Promote and encourage the restoration of dune vegetation on dune systems used for arable farming or improved grassland
  - Monitor, regulate and promote remedial action where water tables in dune systems may be affected by water abstraction and drainage
  - Discourage unnecessary stabilisation of dunes
  - Support beach management strategies which encourage the protection of seaward fronts of dune systems from vehicles and pedestrians; discourage mechanical beach cleaning
- Advisory
  - Promote and develop demonstration sites for restoration of dune vegetation
  - Encourage appropriate dune management by providing guidance material
  - Advise agri-environment project officers and regional agri-environment conservation groups of the location, importance and management requirements of dune systems
  - Utilise existing estuary management partnerships to carry forward this plan
- International
  - Promote information exchange among European maritime states
  - Ensure lessons from EU *Life* projects are widely disseminated

- Monitoring and research
  - By 2002 compile an inventory of the desirability, feasibility and priority of sites for sand dune restoration from forestry and agriculture, and for the development of Atlantic woodland
  - Identify suitable locations and methods for dune activation
  - Coordinate information on changes in extent and quality of sand dunes resource in UK for monitoring purposes
  - Continue research on the use of remote sensing in monitoring
  - Promote research on the causes of falling water tables in dune systems
  - Promote research on the effects of nitrogen deposition, climate change and sea level rise on sand dunes
- Communications and publicity
  - Raise public awareness
  - Promote awareness of this plan among decision makers

Costing
- Total expenditure to 2005 (first 5 years):   £980 000
- Total expenditure 2005–2014 (next 10 years):   £4 100 600

Source: UK Biodiversity Action Plan (*www.ukbap.org.uk*)

Figure 7.3 Sand dunes at Ainsdale, near Southport. Some erosion is evident as a result of human activity.

The international community's response to that request was adopted at the United Nations Conference on Environment and Development on 14 June 1992, and is called 'Agenda 21'. Agenda 21 is a comprehensive programme of action by governments, development agencies, UN organisations, and independent sectors in every area where human economic activity affects the environment, intended to extend into the twenty-first century.

PANEL 7.10   LOCAL ACTION PLAN

Local Action Plan   Bolton

Plan name
A Biodiversity Action Plan for Bolton

Plan coverage
Bolton

Region/country
North-west

Partners involved
- Bolton Institute
- Bolton Metropolitan Borough Council
- English Nature
- Greater Manchester Ecology Unit
- Lancashire Wildlife Trust
- Red Rose Forest
- Royal Society for the Protection of Birds

Funding
- Source: Bolton Metropolitan Council
- Amount: £2500
- Duration: 1 year

Stage and status of initiative

Stage	Status	Date
Partnership building	Steering group established	2000
Introductory document	In progress	March 2001
Information collation/audit	On going	–
Action plan	In progress	March 2001

Species for which action plans prepared (or are in preparation)

Local/UK species	Species
Local species	Bats (group plan)
Local species	Bluebell
Amphibians	Great crested newt
Mammals	Water vole

Habitats for which action plans prepared (or are in preparation)

Local/UK habitat	Habitats
Local habitat	Lodges
Local habitat	Rivers and floodplains
Local habitat	Semi-improved grassland
Broad habitats	Boundary and linear features
Broad habitats	Broadleaved, mixed and yew woodland
Broad habitats	Inland rock

Broad habitats	Neutral grassland
Broad habitats	Standing open water and canals
Priority habitats	Lowland calcareous grassland
Priority habitats	Lowland dry acid grassland
Priority habitats	Lowland heathland
Priority habitats	Lowland raised bog
Priority habitats	Reedbeds
Priority habitats	Upland heathland
Priority habitats	Upland oakwood
Priority habitats	Wet woodland

Information last updated
01/02/2001

Source: UK Biodiversity Action Plan (*www.ukbap.org.uk*)

The underlying principle of Agenda 21 is a recognition that man has reached a defining moment in history where we are faced with a choice. Either we continue with present policies which increase global poverty, hunger, sickness and cause a progressive deterioration of the global ecosystem that supports life on Earth. Or we improve our management and protection of the global ecosystem and bring about a more secure and prosperous future for mankind. The basis of Agenda 21 is a global partnership for sustainable development within which local actions will collectively have a beneficial global impact.

Agenda 21 does not have any statutory basis within UK domestic law but the government wanted all local authorities to adopt a Local Agenda 21 strategy by December 2000, working towards the principles agreed at Rio.

Many of the chapters within Agenda 21 address problems relating to wildlife, nature conservation and development:

Section I
Chapter 7   Promoting sustainable human developments

Section II
Chapter 10   An integrated approach to land-resource use
Chapter 11   Combating deforestation
Chapter 13   Protecting mountain ecosystems
Chapter 14   Meeting agricultural needs without destroying land
Chapter 15   Sustaining biological diversity
Chapter 16   Environmentally sound management of biotechnology
Chapter 17   Safeguarding the ocean's resources
Chapter 18   Protecting and managing freshwater resources
Chapter 19   Safe use of toxic chemicals

Many local authorities have embraced Local Agenda 21, and some have appointed officers specifically to deal with Agenda 21 issues. Some authorities produce environment reports which include 'sustainability indicators', but in many cases these are simplistic statistics that say little about sustainability *per se*. They may contain data on the total length of cycle paths constructed (with no data on usage), crime, employment and health statistics, drinking water quality and recycling. Biological indicators are some of the best indicators of changes in environmental quality. Surveys of lichens provide information about air quality, and studies of songbird diversity are good indicators of the general state of the environment. However, local authorities do not have the resources to invest in this type of long-term quality monitoring.

REFERENCES

Anon. (1995) *Biodiversity: The UK Steering Group Report Meeting the Rio Challenge (Volume I)*. HMSO, London.

Anon. (1999) Factfile: Wildlife Trade in the UK. *www.traffic.org/factfile/wildlife tradeinuk.html*, accessed 24 January 2002.

Anon. (2000) *The State of the Environment of England and Wales: The Land*. Environment Agency, The Stationery Office Ltd, London.

Anon. (2001a) Blackpool boost, *The Times*, 7 November 2001, p. 14

Anon. (2001b) Jail sentence for bushmeat sellers. *Care for the Wild News*, No.17, p. 19. Care for the Wild, Rusper, W. Sussex.

Anon. (2001c) *'Sustaining the variety of life': 5 years of the UK Biodiversity Action Plan*. The Stationery Office Ltd, London.

Anon. (2002a) Recent prosecutions. Partnership for Action Against Wildlife Crime, DEFRA. *www.defra.gov.uk/paw/prosecutions*, accessed 26 April 2002.

Anon. (2002b) 'Green' police to target trade in wildlife. *Sunday Times*, 21 April 2002, p. 32.

Bell, S. and McGillivray, D. (2000). *Environmental Law. The Law and Policy Relating to the Protection of the Environment*, 5th edn. Blackstone Press Ltd, London.

Gould, D. (2001) Editorial. *International Zoo News*, 48(6): 358–359.

Horsnell, M. (2002) Suitcases full of birds land dealer in jail. *The Times*, 19 January 2002, p. 8.

Lawson, T. (2002) *Traded Towards Extinction? The role of the UK in wildlife trade*. A WWF Report. WWF–UK, Godalming, Surrey.

PART III

Planning, Urban Environments and Environmental Impact Assessment

Part III describes the role of the planning system and environmental impact assessment in the protection of nature. It considers the impact of highways and the construction of modern buildings, and concludes with a brief look at possible changes to the law in the near future.

8   Planning, highways and wildlife   311

9   Environmental impact assessment   337

10 The future   356

8

Planning, highways and wildlife

8.1 PLANNING AND DEVELOPMENT

Local and national government may, by the use of the planning system, significantly affect the extent to which wildlife and natural habitats are protected. The planning system allows the attachment of conditions to planning permissions and requires the assessment of the environmental impact of certain developments, including their effect upon the natural environment.

8.1.1 Planning, development control and conservation

Planning control is the process by which changes in land use are controlled in the UK and is a statutory duty of local authorities. Developers must apply to local planning authorities to change the use of land.

The principles by which development in a particular area will be controlled are set out in development plans. After a technical consideration of an application the planning authority will consider whether or not the proposal conforms to the current development plan for the area. The planning authority may then:

- approve the application;
- reject the application;
- approve the application with conditions.

Planning departments may attach any conditions they like to the granting of planning permission. In Orkney the council required that the construction of a wind turbine was completed before the onset of the breeding season of the red-throated diver (*Gavia stellata*) which would otherwise have been disturbed.

Applicants for planning permission cannot be expected to wait indefinitely for their applications to be determined. In *Lafarge Redland Aggregates Ltd* v. *The Scottish Ministers* [2001] the court found that there was a legitimate expectation by the claimant that its application for planning permission would be determined within a reasonable time. The refusal of the respondents to determine the application pending the outcome of a reference to Scottish Natural Heritage was deemed to be *ultra vires*.

If an application is refused the applicant normally has the right of appeal to the Secretary of State. Where planning applications are particularly problematical the Secretary of State may 'call them in' for a decision. This may happen, for example, if a proposal is potentially damaging to the environment.

8.1.2   The Planning Inspectorate

The Planning Inspectorate processes planning and enforcement appeals and holds inquiries into local development plans. It also deals with other planning-related casework, including reporting on planning applications that have been 'called in' for decision by the Secretary of State, and cases arising from the Environmental Protection and Water Acts.

Even relatively small developments may be considered by the Planning Inspectorate. The Inspectorate gave the mobile phone company One-2-One permission to erect a mast in Lostock, near Bolton, after it had been refused by the local planning authority. At the appeal, the company claimed that its engineers would make the mast look like a 'pretend tree' so that it would not spoil the appearance of the landscape. In fact, the mast simply had numerous antennae attached along its length and looked nothing at all like a tree (Bullock, 2001).

The legitimacy of the Planning Inspectorate as the final arbiter of planning disputes has recently been challenged since it is an agency of the government and as such cannot make independent judgements.

8.1.3   Planning Policy Guidance Notes

The government provides statements of national planning policy to planning authorities through a series of Planning Policy Guidance Notes. Those containing information which specifically concerns the protection of the natural environment and heritage are:

- PPG02   Green Belts
- PPG07   The Countryside: Environmental Quality and Economic and Social Development
- PPG09   Nature Conservation
- PPG15   Planning and the Historic Environment
- PPG16   Archaeology and Planning
- PPG20   Coastal Planning

It should be remembered that the details contained within individual PPGs may lag behind changes in the law.

In addition to the PPGs listed above, Minerals Policy Guidance Notes contain detailed recommendations relating to environmental protection for example:

- MPG07   Reclamation of Mineral Workings
- MPG13   Guidelines for Peat Provision in England, including the place of Alternative Materials

8.1.4   The National Land Use Database

The government is seeking to encourage the greater use of previously developed brownfield sites for new developments (Figure 8.1), as indicated in two White Papers published in November 2000 (*Our Countryside: The Future*, and *Our Towns and Cities: The Future*). This should help to achieve the dual aim of reviving towns and cities, while reducing the loss of land to development in the countryside. To this end the government has recognised the need for a database of available brownfield sites and is currently engaged in the creation of a National Land Use Database (NLUD).

The NLUD is a geographical inventory of brownfield sites that is being developed to facilitate the re-use of land and buildings, but will eventually be extended to cover other land uses. The pilot database

Figure 8.1 A brown field development in Bolton, where housing is being built on the site of an old mill.

included sites identified by local authorities as vacant or derelict in 1998. By June 2002 information had been included for over 10 000 sites in 171 local authorities in England. It shows the stock of previously developed vacant and derelict land and information about the land and buildings which have been redeveloped. The database shows sites on a map and gives details of previous land use and planning status. The database is accessible via the internet and it is intended that it will be updated annually.

8.2   HIGHWAYS AND WILDLIFE

8.2.1   The Highways Agency and conservation

The Highways Agency's Environmental Committee was established in 1999 and its purpose is to help the agency to view its work from the point of view of official and non-governmental organisations. The groups represented on the committee are:

■ Environment Agency
■ English Nature
■ English Heritage
■ Countryside Agency
■ National Air Quality Forum
■ Civic Trust
■ National Trust
■ Royal Town Planning Institute
■ RSPB
■ Association of National Park Authorities
■ Transport 2000 Trust
■ Friends of the Earth
■ Council for the Protection of Rural England

The Agency recognises that new and existing roads can have serious implications for wildlife and nature conservation and that care must be taken over route planning, construction, operation and maintenance.

The actions that the Highways Agency proposes to take to protect biodiversity are an integral part of its strategic plan *Towards a Balance with Nature*:

■ Developing a biodiversity action plan for the management of the 'soft estate' – the land surrounding the road network.
■ Working with English Nature and other partners on a programme of research on the relationship between biodiversity and highways management.
■ Implementing a programme of engineering measures to protect wildlife where their habitats are crossed by highways.

Examples of conservation measures taken by the Agency include:

- Restoration of the site of the old Winchester Bypass, Twyford Down.
- A reduction in light pollution on the M62 over Saddleworth Moor.
- Construction of badger tunnels under roads in Cumbria.
- Creation of Bathampton Meadows as part of the A4 Batheaston Bypass.
- Creation of a wildlife habitat from a balancing pond which filters runoff water from the M6 at Tebay, Cumbria.
- Promotion of tree planting in collaboration with the Urban Forestry Unit along motorways, e.g. at Junction 7 of the M6.

The Highways Agency also recognises its responsibility to protect heritage features which may be affected by its activities and the need to respect the historic fabric of the landscape. Roads are a part of the historic fabric of Britain and many trunk roads have been developed from Roman roads and other ancient trackways. While some road developments may have a detrimental effect on the survival and setting of some heritage features, other road improvements may enhance the setting of historical features. The Agency has listed the following actions it proposes to take to ensure the protection of heritage features:

- Review with English Heritage, English Nature and other interested parties the Agency's guidance on the treatment of heritage issues (including geological heritage) within its network management activities.
- Establish a database of current and previous cultural heritage studies by the Agency.
- Work with English Heritage and others in considering joint initiatives to protect and enhance the historic environment.

The process the Highways Agency uses to assess the effects of a proposed new trunk road is shown in Figure 8.2.

8.2.2 Preventing roadkills

Large numbers of wild animals are killed on the roads each year. Much of this destruction of wildlife is unnecessary and can be avoided if highways are designed with the protection of wildlife in mind.

Measures that have been used to reduce road kills include:

- fences, tunnels and bridges;
- reflecting devices (that bounce light from headlights into the verges and woodlands);
- road signs.

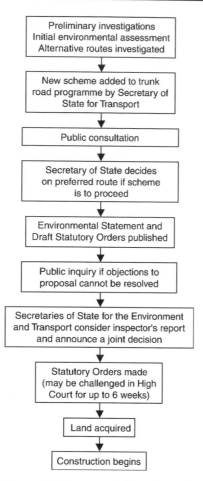

Figure 8.2 Approval of trunk road proposals (adapted from a diagram produced by the Highways Agency).

Fences, tunnels and bridges

It has been estimated that between 30 000 and 50 000 road accidents involving deer occur every year. Deer fences may be used to keep deer off roads or out of forests containing trees vulnerable to deer damage. The lower high tensile strainer-wire must be as close to the ground as possible and any gaps under the fence produced by uneven ground should be blocked by rails or additional netting. Where access gates are necessary they should be designed to swing closed if left open (Springthorpe and Myhill, 1985). New deer fencing types are described by Pepper (1999).

Fences may be used to control the paths used by badgers. Badgers can climb over low fences and may dig underneath them. Fences should be designed so that the mesh at the base is buried under the ground (see Section 4.4.4).

Where culverts pass under roads they should pass through a tunnel which incorporates a ledge well above the water level so that wildlife can safely pass from one side to the other. Tunnels should also be provided where roads cross traditional badger paths. In some situations bridges may be more appropriate to allow the safe passage of wildlife.

Wildlife warning reflectors

Warning reflectors (Figure 8.3) may be used to deter wildlife, especially deer, from crossing roads at night when vehicles are passing, reducing road kills and saving human lives. The reflectors are mounted on short poles (approximately 600 mm tall) and placed at intervals along both sides of the road. When a vehicle approaches the reflectors create an optical warning fence of red light by bouncing the light from headlights into the areas on either side of the road. This deters wildlife

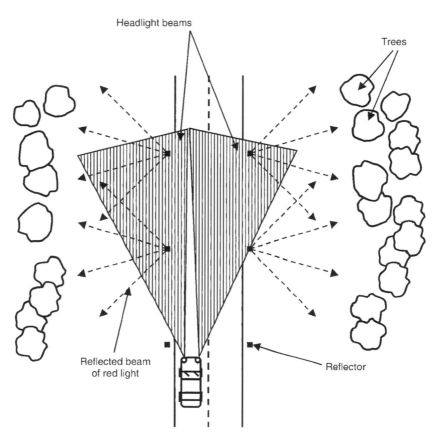

Figure 8.3 Wildlife warning reflectors deter animals from crossing a road by reflecting light from vehicle headlights into the adjacent areas.

from crossing the road until the vehicle has passed. The driver cannot see the optical fence because the beams of light are directed away from the road. When the vehicle has passed the reflectors become inactive and animals can cross safely.

These reflectors cannot prevent all road kills as they do not operate during daylight and some animals may be already on the road before the approach of a vehicle. Nevertheless, they have been used successfully in various countries (in Europe, in Australia, and in the United States) and studies have shown that reflectors may significantly reduce the total number of animals killed.

Swareflex wildlife warning reflectors are manufactured by Swarovski AG of Austria in two designs:

■ Standard reflector – for level or slightly sloping roadsides.
■ Slope reflector – for steep roadsides.

The number, height and positioning of reflectors on any stretch of road will vary depending upon local conditions such as road width and the nature of the adjacent land. Once installed, maintenance is generally low, but reflectors need to be regularly cleaned, realigned and replaced if damaged.

Road signs

A variety of road signs warn of the possibility that wild animals may cross the road. These include signs comprising a red warning triangle and a black silhouette of an appropriate animal as follows:

■ Wildlife (stag)
■ Waterfowl (duck)
■ Frogs or toads (toad)
■ Otter (otter)

The majority of motorists probably pay little attention to these signs (Figure 8.4) since most of our wildlife species are relatively small and unlikely to cause a serious accident if hit. However, where the larger species of deer are present there is clearly the potential for serious accidents. In Scotland even motorways are no barrier to deer dispersal.

Many birds are killed every year by vehicles. Owls often fly low over roads and may be hit by vehicles. One road sign in a rural area in Avon warns of 'Low-flying owls', but it is difficult to see what the motorist could do to avoid striking an owl, even if warned of the possibility in advance (Figure 8.4(c)). The Environment Agency has erected signs to warn drivers of otters crossing roads and in places voluntary groups organise temporary 'toad crossings' to prevent migrating toads from being killed by vehicles.

(a)   (b)   (c)

Figure 8.4 A variety of road signs warn drivers of the possible presence of wildlife: (a) wild animals; (b) wildfowl; (c) owls.

8.2.3   The Highways Agency's *Design Manual for Roads and Bridges*

The *Design Manual for Roads and Bridges* was first introduced in England and Wales in 1992 and subsequently in Scotland and Northern Ireland. It provides a comprehensive loose-leaf manual system of documents relating to the design, assessment and operation of trunk roads, including motorways. The design manual is available on the Highways Agency's website.

Volume 10 of the manual gives information and guidance on environmental design and management, including:

- Landscape management
- Nature conservation
- Archaeology

The nature conservation section contains specific advice in relation to:

- Biodiversity
- Badgers
- Bats
- Otters
- Dormice
- Amphibians

The section on landscape management gives advice on growing wild flowers, including the selection of suitable sites, sowing and growing methods, and choosing suitable mixtures of species.

8.2.4   Trunk road Biodiversity Action Plan for Scotland

The Scottish Executive has produced a Biodiversity Action Plan for the trunk road network in Scotland. It is based on a seven-point plan:

- Work to identify and understand the variety of species and habitats that exist along the network and the interrelationships between them and the activities of the Trunk Road Division.

■ Develop sustainable, cost-effective management practices that work with natural processes.
■ Formulate best practice guidance on maintaining and enhancing areas of value, setting targets for action which are compatible with local, regional and national biodiversity objectives.
■ Work in partnership with other organisations.
■ Promote biodiversity awareness and understanding of locally important species and habitats.
■ Develop and publicise progress and examples of best practice.

The key issues identified by the BAP are:

■ Habitat fragmentation.
■ Use of chemicals, e.g. weedkillers.
■ Accidental pollution during road construction.
■ Changes to hydrology and drainage.
■ Disturbance to wildlife.
■ Road crossing – provision of suitable deterrents and crossing places, such as underpasses or bridges.

The BAP is a comprehensive document which includes a number of useful case studies. It is available on the website of the Scottish Executive.

8.2.5   Managing roadside verges

Grasslands can be actively managed to encourage the growth and survival of wild plant species. This management may simply involve altering the grass-cutting regime to avoid destroying annual plants until after they have flowered and set seed.

One aspect of the nature conservation value of grassland can be measured by the species richness of the sward. This is usually highest in soil of moderate fertility where the turf has been closely grazed. The highest species richness is produced by sheep grazing because it creates a short turf that favours dwarf perennial herbs. Where sheep are not present their effect may be mimicked by appropriate mowing. The following factors may all influence the species composition of grassland:

■ type of mowing machine used;
■ timing and frequency of the cut;
■ fate of cut grass (i.e. removed, shredded or left).

Grassland management also needs to take into account the ecological requirements of animals. The nesting and breeding behaviour of the corncrake (*Crex crex*) has been disrupted by the early cutting of grass and its populations have declined as a result. The populations of the skylark (*Alauda arvensis*) and the lapwing (*Vanellus vanellus*) have also been affected by changes to the management of grasslands. Grassland butterfly populations may be adversely affected if the flowers upon which they depend for their nectar and the host plants of their caterpillars are damaged (Alma, 1993).

Kent County Council has designated 140 roadside sites that are managed as Roadside Nature Reserves. They have been particularly selected for their importance as refuges for wild plants, with the presence of orchid species being used as an indicator of site quality. The sites are actively managed with the cooperation of the council's Highways Management Unit, which may selectively cut back some plant species in order to create favourable habitats for others. The verges are patrolled by voluntary wardens who liaise with grass cutters to prevent the destruction of rare or important plants.

Roads may have subtle effects on the ecology and genetics of plants. Lead-tolerant strains of bent grass (*Agrostis capillaries*) appeared in soil around the convergence of the M5 and M6 motorways in Birmingham ('Spaghetti Junction') as a result of the selection pressure cause by lead compounds that were previously widely used in petrol. Some roadside soils contain such high concentrations of salt that over 20 saltmarsh plants have spread along road networks into the centre of the UK including thrift (*Armeria maritima*), saltmarsh rush (*Juncus gerardii*) and sea plantain (*Plantago maritima*) (Alma, 1993).

Roadside verges provide habitats for small mammals that attract kestrels (*Falco tinnunculus*). These birds can frequently be seen hovering over roadsides and nesting under motorway bridges.

Where highway works are likely to affect trees, construction work should identify a precautionary zone around each tree to avoid damage to the roots (see Figure 6.4).

8.2.6   Case studies

Badger deaths on roads

A very large number of badgers are killed on British roads each year. In a study conducted at Woodchester Park, Gloucestershire, Cheeseman *et al.* (1987) estimated that 25–30 per cent of annual adult badger deaths were the result of road accidents. Neal and Cheeseman (1996) have extrapolated this figure for the whole of mainland Britain and estimate that 37 500 badgers are killed annually on the roads.

Badgers are also frequently killed on railway lines where the track separates one sett from another and ancestral paths connect them. Badgers are killed by being struck by trains but also by electrified lines. During the few months following the electrification of the line between Ashford and Deal in Kent 68 badgers were electrocuted, and over a longer period the total rose to some 200 (Neal and Cheeseman, 1996).

In some areas badgers and foxes appear to have learned to avoid electrocution by walking between the gaps in the live rails.

Otters and the Highways Agency

Research has found that 57 per cent of 673 otter road deaths recorded between 1971 and 1996 occurred on trunk roads and A-roads, and 67 per cent of these accidents happened within 100 m of either freshwater or the sea. A high proportion of deaths occurred where road bridges cross waterways. The Highways Agency is now planning to add ledges to existing bridges and small culverts, and to build otter underpasses to reduce the need for otters to cross roads. Work on modifying some 600 crossings on the road network in England started in 2002. These measures are expected to be relatively low-cost because they will be combined with other bridgeworks.

In the first six months of 2000 four times as many otters were killed on Welsh roads than in the whole of the 1980s. A study funded by the Environment Agency Wales has identified 30 priority sites and 10 top priority sites scattered throughout Wales where improvements to highways are necessary to reduce the danger to otters. The study showed that deaths were not random, but associated with specific features such as culverts and bridges. Otters tend to cross roads when they are unable to pass through bridges or culverts due to floodwater or poor design. Otters will also take shortcuts across roads (Anon., 2000). The location of dead otters found on roads should be reported to the Environment Agency.

Voles halt road improvements

Improvements to the A259 near Rye, West Sussex, were delayed to ensure that construction work would not damage the nearby marshlands, which are an important habitat for voles (Anon., 2001a).

Toads and roads

Froglife maintains a register of toad road-crossing sites on behalf of DEFRA. Once approved and registered by Froglife, a crossing point may be marked by road signs by the local authority. Further information on this scheme is available from Froglife in *Froglife Advice Sheet 3, Amphibians and Roads: Guidelines to Help Reduce the Risk of Injury and Death to Amphibians from Vehicles and Road Drains.*

8.2.7   Translocations

Translocation may be a useful conservation tool where proven techniques are available, but planning inspectors will not simply assume that all such proposals will automatically succeed (see the results of the inquiry into the proposed translocation of Brock's Farm Site of Special Scientific Interest below).

Translocating ancient woodland: A2/M2 road-widening scheme

The Highways Agency has funded the translocation of ancient woodland as part of the A2/M2 road-widening scheme in Kent. The £1.5 million scheme will link the Channel Tunnel and ports with London and the South East. Frith Wood bounds one section of the M2 and the project has involved moving about 100 ancient hazel trees that would otherwise have been felled. The trees have been moved using specialised machines to a new location on former agricultural land at Cressington Fields near Gillingham. The original farmland soil was removed from this eight hectare site before some 10 000 tonnes of ancient woodland topsoil were applied as a 25 cm thick layer. This was collected from Frith Wood and other sites nearby. This translocation of soil has ensured that there will be a supply of woodland plant seeds to help speed up the re-establishment of the ecosystem.

In addition to the ancient trees moved from Frith Wood a further 60 000 trees grown from the seeds of Kentish trees have been planted on the site. The trees are of varying ages and link the newly planted area with the remaining fragment of Frith Wood and nearby Malling and Tunbury Woods. Piles of rotting wood have been artificially created within the new woodland to encourage fungi and insects. Such decomposing wood would normally be absent from newly planted woodland for many years.

Great care has been taken to consider the life cycles of woodland animals in the timing of operations at the site. In November 1998 trees were coppiced by hand to encourage dormice (protected under WCA 1981) emerging from hibernation in the ground to move to taller trees during the spring, away from the road-widening. The remaining dormice were trapped and moved by hand. The digging machines used to move the ancient trees were brought onto the site in September 1999, after the birds had finished breeding.

Brock's Farm Site of Special Scientific Interest

Ecosystem translocation was rejected as a conservation tool in a recent planning inquiry. In July 1998 a planning inspector rejected a proposal by English China Clay International (ECCI) to move Brock's Farm Site of Special Scientific Interest to make way for a waste dump. At the inquiry English Nature was able to show that ECCI's attempt to translocate part of the SSSI several years earlier had failed.

8.3   CASE STUDY: PLANNING AND BIODIVERSITY IN CAMBRIDGESHIRE

The best way of appreciating the role of the local authority planning department in nature conservation is to consider the activities of a single department in detail.

8.3.1   Cambridgeshire's Biodiversity Partnership

Cambridgeshire's Biodiversity Partnership (CBP) promotes the importance of species and habitat conservation within the area governed by Cambridgeshire County Council and the new Peterborough Unitary Authority. Work on writing Local Biodiversity Plans was begun in July 1996 and completed before the end of 2001. There were three phases of consultation involving over 100 people. The partnership includes all local authorities, English Nature, the RSPB, the Wildlife Trusts, the Environment Agency and Anglian Water.

A total of 45 Biodiversity Action Plans (Table 8.1) were written for Cambridgeshire and Peterborough, including:

■ 26 Habitat Action Plans (HAPs); and
■ 19 Species Action Plans (SAPs).

These are grouped into five broad habitat types:

■ Rivers and wetlands
■ Trees and woodlands
■ Farmland
■ Cities, towns and villages
■ Dry grasslands

In Cambridgeshire, applicants for planning permission receive information about biodiversity. The authority produces a biodiversity guidance document, a leaflet entitled *Biodiversity and Householder – Planning Applications*, and information on barn owls and bats where a barn conversion is involved. It is extremely important that discussions between the applicant and the planning authority begin at a very early stage so that ecological concerns may be raised at the beginning of the process.

The CBP lists five objectives that can be achieved through the planning process to benefit biodiversity:

■ **Protect** current key habitats
  – Use development plans policies and restrictive conditions to amend plans and working methods.
  – Exclude important biodiversity areas from permission, especially those with Biodiversity Action Plans.

Table 8.1 Habitat and Species Action Plans for Cambridgeshire

Habitat Action Plans

Fens
Reedbeds
Wet grassland
Rivers and streams
Lakes and irrigation reservoirs
Mineral restorations
Ditches
Wet woodlands
Woodland
Scrub
Old orchards
Parklands and veteran trees
Urban forest
Cereal field margins

Hedgerows
Arable land
Ponds
Gardens, churchyards and cemeteries
Parks/shelterbelts and open spaces
Brownfield sites
Built environment
Allotments
Lowland calcareous grassland
Road verges
Meadows and pasture
Heathland and acid grassland

Species Action Plans

Otter
Water vole
Bittern
Large copper butterfly
Desmoulin's whorl snail
Glutinous snail
Shining ram's-horn snail
Freshwater white-clawed crayfish
Ribbon-leaved water plantain
Dormouse

Black hairstreak butterfly
Brown hare
Grey partridge
Skylark
Pipistrelle bat
Song thrush
Great crested newt
Stone curlew
Pasqueflower

- **Enhance** existing habitats or create new areas
  - Routinely look for opportunities to improve habitats for nature conservation purposes, rather than simply using compensatory measures.
  - Create habitats.
  - Reintroduce species.
  - Reduce fragmentation through wildlife corridor development.
  - Use more appropriate management for existing or new sites.
  - Introduce appropriate afteruse of mineral restoration sites.
- **Mitigate** against potentially damaging impacts
  - Use mitigation measures to make the project acceptable
  - Use planning conditions or agreements on design, methods, timing of operations etc.
  - Obtain adequate information from environmental impact assessments and ecological surveys.
- **Compensate** where damage is unavoidable

  - Only in limited circumstances, where the loss is justified.
  - Recreating quality habitat is very difficult.
  - Loss of habitat continuity as the development progresses results in species loss.
  - Use the precautionary principle.
- **Monitor and enforce** conditions and agreements
  - Enforce planning conditions
  - Assess the success of enhancement, mitigatory and compensatory measures.

The CBP emphasises the legal bases for these measures, listing the relevant provisions in the planning law, planning policy guidance, wildlife laws, the requirements of EIA, the Habitats Directive and the Berne Convention.

For minor proposals, householders are sent an information leaflet which provides advice on improving the environment for wildlife and warning about protected species:

- Plant native trees such as oak, ash and beech in the garden.
- Contact District Council Tree Officer for information on protected trees and hedgerows.
- Create window boxes to attract wildlife.
- Plant native wildflowers such as poppies and cornflowers to attract butterflies and birds, and insects as food for bats.
- Install swift, swallow and house martin boxes and bird feeders.
- Avoid roofing works during the bird breeding season.
- Cut half of the hedgerow each year so that berries are left over the winter.
- Leave long grass under hedges as habitat for small mammals.
- Leave dead wood piles for invertebrates and hibernating hedgehogs.
- Create a pond; check existing ponds for great crested newts; leave long grass around ponds.
- Install bat boxes, bricks or shuttering in the eaves to provide habitats for bats.
- Contact English Nature for advice if bats, great crested newts, grass snakes, slow worms or lizards are likely to be affected by development.

Key points to be considered for major proposals (as defined by the Town and Country Planning (General Development Procedure) Order 1995 (SI 1995/419) are:

- **Survey**
  - Ensure an adequate survey is available from the outset.
  - Some developments require an environmental impact assessment (EIA) under the Town and Country Planning (Environmental

Impact Assessment) (England and Wales) Regulations 1999 (SI 1999/293).
- Advice should be sought from appropriate conservation organisations.

■ **Protect**
- Avoid adverse impacts on designated sites (refer to Structure and Local Plan policies).
- Ensure preservation of statutorily protected species and habitats.
- A landscaping scheme should be provided prior to the planning decision. Site design should retain existing habitats and wildlife features, including Biodiversity Action Plan species and habitats.
- Encourage wildlife corridors and linking habitats to reduce habitat fragmentation.

■ **Mitigate**
- Minimise damage to species and habitats where possible.
- Use a planning condition to require a mitigation strategy.
- English Nature will provide advice on protected species and habitats and may require the issue of a licence for some operations.
- Use planning conditions to ensure works are carried out at appropriate times to avoid disturbance, e.g. outside bird breeding season.

■ **Enhance**
- Planning authorities should be proactive in enhancing existing habitats and creating new habitats where appropriate, in accordance with planning guidance.
- Specific enhancement methods are available for each development type (e.g. highways, housing developments, railway lines etc.).

■ **Compensate**
- Use planning conditions and agreements to ensure re-creation of habitat on- or off-site at the expense of the developer, where damage is unavoidable.
- Where the development could lead to increased pressure on nearby sites it may be appropriate to request a financial contribution from the developer towards the management costs of these sites.

■ **Monitoring and management**
- Provision must be made for the appropriate management of retained, new or enhanced habitats.
- The developer should monitor the site to determine any effects upon wildlife during or after works.
- A long-term management plan should be established with the developer, possibly in collaboration with the Wildlife Trust, a local conservation group or the local authority. Funding from the developer could be secured through planning obligations.

    – Planning agreements may secure the preparation and implemen-
tation of a management plan and long-term site monitoring in
accordance with the objectives of the management plan.

8.3.2   Good practice examples from Cambridgeshire

The following are some examples of conditions used by the planning
authority to ensure the protection or enhancement of wildlife habitats
at specific sites:

■ **Park and ride sites outside Cambridge**
At one site a balancing pond provides a new habitat and a local
wildlife group helped with tree planting. At a second site, native
chalk grassland has been sown in place of the usual amenity mix.
■ **British Sugar site**
Settling ponds at the former British Sugar site in Peterborough are
classified as a County Wildlife Site. Conditions were attached to a
new planning permission for the site requiring safeguards against
pollution, ecological enhancement and the full use of surface water
drainage. A full nature conservation management plan was also
required.
■ **Needingworth Sand and Gravel site**
This site will create 720 ha of wetland habitats over a period of 30
years, following gravel extraction.
■ **Proposed landfill and household waste recycling centre at Kennet**
In collaboration with the Wildlife Trust the planning authority
required the site restoration plan to include the creation of rough
grassland and heathland areas.

8.4   CASE STUDY: THE *LIFE* ECONET PROJECT, CHESHIRE

Ecological projects in the UK are increasingly involving partnerships
between a variety of interested organisations. One such project, the *Life
Econet*, is an attempt to create an ecological network for Cheshire
(James, 1999). The decline in the diversity of organisms and landscapes
within the European Union is evident in the increased fragmentation
and isolation of landscape features, which inhibits their proper ecological functioning. The aim of the econet is to provide a framework for the
integration of environmental issues into land use planning and management by targeting actions and resources, but it will not designate new
areas for protection.

    Ecological networks that are embedded in nature conservation
policy typically comprise four main elements:

■ Core areas representing critical habitat types.

- Corridors or stepping stones linking the core areas, allowing species migration and dispersal.
- Nature restoration areas, adjoining or close to core areas.
- Buffer zones which protect the network from potentially damaging external influences.

The econet project is being funded by the European Commission's *Life* Environment Programme, for the benefit of Cheshire County Council. The UK partners in the project are:

- Vale Royal Borough Council
- English Nature
- Environment Agency
- Sustainability North West
- United Utilities (formerly North West Water)
- University of Salford
- Liverpool John Moores University
- University of Reading

However, the *Life* Econet Project is a multinational venture, which also involves local government and academic partners in Italy and the Netherlands. It will demonstrate how environmental issues can be integrated into the land-use planning and management activities of local authorities in the UK and other EU member states. The key tasks of the project are:

- Technical development of geographical information systems and the application of landscape ecology principles.
- Assessing and influencing land-use policy and instruments.
- Demonstrating integrated land management.
- Engaging stakeholders.
- Dissemination of solutions throughout the EU.

Partners in the *Life* Econet Project hope that the use of regional ecological networks as a targeting mechanism and an aid to decision making will provide a methodology for advancing nature conservation across Europe.

8.5   CASE STUDY: DIBDEN TERMINAL INQUIRY

An examination of a planning inquiry gives a useful insight into the types of conflict which arise between developers and parties seeking to protect nature conservation interests. The development which is the subject of this case study is of particular interest because of the wide range of predicted impacts, and the large number of protected areas involved.

8.5.1   Introduction

Associated British Ports (ABP) proposes to construct a deep-water container terminal at Dibden Bay, on Southampton Water. On 31 January 2001, John Prescott, the then Secretary of State for the Environment, Transport and the Regions, announced an inquiry into the proposal, after being advised by English Nature that it required an 'appropriate assessment' under regulation 48 of the Conservation (Natural Habitats etc.) Regulations 1994, because of its implications for protected areas. The inquiry opened on 27 November 2001 and was expected to last 12 months (Anon., 2001b).

8.5.2   The proposal

Associated British Ports propose to construct a new container terminal consisting of deep-water quays, a rail terminal, a lorry handling area, offices, warehouses and other port infrastructure capable of handling almost one-and-a-half million containers every year. The site includes mudflats, saltmarsh and an area which was reclaimed over 50 years ago and now supports a great diversity of wildlife.

8.5.3   The inquiry

The inquiry was, in effect concurrent inquiries into:

- the Port of Southampton (Dibden Terminal) Harbour Revision Order (Harbours Act 1964 s. 14);
- the Fawley Branch Line Improvements Order (Transport and Works Act 1992 s. 1 and s. 5);
- a Stopping-up Order relating to two areas of highway at the Hyde Road, near Marchwood (Town and Country Planning Act 1990 s. 248);
- three planning applications (called in by the Secretary of State) for improvements to the A326 and Terminal Access Road Junction and for noise barriers along Fawley Branch Line; and
- a proposal to provide land at West Cliff Hall in exchange for open space at the Hythe Marina Bund (Acquisition of Land Act 1981 s. 19).

The inquiry was held by a team from the Planning Inspectorate comprising:

- An inspector
- A deputy inspector
- An assistant inspector (Ecology)
- An assessor (Erosion and Deposition of Sediment)
- An inquiry manager

Evidence to the inquiry was divided into 23 topics:

- Proposals
- Planning background and overview
- Need for the proposed development
- Alternative solutions
- Erosion and deposition of sediment
- Nature conservation
- Fisheries
- Navigation
- Land access to Dibden Bay
- Noise and vibration
- Visual impact
- Air quality and climatic factors
- Countryside issues
- Effect on public rights of way
- Effect on tourism and recreation
- Flooding and drainage
- Archaeology and architecture
- Contaminated land
- Effect on Marchwood Military Port safeguarded area
- Effect on seawater abstraction by local industries
- Effect on outfalls, pipe-lines and other apparatus
- Compulsory purchase issues
- Human rights issues, including the effects on property values

By the beginning of the inquiry, 38 statements of case had been received from individuals and interested organisations, including:

- Local government and parish councils
  - Hythe and Dibden Parish Council
  - New Forest District Council
  - Southampton City Council
  - Marchwood Parish Council
  - Hampshire County Council
  - Totton and Eling Town Council
- Business interests
  - Associated British Ports
  - Southern Water Services
  - Cable and Wireless UK Services Ltd
  - Transco plc
  - Esso Petroleum, ExxonMobil Chemical Ltd and ExxonMobil Olefins Inc.
  - Shanks Chemical Services Ltd
  - EniChem UK Ltd
  - Railtrack plc

- White Horse Ferries Ltd
- Hythe Marina Ltd/Hythe Marina Village Ltd
- English Welsh and Scottish Railway Ltd
- Southampton and Fareham Chamber of Commerce and Industry
■ Conservation interests
- Royal Society for the Protection of Birds
- New Forest Committee
- Hampshire Ornithological Society
- Hampshire and Isle of Wight Wildlife Trust Ltd
- Environment Agency
- English Nature
- Council for National Parks
- Residents Against Dibden Bay Port
- The Ramblers Association
- Solent Protection Society
- Council for the Protection of Rural England (Hampshire Branch)
- Friends of the Earth (Hampshire)
- The National Trust
■ Other interests
- Defence Estates
- Cruising Association
- Royal Yachting Association
- Hythe Marina Association

8.5.4   English Nature's Objections

The proposed terminal at Dibden Bay will affect a number of areas of nature conservation interest including:

■ 42 hectares of mudflat within the Solent and Southampton Water Special Protection Area (SPA) (Wild Birds Directive), and Ramsar site;
■ 34 hectares of mudflat below the mean low water mark which is outside the SPA but considered integral to the bird use of the site;
■ River Itchen candidate Special Area of Conservation (cSAC) (Habitats Directive);
■ Solent Maritime candidate Special Area of Conservation (cSAC) (Habitats Directive);
■ 8 Sites of Special Scientific Interest, including the almost total loss of the recently notified Dibden Bay SSSI.

The SPA is used by 50 000 waterbirds every winter and is one of the most important sites in the UK for wintering wildfowl. Up to one-fifth of the local population of oystercatchers uses the foreshore, along with grey plover, widgeon, curlew, lapwing and an internationally impor-tant population of dark-bellied Brent geese. The site is of European importance due to the presence of two distinct types of saltmarsh: Atlantic Salt Meadows and Cordgrass Swards.

When asked to comment on the proposed development in 1999, English Nature advised the former DETR that the scheme would be likely to have a significant effect on the SPA, cSAC and Ramsar sites. If approved it would result in the immediate loss of important areas of the SPA/Ramsar site and accelerated erosion of saltmarsh and mudflats within the area proposed as a candidate SAC. In the long term this will result in additional foreshore loss, reducing the ability of intertidal habitats to support migratory waterfowl. Bird populations will be disturbed and displaced (Anon., 2001c).

8.5.5 Mitigation measures proposed by Associated British Ports (ABP)

ABP has spent four years researching the environment and ecology of the terrestrial and aquatic ecosystems that would be affected by its proposals. The company commissioned independent experts to study the site and has also held discussions with English Nature and the RSPB.

Almost half of ABP's land (180 hectares) will be used for environmental improvements including:

- landscaping to restrict views of the terminal;
- nature reserves including reedbeds, saltmarshes and woodland;
- public open space;
- an interpretation facility;
- planting approximately a third of a million native trees and shrubs, including many that will be fully grown;
- extending existing mudflats by capping them with freshly dredged sediment;
- creation of a mile-long intertidal creek that meets the water and restores the original shoreline of Dibden Bay.

This last proposal would create new intertidal areas and a restored mudflat providing a new bird-feeding area. ABP claims that its proposals offer improved open space for the public along with secure and sustainable habitats for wildlife.

8.5.6 Objections to the mitigation

In spite of the offsetting measures proposed by ABP, there will be an overall loss of mudflat and saltmarsh habitats from within the SPA/Ramsar site boundary. The new terminal will require regular dredging to prevent the channel from silting up. English Nature believes that the reduced availability of sediment to saltmarsh habitats will prevent them from accreting at the same rate as sea level rise, limiting their ability to maintain themselves in the long term. The proposed artificial creek is considered by English Nature as little more than an experiment which will be disturbed by the port on one side and a marina on the other. ABP's plan to extend three kilometres of mudflats will disturb the equilibrium of the existing mudflats and disrupt the flow of water in the finely balanced saltmarsh which lies above it. The remodelled mudflats would also have an adverse impact on migratory salmon that move up and down Southampton Water to and from their breeding grounds in the Rivers Test and Itchen.

8.6   BUILDING DESIGN AND WILDLIFE

Modern design excludes many animal species which would previously have found shelter in buildings. There are good public health and economic reasons for wishing to prevent many organisms from taking up residence in buildings. However, building designs and modifications that exclude undesirable species may also have a significant impact upon rare species.

Murphy and Todd (1993) have suggested that voids at eaves level in buildings should be controlled to reduce the likelihood of birds and insects entering the roof space. They note that it is common practice to fit a plastic ventilation strip in these gaps with a nominal mesh size of 4 mm (as recommended in BS 5250) in order to maintain adequate air flow to the roof space.

Murphy and Todd draw attention to the potential problems caused by the use of pantiles to finish a roof. Their humped shape can create harbourage underneath the tile and allow access to the roof space. They suggest the use of comb fillers to close these gaps and stress the importance of integral insect screens in roof tile and ridge ventilators. They also recommend that flashings around chimney stacks should be checked for large gaps.

As building design makes dwelling-houses and other buildings less accessible to wildlife, urban populations of some species are likely to suffer. House sparrows and starlings commonly nest in the roof spaces of dwelling-houses and both of these species have experienced significant declines in recent years. It is becoming increasingly common to see bird netting covering the tops of city centre buildings blocking the traditional roosting sites of starlings (Figure 8.5).

Bats commonly use roof spaces for roosting. They also utilise spaces under roofing felt and tiles, under flashings, behind barge boards and between soffits and walls. Many of the measures intended to prevent access by birds undoubtedly also exclude bats. In some areas of the south of England the Building Regulations require all softwood roof

Figure 8.5 Bird netting used to protect buildings is depriving many urban birds such as sparrows and starlings of roosts.

timbers, including ceiling joists, to be treated with a suitable preservative against the house longhorn beetle (*Hylotrupes bajulus*) (Powell-Smith and Billington, 1999). Where wood preservatives are used, because of Building Regulation requirements or otherwise, care must be taken to avoid exposing non-target species like bats to toxins.

Many old buildings have wooden floors at ground level and were designed with large metal grills giving access to the underfloor spaces and basement areas. As a result quite large animals could take up residence underneath these buildings and in the associated heating ducts. Some large Victorian hospitals can support substantial numbers of feral cats living underground. In the early 1980s Winwick Hospital in Warrington was home to a population of around 35 adult feral cats and their kittens (Rees, 1982). These animals sheltered in a network of underground ducts and subways that extended over several miles. Many of the ventilation grids had become broken over the years giving the animals free access to the buildings. In some hospitals feral cats have been blamed for introducing fleas into laundries and even for the closure of infested operating theatres. The solid floors of modern buildings make such access by large mammals impossible. The tendency for modern buildings to have few if any points of access at ground level is likely to affect the populations of urban animals in the long term.

While it would be clearly inappropriate to design buildings so that species regarded as pests could easily penetrate them, designers can nevertheless play an important part in making structures wildlife-friendly. Allowing bats restricted access to roof spaces and cavity walls is unlikely to cause harm. Likewise, small songbirds are unlikely to cause extensive damage during the relatively short time they nest in roof spaces.

REFERENCES

Alma, P.J. (1993) *Environmental Concerns*. Cambridge University Press, Cambridge.

Anon. (no date) *Towards a Balance with Nature. Highways Agency Environmental Strategic Plan*. Traffic Safety and Environmental Division, Highways Agency, London.

Anon. (2000) Action for otters on Welsh roads. Environment Agency document SM224/00LR. *www.environment-agency.gov.uk/modules/MOD44.2300*. html, accessed 19 April 2001.

Anon. (2001a) Voles halt work. *The Times*, 3 January 2001, p. 10.

Anon. (2001b) Dibden Terminal Inquiry, Planning Inspectorate, *www.planning-inspectorate.gov.uk/dibden/index.htm*, accessed 28 November 2001.

Anon. (2001c) Public enquiries. Dibden Bay – protecting it for nature is a must. *www.english-nature.org.uk/news/enquiry/dibden.asp*

Bullock, J. (2001) Loggerheads over 'pretend tree' mast. *Bolton Evening News*, 20 December 2001, p. 8.

Cheeseman, C.L., Wilesmith, J.W., Ryan, J. and Mallinson, P.J. (1987) Badger population dynamics in a high-density area. *Symp. Zool. Soc. Lond.*, 58: 279–294.

Design Manual for Roads and Bridges. Highways Agency, Scottish Office Development Department, Welsh Office, Department of the Environment for Northern Ireland. Available at *www.highways.gov.uk/info/techinf/index*

James, P. (1999) Ecological networks: creating landscapes for people and wildlife. *Journal of Practical Ecology and Conservation*, 3(2): 3–10.

Murphy, R.G. and Todd, S. (1993) Pest implications in housing design – research and practices in the UK. *International Journal of Environmental Health Research*, 3: 67–72.

Neal, E. and Cheeseman, C. (1996) *Badgers*. T. & A.D. Poysner Ltd, London.

Pepper, H. (1999) *Recommendations for Fallow, Roe and Muntjac Deer Fencing: New Proposals for Temporary and Reusable Fencing*. Forestry Commission, Edinburgh.

Powell-Smith, V. and Billington, M.J. (1999) *The Building Regulations Explained and Illustrated*, 11th edn. Blackwell Science, Oxford.

Rees, P.A. (1982) The ecology and management of feral cat colonies. Unpublished PhD thesis, University of Bradford.

Springthorpe, G.D. and Myhill, N.G. (1985) *Wildlife Rangers Handbook*. Forestry Commission.

Trunk Road Biodiversity Action Plan. Scottish Executive. *www.scotland.gov.uk/library2/doc/doc11/tbap-00.asp*

9

Environmental impact assessment

Environmental impact assessment (EIA) is a procedure designed to assess the likely effects of a new development before it is allowed to proceed. The subject of EIA is deserving of a book in itself. The intention here is to provide an overview of the EIA requirements with respect to conservation interests affected by developments. The principles of EIA and the approaches adopted by various countries have been discussed by Wathern (1988). A detailed guide to EIA procedures in England and Wales has been produced by the former DETR (Anon., 2000).

9.1 INTRODUCTION

EIA has been required by European law for certain types of projects in the United Kingdom since July 1988, when Directive 85/337/EEC on the assessment of the effects of certain public and private projects on the environment came into effect. The original Directive has since been amended by Directive 97/11/EC, which came into effect in March 1999.

9.1.1 The EIA Directive and its implementation

The Directive requires EIA to be carried out for certain types of major projects which are likely to have significant environmental effects, before development consent is granted. In England and Wales the Directive was given legal effect through the Town and Country Planning (Environmental Impact Assessment) (England and Wales) Regulations 1999 (SI 1999/293), for those projects requiring planning permission. These regulations came into force on 14 March 1999 and apply to all relevant planning applications lodged on or after that date. The Town and Country Planning (Environmental Impact Assessment) (England and Wales) (Amendment) Regulations 2000 (SI 2000/2867) amend the 1999 regulations and implement the Directive in relation to applications to mineral planning authorities.

Environmental Impact Assessment Directive 1985

Article 1
1. This Directive shall apply to the assessment of the environmental effects of those public and private projects which are likely to have significant effects on the environment.

2. For the purposes of this Directive:
    'project' means:
    >   the execution of construction works or of other installations or schemes, other interventions in the natural surroundings and landscape including those involving the extraction of mineral resources;
    'developer' means:
    >   the applicant for authorisation for a private project or the public authority which initiates the project;
    'development consent' means:
    >   the decision of the competent authority or authorities which entitled the developer to proceed with the project.

Article 2
1. Member States shall adopt all measures necessary to ensure that, before consent is given, projects likely to have significant effects on the environment by virtue, *inter alia*, of their nature, size or location are made subject to a requirement for development consent and an assessment with regard to their effects.

9.1.2   Town and Country Planning (Environmental Impact Assessment) (England and Wales) Regulations 1999

Under the 1999 regulations, planning permission may not be granted for development requiring EIA without first taking environmental information into account (reg. 3(2)). EIA development is defined by regulation 2(1) as development which is either:

(a) Schedule 1 development; or
(b) Schedule 2 development likely to have significant effects on the environment by virtue of factors such as its nature, size or location.

Scoping and screening

Developers should consult the planning authority where there is a possibility that a development may require EIA. Developers may decide for themselves that EIA is required, or they may apply to the

---

Town and Country Planning (Environmental Impact Assessment) (England and Wales) Regulations 1999 (SI 1999/293) reg. 3

---

(2) The relevant planning authority or the Secretary of State or an inspector shall not grant planning permission pursuant to an application to which this regulation applies unless they have first taken the environmental information* into consideration, and they shall state in their decision that they have done so.

* The environmental statement, any further information supplied, any representations received (reg. 2(1)).

---

local planning authority (LPA) for an opinion ('screening opinion') on whether EIA is needed (reg. 5).

If the planning authority decides that EIA is required but the developer is not satisfied with this decision, application may be made to the Secretary of State (or for projects in Wales, the National Assembly for Wales) for a 'screening direction' (reg. 6) (see Figure 9.1). Under regulation 6(7) the Secretary of State may make a screening direction irrespective of whether he has received a request to do so.

If EIA is required the developer must produce an environmental statement. The developer may ask the planning authority what is required in the environmental statement ('scoping opinion') (reg. 10), or this information may be provided by the Secretary of State ('scoping direction') (reg. 11). Before adopting a scoping opinion or a scoping direction, the local planning authority or the Secretary of State must consult bodies with environmental responsibilities (defined in reg. 2(1)), including the Environment Agency, English Nature and the Countryside Council for Wales.

After an environmental statement has been submitted a planning authority, the Secretary of State or a planning inspector may request additional information under regulation 19. Copies of the environmental statement and any 'further information' must be made available to the public (regs 15, 17 and 19). A reasonable charge may be made for copies.

Consultation and publicity

The local planning authority must notify the consultation bodies (defined in reg. 2(1)) of any application for planning permission which

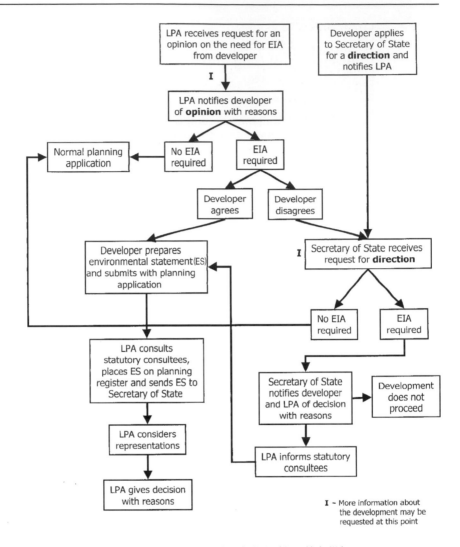

In Wales references to the Secretary of State apply to the National Assembly for Wales

Figure 9.1 Pre-application requests for a screening opinion to a local planning authority (LPA) or a screening direction to the Secretary of State.

is accompanied by an environmental statement (reg. 13). Equivalent provisions apply to the Secretary of State in relation to an appeal or a called-in application (reg. 16). Regulation 20 provides for documents to be placed on the planning register or otherwise made available for public inspection, including screening opinions and directions, scoping opinions and directions, environmental statements and reasons for decisions. Information on decisions made by the local planning authority or the Secretary of State must be made available under regulation 21.

Other provisions

Regulations 23 and 24 restrict the grant of planning permission by simplified planning zone schemes or enterprise zone orders, and regulation 25 establishes procedures for EIA in relation to the enforcement of planning control.

If development proposed to be carried out in England or Wales is the subject of an EIA application and is likely to have significant effects on the environment of another member state, regulation 27 requires the Secretary of State to send to the member state:

■ a description of the development;
■ information on its possible significant effects on the environment in another member state;
■ information on the nature of the decision which may be taken.

Regulation 28 makes provisions for consultations where development in another member state may have significant effects on the environment in England and Wales. Regulations 27 and 28 implement Article 7 of the Directive.

9.1.3 Types of project that require EIA

The types of project which shall be made subject to an EIA are described in Annexes I and II to the Directive, and Schedules 1 and 2 of the regulations (SI 1999/293). All development in Schedule 1 requires EIA. This includes large-scale projects such as oil refineries, long distance railway lines and airports (see Appendix 5). Schedule 2 lists small-scale projects such as golf courses and motorway services (see Appendix 6). Development in Schedule 2 requires EIA if it is either:

■ to be carried out in a sensitive area; or
■ satisfies a threshold or criterion listed in Schedule 2.

The selection criteria for screening Schedule 2 developments are listed in Schedule 3 of the regulations. In relation to considerations of location, they include the presence of protected areas and certain types of habitats.

9.1.4 Exemptions of certain projects from EIA

In exceptional circumstances, member states may exempt a specific project in whole or in part from the provisions of the Directive (Art. 2 (3), implemented by regulation 4(4)). In this event, the member state must consider whether another form of assessment is appropriate and whether the information collected by such an assessment should be

> ### Town and Country Planning (Environmental Impact Assessment) (England and Wales) Regulations 1999 (SI 1999/293) Schedule 3
>
> 2. Location of development
>   The environmental sensitivity of geographical areas likely to be affected by development must be considered, having regard, in particular, to –
>   (a) the existing land use;
>   (b) the relative abundance, quality and regenerative capacity of natural resources in the area;
>   (c) the absorption capacity of the natural environment, paying particular attention to the following areas –
>   (i) wetlands;
>   (ii) coastal zones;
>   (iii) mountain and forest areas;
>   (iv) nature reserves and parks;
>   (v) areas classified or protected under Member States' legislation; areas designated by Member States pursuant to Council Directive 79/409/EEC on the conservation of wild birds and Council Directive 92/43/EEC on the conservation of natural habitats and of wild fauna and flora;
>   (vi) areas in which the environmental quality standards laid down in Community legislation have already been exceeded;
>   (vii) densely populated areas;
>   (viii) landscapes of historical, cultural or archaeological significance.

made available to the public. It must also make available to the public and the Commission the reasons for granting an exemption.

The Directive exempts from the requirement for an EIA 'projects serving national defence purposes' (Art. 1 (4)), and projects which are 'adopted by a specific act of national legislation' (Art. 1 (5)). In the latter case, Article 1 (5) acknowledges that the objectives of the Directive are achieved through the legislative process, including that of supplying information. An example of a recent infrastructure project which was authorised by Act of Parliament is the construction of the Channel Tunnel Rail Link (Channel Tunnel Rail Link Act 1996) (Panel 9.1).

9.1.5   EIA of projects not approved under the planning system

The projects covered by Schedules 1 and 2 of the regulations are subject to planning control in England and Wales. However, where projects are

---

PANEL 9.1   CHANNEL TUNNEL RAIL LINK

The construction of the Channel Tunnel Rail Link was authorised by Act of Parliament in 1996. Approximately 85 per cent of the route of the rail link is designed to run within tunnels or existing transport corridors, next to railways, motorways or major roads. Before the completion of Section 1 of the project the following will have been created or developed:

- 55 hectares of mixed broad-leaved woodland, including about 30 hectares on translocated ancient woodland soils, and the remainder on low nutrient soil
- 12 kilometres of new hedgerow
- 80 hectares of amenity woodland and tree and shrub planting
- 180 hectares of grassland seeding, including 15 hectares of wildflower grassland
- 9 hectares of reedbeds, alder carr (a fen wood) and wet grassland
- 1 kilometre of ditches
- 8 new ponds for amphibians
- 5 new artificial badger setts and specially designed track crossing points
- 50 new artificial bat roosts
- New habitat for the grey mouse-ear (*Cerastium brachypetalum*), a rare flowing plant
- Reintroduction of over 100 dormice from Kent in two other woodlands
- Translocation of great crested newts, slow worms and common lizards
- Mitigation for water vole populations
- A monitoring programme for surface water quality and fisheries on major water courses

In addition, over 30 detailed investigations of archaeological sites have been undertaken.

Source: *www.ctrl.co.uk/ecology*

(In July 2001 Paignton Zoo released four pairs of captive-bred common dormice into a site in Cambridgeshire. The zoo's captive breeding programme is run in collaboration with English Nature and started with a breeding pair saved from building work on the Channel Tunnel in 1998.)

---

not approved under the planning system specific regulations apply. Examples of these regulations include:

- The trunk road network (including most motorways) – Highways (Assessment of Environmental Effects) Regulations 1999 (SI 1999/369).
- Oil and gas pipe-lines – Pipe-line Works (Environmental Impact Assessment) Regulations 2000 (SI 2000/1928)*.
- Public gas transporter pipe-line works – Public Gas Transporter Pipe-line Works (Environmental Impact Assessment) Regulations 1999 (SI 1999/1672)*.

- Offshore oil and gas projects – Offshore Petroleum Production and Pipe-lines (Assessment of Environmental Effects) Regulations 1999 (SI 1999/360)**.
- Nuclear power stations – Electricity Works (Environmental Impact Assessment) (England and Wales) Regulations 2000 (SI 2000/1927); Nuclear Reactors (Environmental Impact Assessment for Decommissioning) Regulations 1999 (SI 1999/2892)*.
- Other power stations and overhead power lines – Electricity Works (Environmental Impact Assessment) (England and Wales) Regulations 2000 (SI 2000/1927).
- Forestry projects – Environmental Impact Assessment (Forestry) (England and Wales) Regulations 1999 (SI 1999/2228).
- Land drainage improvements – Environmental Impact Assessment (Land Drainage Improvement Works) Regulations 1999 (SI 1999/1783).
- Ports and harbours – Harbour Works (Environmental Impact Assessment) Regulations 1999 (SI 1999/3445)*.
- Marine fish farming – Environmental Impact Assessment (Fish Farming in Marine Waters) Regulations 1999 (SI 1999/367)*.
- Construction or operation of railways, tramways etc., inland waterways, interference with rights of navigation (works authorised under the Transport and Works Act 1992 – Transport and Works (Applications and Objections Procedure) (England and Wales) Rules 2000 (SI 2000/2190).

The regulations listed above all apply to England and Wales but those marked* also apply to Scotland. Those marked** apply to the whole of the UK.

9.1.6   EIA of projects in Scotland and Northern Ireland

The arrangements for EIA in Scotland and Northern Ireland are broadly similar to those in England and Wales. In Scotland, projects requiring planning permission and certain trunk road projects and land drainage works are covered by the Environmental Impact Assessment (Scotland) Regulations 1999 (SSI 1999/1). The Directive is implemented in Northern Ireland by the Planning (Environmental Impact Assessment) Regulations (Northern Ireland) 1999 (SR 1999/73), for projects requiring planning permission.

Projects outside the planning systems in Scotland and Northern Ireland are included in some of the regulations listed above, as indicated. In addition, the following regulations apply to Scotland or Northern Ireland alone:

- Environmental Impact Assessment (Forestry) (Scotland) Regulations 1999 (SSI 1999/43).

- Environmental Impact Assessment (Forestry) Regulations (Northern Ireland) 2000 (SR 2000/84).
- Environmental Impact Assessment (Fish Farming in Marine Waters) Regulations (Northern Ireland) 1999 (SR 1999/415).
- Roads (Environmental Impact Assessment) Regulations (Northern Ireland) 1999 (SR 1999/89).

9.1.7 What impacts must be assessed?

The Directive requires developers to provide a wide range of information on environmental impacts (Art. 3). This takes the form of an 'environmental statement' which sets out the developer's own assessment of a project's likely environmental effects, and is submitted in conjunction with the application for consent.

Environmental Impact Assessment Directive 1985 Art. 3

The environmental impact assessment shall identify, describe and assess in an appropriate manner, in the light of each individual case and in accordance with Articles 4 to 11, the direct and indirect effects of a project on the following factors:

human beings, fauna and flora;
soil, water, air, climate and the landscape;
material assets and the cultural heritage;
the interaction between the factors mentioned in the first, second and third indents.

This information must include:

- a detailed description of the project;
- an outline of the main alternatives to the project;
- a description of likely environmental effects of the project;
- a description of the measures envisaged to mitigate adverse environmental effects;
- a non-technical summary of the information provided above;
- an indication of any difficulties and technical deficiencies encountered by the developer in compiling the information.

These requirements are listed in Annex IV to the Directive and Schedule 4 of the regulations (see Panel 9.2).

The former DETR has produced a checklist of matters which should be included in an environmental statement, including specific requirements in relation to land use and the natural environment (Anon., 2000, appendix 5).

PANEL 9.2   INFORMATION FOR INCLUSION IN
ENVIRONMENTAL STATEMENTS

SI 1999/293 SCHEDULE 4

Regulation 2(1)

INFORMATION FOR INCLUSION IN ENVIRONMENTAL STATEMENTS

PART I

1. Description of the development, including in particular –

    (a) a description of the physical characteristics of the whole
        development and the land-use requirements during the construction
        and operational phases;
    (b) a description of the main characteristics of the production process,
        for instance, nature and quantity of materials used;
    (c) an estimate, by type and quantity, of expected residues and
        emissions (water, air and soil pollution, noise, vibration, light, heat,
        radiation, etc.) resulting from the operation of the proposed
        development.

2. An outline of the main alternatives studied by the applicant or appellant
    and an indication of the main reasons for his choice, taking into account
    the environmental effects.

3. **A description of the aspects of the environment likely to be
    significantly affected by the development, including, in
    particular, population, fauna, flora, soil, water, air, climatic
    factors, material assets, including the architectural and
    archaeological heritage, landscape and the interrelationship
    between the above factors.**

4. **A description of the likely significant effects of the
    development on the environment which should cover the direct
    effects and any indirect effects and any indirect, secondary,
    cumulative, short, medium and long-term, permanent and
    temporary, positive and negative effects of the development,
    resulting from:**

    (a) the existence of the development;
    (b) the use of natural resources;
    **(c) the emission of pollutants, the creation of nuisances and
        the elimination of waste;**

    **and the description by the applicant of the forecasting methods
    used to assess the effects on the environment.**

5. **A description of the measures envisaged to prevent, reduce and
    where possible offset any significant adverse effects on the
    environment.**

6. A non-technical summary of the information provided under paragraphs 1 to 5 of this Part.

7. An indication of any difficulties (technical deficiencies or lack of know-how) encountered by the applicant in compiling the required information.

PART II*

1. A description of the development comprising information on the site, design and size of the development.

2. **A description of the measures envisaged in order to avoid, reduce and, if possible, remedy significant adverse effects.**

3. **The data required to identify and assess the main effects which the development is likely to have on the environment.**

4. An outline of the main alternatives studied by the applicant or appellant and an indication of the main reasons for his choice, taking into account the environmental effects.

5. A non-technical summary of the information provided under Paragraphs 1 to 4 of this Part.

* Part II lists the minimum information required by the regulations.

Sections indicated in **bold** are likely to contain information relating to wildlife and nature conservation issues.

The statement should include:

■ *Information describing the project*
  – Land use requirement and other physical features of the project:
    – during the operation;
    – when operational;
    – after land use has ceased.
  – Emissions to air, discharges to water, deposits to land and soil, noise, heat etc.
  – Main alternative sites and processes considered and reasons for final choice.

■ *Information describing the site and its environment*
  – Flora and fauna, including habitats and species, in particular, protected species and their habitats.
  – Soil:
    – agricultural quality;
    – geology;
    – geomorphology.
  – Water:
    – aquifers;

- water courses;
- shoreline;
- including the type, quantity, composition and strength of existing discharges.
- Architectural and historic heritage, archaeological sites and features, and other material assets.
- Landscape and topography.
- Recreational uses.
- Policy framework:
  - national policies, regional and local plans, approved or emerging development plans;
  - statutory designations: national parks, SSSIs, NNRs, AONBs, TPOs, green belt, country parks, heritage coasts, water protection zones, conservation areas, listed buildings, scheduled ancient monuments etc.
- European and international designations e.g. SPA, SACs, Ramsar sites etc.

■ *Assessment of effects*
  - Visual effect on the surrounding area and landscape.
  - Levels and effects of emissions during normal operation.
  - Effects on:
    - architectural and historic heritage;
    - archaeological features;
    - other human artefacts.
  - Loss of, and damage to:
    - habitats;
    - animal species;
    - plant species.
  - Loss of, and damage to:
    - geological features;
    - palaeontological features;
    - physiographic features.
  - Other ecological consequences.
  - Physical effects on land:
    - soil stability;
    - soil erosion;
    - local topography.
  - Effects of chemical emissions and deposits on:
    - the soil of the site;
    - surrounding land.
  - Effects on land use and resources:
    - quality and quantity of agricultural land to be taken;
    - sterilisation of mineral resources;
    - alternative uses of the site, including 'do nothing' option;
    - effect on surrounding land uses, including agriculture.

 – Effects on water:
   – drainage pattern;
   – other hydrology, including groundwater level and water courses;
   – coastal or estuarine hydrology;
   – effects of pollutants on water quality.
 – Effects of chemical emission on the environment, including climatic effects.
 – Other indirect and secondary effects:
   – traffic etc.;
   – effects arising from extraction and consumption of materials, energy, water and other resources;
   – effects of associated developments, e.g. new roads, sewers, housing, power lines etc.;
   – effects of association of the development with other existing or proposed developments;
   – secondary effects resulting from the interaction of direct effects listed above.

■ *Mitigating measures*
 – Site planning.
 – Technical measures e.g. pollution control.
 – Aesthetic and ecological measures:
   – mounding;
   – colour and design of buildings etc.;
   – landscaping;
   – tree planting;
   – habitat preservation measures;
   – creation of new habitats;
   – recording of archaeological sites;
   – safeguarding of historic buildings and sites;
 – Assessment of the likely effectiveness of mitigating measures.

It should be noted that the above list contains only those matters which are likely to have an effect on species, natural habitats and heritage sites. Other considerations, such as the general effects of traffic or human population increase, have been omitted.

9.2   NATURE CONSERVATION AND EIA

9.2.1   Identifying impacts

Environmental impact assessment aims to provide decision makers with an indication of the likely consequences of their proposed actions. While the focus of many environmental statements is to assist a project to successful authorisation the process also has an important part to play in reducing adverse impacts before the authorisation phase is reached.

A development may have no significant effect on air or water but it is inconceivable that it could have no effect on the land. Soil will be both removed and disturbed. It is, therefore, inevitable that soil organisms and plants will be destroyed. If a development project has no other environmental effect, it will always have an impact on nature.

It would be possible to quantify the amount of soil removed from a development site and to estimate the damage done to plant and animal life in the process. However, unless the species that will be lost are rare or of great public interest, they are unlikely to be mentioned in an environmental statement.

For a development such as a large industrial complex it is possible to model the likely dispersion of air pollutants, using estimates of likely emissions along with data on wind speed and direction. It is also possible to estimate the amount of waste that would be produced, increases in traffic volumes, the effect on the landscape and many other environmental impacts. Similarly, attempts can be made to quantify the possible effects of development upon the ecology of a site.

Some early environmental statements made little serious attempt at ecological surveys. They stated simply that there were no nature reserves or SSSIs in the vicinity and that an ecological survey had not identified the presence of any rare species. However, it is in the nature of rare species that they are difficult to find. If the 'ecological survey' consisted of a biologist spending no more than a day walking around the site noting each species he observed, it is unlikely that he would record anything other than the most common species of plants and birds. Mammals and insects are unlikely to be observed unless traps are used. Furthermore, if species lists contain only scientific names they are incomprehensible to the layperson. From such lists, it may be impossible to distinguish a tree species from a grass. Such 'surveys' have little value other than to satisfy the local planning department that an ecological study has been undertaken.

Ecological surveys should consider all forms of organisms from large mammals and birds to the smallest invertebrates and fungi. They should be conducted over an extended period as many flowering plants (especially annuals) are difficult to find and identify when not in flower, and many animal species exhibit seasonal variations in both distribution and abundance. Plant surveys should be conducted in a systematic fashion, collecting samples from quadrats or transects as appropriate to the habitat. Such surveys may be usefully supplemented by data from other sources such as local museums and natural history societies. Local experts (both professional and amateur) may be able to provide valuable information on the location of rare or interesting species or communities. Vegetation classification systems are described in Appendix 8.

The EIA process has tended to concentrate on spatial impacts rather than those that have a temporal dimension. In the context of nature conservation temporal impacts are particularly important. A project may initially have a limited impact on an animal population and yet lead to its eventual decline. Damage to tree roots is unlikely to lead to the sudden death of the tree. Wathern (1988) distinguishes between direct (primary) and indirect (secondary) impacts. The construction of a dam on a river will prevent upward migration of fish (primary impact) while long-term changes in oxygen levels will affect survival and increase siltation, making conditions unsuitable for breeding (secondary impacts).

It is extremely difficult to predict accurately the likely effects of pollutants on individual species or ecosystems. In particular, the long-term effects of exposing ecosystems (and humans) to low levels of chemicals, has not been well studied. Although the precautionary principle should be applied when known risks are involved, at least in the short term, it is likely that many potentially damaging projects will be authorised simply through lack of scientific knowledge.

9.2.2 Mitigation of ecological damage

Where considerable ecological damage is likely to be sustained by a large development site, it is essential that developers undertake a comprehensive programme of environmental mitigation. The detail of these measures may be important in persuading a planning authority to authorise a development. If 20 trees are to be removed but a hundred saplings will be planted to replace them, there would appear to be an environmental gain, provided that the lost trees were not of particular ecological value. In some circumstances, it may even be possible to translocate individual animals or even entire ecosystems. However, there is no guarantee that tranlocations will be successful and the courts may not always be satisfied that such a proposal will necessarily mitigate ecological damage (see Section 8.2.7).

9.3 CASE STUDIES

The following two case studies provide examples of how organisations involved with large-scale construction projects have attempted to mitigate the environmental damage caused.

9.3.1 The second runway at Manchester International Airport

Major infrastructure projects such as motorways and runways are Schedule 1 projects under the Town and Country Planning

Figure 9.2 Manchester International Airport.

(Environmental Impact Assessment) (England and Wales) Regulations 1999 and as such *must* be the subject of an environmental assessment. In 1997 work began on a second runway at Manchester Airport (Figure 9.2). When completed the runway will be 3050 metres long and is expected to cost £172 million. This is the first new runway to be built in the UK for 20 years and it has met with considerable opposition from local communities and environmental pressure groups.

Over half of the new runway site is within the existing perimeter of the airport. Of the new land being used, more than half has been set aside for environmental works. In total the airport plans to spend £17 million on a detailed environmental mitigation package. This was developed as part of a Green Chapter for the second runway and a management plan was developed to protect and improve over 350 ha of countryside around the site. Progress is monitored and reviewed by the Nature Conservation and Landscape Steering Group which brings together the airport company, local authorities, representatives of the local community, English Nature, and other specialist bodies. Ecologists have worked closely with the project engineers to conserve habitats and minimise intrusion into the most ecologically valuable areas.

It is inevitable with a project of this scale that considerable environmental damage will result. However, the mitigation measures planned are also large scale. Large areas of woodland will be planted (the equivalent of 50 football pitches) amounting to six times the area of woodland that will be lost. In addition, over 36 km of hedgerow will be planted or restored and new wild flower grassland will be created.

For each pond that is lost at least two new ponds will be created. As part of this project 30 000 newts, frogs and toads have been translocated to over 90 new or restored ponds. A rare mud snail will be captive-bred at Chester Zoo and reintroduced after construction is complete. Three new bat barns have been constructed specifically for the pipistrelle bat, and a family of badgers has been moved to a new artificial sett.

The River Bollin runs through the construction site and has had to be diverted into a new 260 metre long channel that passes through a tunnel under the new runway. The Environment Agency has transferred fish from the redundant channel into the new channel of the river. The new tunnel is sufficiently wide to accommodate the river's meanders and incorporates bat roosts and nesting boxes for wagtails.

During construction work evidence of a Bronze Age settlement was discovered by archaeologists. Manchester Airport interrupted work at this site and funded an excavation project that has allowed the site to be documented and artefacts to be removed for storage with Cheshire and Manchester museum services. In addition, two historically important buildings (Hill House and Hanson House) have been carefully dismantled and it is intended that the seventeenth century timber cores of the original buildings will be re-erected on new sites.

9.3.2 The Samlesbury to Helmshore Natural Gas Pipe-line

Transco, the company that owns and operates the national gas pipe-line system, is building a new high pressure gas pipe-line in Lancashire, between Samlesbury near Preston to Helmshore, near Haslingden. The pipe-line will be approximately 30 km long and a metre in diameter, and was subject to an environmental impact assessment.

Transco carried out a detailed environmental review before the project began, and produced an Environmental Statement. The route of the pipe-line was pegged out before construction began and chosen:

- to avoid population centres;
- to avoid existing and future development areas;
- to avoid recorded archaeological sites;
- to minimise impact on woodland, hedgerows, rivers and other wildlife habitats;
- to minimise impact on peat moorland areas.

In planning the project, Transco consulted with the following organisations, and refined its route as a result:

- English Nature
- Environment Agency
- Countryside Agency

- English Heritage
- Lancashire Wildlife Trust
- Lancashire County Council
- Chorley Borough Council
- South Ribble Borough Council
- Blackburn with Darwin Borough Council
- Rossendale Borough Council
- RSPB
- Lancashire Bat Group
- Lancashire Badger Group
- United Utilities (formerly North West Water)
- Lancashire Tourist Board
- Lancashire and Cheshire Bird Club
- Lancashire Aircraft Investigation Team
- Chorley and District Natural History Society
- Volunteer Otter Survey

By working with this wide range of organisations Transco was able to identify existing and potentially sensitive sites. As a result, 66 per cent of the land crossed is agricultural. However, the pipe-line crosses seven Biological Heritage Sites (BHSs) which have been designated by Lancashire County Council. These are:

- Darwen Moors
- Cranberry Moss
- Grimehills Moor
- Hoddlesden Moss
- Broadhead Valley
- Holcombe Moor
- Musbury Valley

The construction of the pipe-line will do damage to these areas but Transco has undertaken to minimise the impact on wildlife by laying it as quickly as possible. The pipe-line has been routed so as to avoid large areas of woodland and isolated large trees. Only one tree protected by a TPO stands within 10 metres of the path of the pipe-line. The pipe-line will cross 62 hedgerows but at these points the working width will be reduced to 20 metres (from the normal width of 38 metres). Where possible, the route uses existing gaps between hedgerows and trees. Hedgerows and trees in the path of the pipeline will be removed before the main breeding season for birds, and replaced with indigenous species.

Thirty-six ponds lie within 50 metres of the pipe-line. These have been surveyed for the presence of great crested newts, and three ponds showed signs of habitation. A licensed handler will carry out detailed searches and erect newt fencing before the start of construction work.

The pipe-line will cross 3.5 km of peat moorland, but avoids the deepest peat and the most ecologically sensitive areas. A specialist subcontractor will remove the heather or grass layer along with the top layer of peat, and re-lay it a few days later, once the construction work has been completed. Non-essential machinery and plant will not be allowed to use the working width of sections that cross peat moorland as a highway to other areas of the pipe-line. A special membrane will cover the heather and peat areas to prevent damage from machinery, and timber mats will be laid on this surface forming a running track for machinery and distributing the load on the peat. When the peat has been reinstated, stockproof fencing will be installed along the working width to prevent grazing for a period of five years. Areas of bare peat will be seeded with a suitable mix of heather or grass seed, or seed collected locally. The restored areas will be monitored annually by an ecologist.

Where possible the pipe-line has been routed away from known archaeological sites. An Archaeological Management Plan has been drawn up in consultation with Lancashire County Council and an archaeologist will monitor the sites as the topsoil is removed.

The site of a Second World War aircraft crash on Wives Hill is legally protected and may be affected by construction work, as some debris from the crash may be found along the route. The Manchester to Ribchester Roman road and the Leeds–Liverpool Canal will be crossed by the pipe-line. In both cases a tunnel will be constructed to carry the pipe-line beneath them. The route will also cross an avenue of trees at Hoghton Tower Park, and this will be reinstated in its original form.

REFERENCES

Anon. (2000) *Environmental Impact Assessment: A Guide to Procedures.* National Assembly for Wales, Department of the Environment Transport and the Regions, London.

Wathern, P. (1988) *Environmental Impact Assessment. Theory and Practice.* Routledge, London.

10
The future

Concerns about the oceans, marine wildlife and marine habitats seem likely to result in new legislation in the near future. Some authorities have called for an 'Ocean Act' but as this book goes to press (June 2002) there has been little development of this concept. However, the Marine Wildlife Conservation Bill 2001 has made progress and will hopefully be adopted in 2002.

The Marine Wildlife Conservation Bill 2001 is a Private Member's Bill which was introduced by John Randall, MP for Uxbridge, and was given a second reading in the House of Commons in October 2001. The purpose of the Bill is to provide improved protection of nationally important marine sites around the coasts of England and Wales, and to ensure that this protection is properly enforced.

If passed, the Bill will create a new designation to protect and manage nationally important marine sites (Marine Sites of Special Interest, MSSIs). These sites will be designated by English Nature or the Countryside Council for Wales (the 'appropriate nature conservation bodies') where there is a special interest by reason of any of its flora, fauna or geological or physiographical features. The marine areas that may be included may extend from the mean low water mark to the seaward limit of territorial waters (12 nautical miles) and the existence of any protected site may be indicated by markers. Registers of designated sites will be kept by the appropriate nature conservation body.

The Bill would impose marine conservation duties upon 'competent marine authorities', which include ministers of the Crown, government departments, the National Assembly for Wales, local authorities, statutory undertakers and other public bodies which have functions in relation to marine areas. These would include the Environment Agency, HM Coastguard and Sea Fisheries Committees. A management scheme may be established by the relevant marine authorities for a MSSI and the appropriate nature conservation body may make byelaws to protect the site.

Some of the important sites which could be protected if the Marine Wildlife Conservation Bill becomes law are:

- *Lyme Bay* (Devon and Cornwall)
  One of the few areas off UK shores where the exotic southern sunset coral and pink sea fan occur. The sandy beaches support populations of sand eels that provide food for little and common terns.

- *The Skerries* (3 km off the north-west coast of Angelsey)
  Marine habitats here are affected by tidal rapids, and support a diverse range of wildlife, including sea firs, sea mats, brittlestars and sub-tidal mussels. The area is an important feeding ground for four species of tern.

- *The Cumbrian coast*
  This area includes extensive reefs produced by the sandy tubes of the honeycomb worm (*Sabellaria alveolata*). The coast is also a fishing ground for seabirds, including the only breeding black guillemots in England.

- *Seven Sisters* (East Sussex)
  Underwater chalk habitats here include reefs which support a rich diversity of sponges, anemones, molluscs and marine worms. The area is also important to breeding colonies of fulmars, herring gulls and kittiwakes, along with occasional wintering auks and divers.

In the planning Green Paper entitled *Planning: Delivering a Fundamental Change* the Minister of State for Housing, Planning and Regeneration, Lord Falconer of Thoroton, has proposed that Parliament rather than local councils should decide if large developments should proceed. The proposals would affect large infrastructure projects such as main roads and airports, and have resulted from cabinet frustration over delays to projects such as the Channel Tunnel Rail Link and Heathrow Airport's Terminal Five. The inquiry into the latter project lasted for six years.

If the proposals were to be implemented, inquiries would be held after projects had been agreed in principle by Parliament and compulsory purchase of property would be made easier. Conservation organisations such as Friends of the Earth and the Council for the Protection of Rural England have expressed concern that the proposed changes to the planning system would prevent detailed scrutiny of proposed projects by local people.

As this book goes to press (June 2002), the Countryside Agency has just begun the process of consulting with the public regarding the new rights of access to the countryside given by the Countryside and Rights of Way Act 2000. It is to be hoped that this will result in a compromise that will satisfy the needs of walkers, farmers and other landowners and the requirements of nature conservation. Early indications suggest that there is much work to do in allaying the concerns of the farming community.

On 29 January 2002 the Policy Commission on the Future of Farming and Food presented its report to the government, following the 2001 foot-and-mouth disease outbreak. The Commission has emphasised the need for the farming industry to concentrate on food quality rather than quantity, and has also stressed the need for the farming community to pay a more significant role in the protection of the natural environment. The Commission strongly criticises the Common Agricultural Policy in its current form and makes a strong case for putting more resources into agri-environment schemes. It seems likely that in the long term UK and European policies will favour a move towards increased subsidy of nature protection in the countryside rather than the unsustainable subsidy of food production.

The Environmental Audit Bill 2002 will require the Comptroller and Auditor-General to examine and report to the House of Commons, at least annually, on the environmental impact of public expenditure, and the environmental performance of government departments and certain other public bodies against targets set by ministers, including contributions to sustainable development. The Bill extends to England and Wales.

On 1 May 2002 the government published the first Marine Stewardship Report – *Safeguarding Our Seas: A Strategy for the Conservation and Sustainable Development of our Marine Environment*. The report was launched by the DEFRA Secretary Margaret Beckett, and sets out a framework for action to save the marine environment, including:

■ a pilot conservation scheme in the Irish Sea;
■ the extension of the Wild Birds and Habitats Directives to cover all UK waters;
■ exploring with other countries the case for marine protected areas on the high seas;
■ an integrated assessment of the state of the seas by 2004;
■ working with the European Commission to review the Common Fisheries Policy and implement its fisheries biodiversity plan.

The report suggests that its proposals will help to conserve over 44 000 different species in UK waters.

Michael Meacher, the Environment Minister, is currently reviewing the law concerning trade in endangered species. It is likely that penalties will be increased and that anomalies will be addressed in an attempt to tackle organised criminal gangs. At present the police can arrest a poacher who illegally takes a relatively common bird yet have no power to arrest someone selling a tiger skin worth thousands of pounds.

The government is also considering updating the law on cruelty to animals (Protection of Animals Act 1911). However, as this book goes to press fox hunting is still legal in England, Wales and Northern Ireland, although it will be unlawful in Scotland from August 2002. It remains to be seen whether or not the rest of the UK will follow Scotland's lead in banning the hunting of wild mammals with dogs.

A number of challenges to SSSI designations have been made recently claiming breaches of the Human Rights Act 1998. In a judicial review of the decision by English Nature to confirm the notification of land at Bramshill plantation in Hampshire as a SSSI, the High Court ruled that the agency had acted in the public interest and that the procedure laid down by s. 28 of the Wildlife and Countryside Act 1981 was compliant with Art. 6(1) of the European Convention on Human Rights (*R* v. *English Nature and Secretary of State for the Environment, Food and Rural Affairs, ex parte Aggregate Industries UK Ltd* [2002]). Aggregate Industries UK Ltd, a quarrying company, claimed that it had been deprived of its right to a fair and independent hearing, and that English Nature's decision had breached the 'legitimate expectation' of the company. The court held that English Nature's decision-making process, taken together with the High Court's powers of judicial review, could be relied upon to produce a fair and reasonable decision, so s. 28 of the 1981 Act did not give rise to any incompatibility with the Convention. The site in question is important for its rare birds, including the nightjar, woodlark and Dartford warbler, and is part of a larger area, the Thames Basins Area, which is of European importance.

It is extremely likely that landowners will continue to make attempts to use the Human Rights Act to challenge, not only SSSI designations, but also the greater rights of access given to the public to the countryside by the Countryside and Rights of Way Act 2000.

APPENDIX 1

History of wildlife and conservation law in the UK

A chronology of major legislation affecting wildlife and nature conservation law in UK.

1772	Game (Scotland) Act
1822	Animal Protection Act
1828	Night Poaching Act
1831	Game Act
1832	Game (Scotland) Act
1844	Night Poaching Act
1848	Hares Act
1848	Hares (Scotland) Act
1860	Game Licences Act
1861	Salmon Fishery Act
1862	Poaching Prevention Act
1865	Salmon Fishery Act
1868	Salmon Fisheries (Scotland) Act
1869	Sea Birds Preservation Act
1872	Wild Birds Protection Act
1876	Cruelty to Animals Act
1876	Wild Fowl Protection Act
1880	Wild Birds Protection Act
1880	Ground Game Act
1881	Wild Birds Protection Act
1884	Access to Mountains (Scotland) Bill
1888	Sand Grouse Protection Act
1892	Hares Preservation Act
1894	Wild Birds Protection Act
1896	Wild Birds Protection Act
1900	Wild Animals in Captivity Protection Act
1902	Wild Birds Protection Act
1902	Freshwater Fisheries (Scotland) Act
1904	Wild Birds (St Kilda) Act
1904	Wild Birds Protection Act

1906	Ground Game (Amendment) Act
1907	National Trust Act
1908	Wild Birds Protection Act
1911	Protection of Animals Act
1912	Protection of Animals (Scotland) Act
1914	Grey Seals (Protection) Act
1919	Forestry Act
1921	Importation of Plumage (Prohibition) Act
1921	Captive Birds Shooting (Prohibition) Act
1925	Protection of Birds Act
1928	Protection of Lapwings Act
1928	Game Preservation Act (Northern Ireland)
1932	Grey Seals (Protection) Act
1933	Wild Birds Protection Act
1933	Trout (Scotland) Act
1934	Whaling Industry (Regulation) Act
1937	Quail Protection Act
1938	National Trust for Scotland Confirmation Act
1939	Wild Birds (Ducks and Geese) Protection Act
1939	Access to Mountains Act
1945	Forestry Act
1945	Wild Birds Protection Act
1947	Agriculture Act
1948	Agriculture (Scotland) Act
1949	Countryside Act
1949	National Parks and Access to the Countryside Act
1951	Salmon and Freshwater Fisheries (Protection) (Scotland) Act
1954	Protection of Birds Act
1954	Pests Act
1959	The Deer (Scotland) Act
1960	Game Laws (Amendment) Act
1964	Restriction of Import (Animals) Act
1964	Protection of Birds Act
1965	Amenity Lands Act (Northern Ireland)
1966	Sea Fisheries Regulation Act
1967	Forestry Act
1967	Wild Birds Protection Act
1967	Antarctic Treaty Act
1968	Countryside Act
1970	Conservation of Seals Act
1971	Wild Creatures and Forest Laws Act
1972	Welfare of Animals Act (Northern Ireland)
1973	The Badgers Act
1975	Conservation of Wild Creatures and Wild Plants Act
1975	Salmon and Freshwater Fisheries Act
1976	Dangerous Wild Animals Act
1976	Endangered Species (Import and Export) Act
1979	Forestry Act
1979	Ancient Monuments and Archaeological Areas Act
1981	Wildlife and Countryside Act
1981	Export of Animals (Protection) Order

1981	Zoo Licensing Act
1982	Wildlife and Countryside (Registration and Ringing of Certain Captive Birds) Regulations
1985	Wildlife (Northern Ireland) Order
1985	Nature Conservation and Amenity Lands (Northern Ireland) Order
1985	Control of Trade in Endangered Species Regulations
1985	Local Government Act
1985	Wildlife and Countryside (Amendment) Act
1985	Wildlife and Countryside (Service of Notices) Act
1986	Salmon Act
1988	Norfolk and Suffolk Broads Act
1988	Town and Country Planning (Assessment of Environmental Effects) Regulations
1989	Water Act
1990	Town and Country Planning Act
1990	Environmental Protection Act
1990	Planning (Listed Buildings and Conservation Areas) Act
1991	Wildlife and Countryside (Amendment) Act
1991	Deer Act
1991	Badgers Act
1991	Natural Heritage (Scotland) Act
1991	Water Resources Act
1991	Water Industry Act
1992	Protection of Badgers Act
1992	Sea Fisheries (Wildlife Conservation) Act
1993	Protection of Animals (Scotland) Act
1994	Conservation (Natural Habitats etc.) Regulations
1994	Game Birds Preservation Order (Northern Ireland)
1995	Environment Act
1996	Salmon Act
1996	Wild Mammals (Protection) Act
1996	Deer (Scotland) Act
1996	Deer (Amendment) (Scotland) Act
1996	The Import of Seals Regulations
1997	The Control of Trade in Endangered Species (Enforcement) Regulations
1997	Hedgerows Regulations
1999	Town and Country Planning (Environmental Impact Assessment) (England and Wales) Regulations
1999	Game Birds Preservation Order (Northern Ireland)
1999	Town and Country Planning (Trees) Regulations
2000	Countryside and Rights of Way Act
2000	National Parks (Scotland) Act
2001	Salmon Conservation (Scotland) Act
2002	Protection of Wild Mammals (Scotland) Act

Not all of this legislation is currently in force.

This is *not* a comprehensive list of all of the legislation affecting wildlife and nature conservation in the UK.

APPENDIX 2
Selected legislation, consultation and policy documents affecting nature conservation

Selected primary legislation, consultation and policy documents affecting nature conservation issued by the former DETR between April 1997 and July 2001.

Primary legislation

Countryside and Rights of Way Act 2000

White Papers

A Better Quality of Life: A Strategy for Sustainable Development for the UK, 1999

Our Countryside: The Future – A Fair Deal for Rural England, 2000

Our Towns and Cities: The Future – Delivering an Urban Renaissance, 2000

Green Papers and major policy documents

Modernising Planning: A Policy Statement by the Minister for the Regions, Regeneration and Planning, 1997

What Role for Trunk Roads in England: Vol. 1, Consultation Paper, 1997

Access to the Open Countryside in England and Wales: A Consultation Paper, 1998

A New Deal for Trunk Roads in England, 1998

Sites of Scientific Interest: Better Protection and Management – A Consultation Document for England and Wales, 1998

High Hedges: A Consultation Document, 1999

Identification of Marine Environmental High Risk Areas in the UK: A Consultation Paper, 1999

Planning Policy Guidance Note 12: Development Plans, 1999

Rural England: A Discussion Document, 1999

Sites of Scientific Interest: Better Protection and Management – Government Framework for Action, 1999

Unlocking the Potential: A New Future for British Waterways, 1999

Greater Protection and Better Management of Common Land in England and Wales: Consultation Paper, 2000

Guidelines on Management Agreement Payments: Public Consultation Paper, 2000

Improving Rights of Way in England and Wales – Consultation Paper, 2000

Planning Policy Guidance Note 11: Regional Planning, 2000

Sites of Special Scientific Interest: Guidance on New Legislation, 2000

Waterways for Tomorrow, 2000

APPENDIX 3

Egg-laying times and typical nest locations for some British birds found in urban areas

Species	Main egg-laying period	Typical nest location
Barn owl	April–May (every month except January)	Barns, ruins, church towers, hollow trees, quarry faces
Black-headed gull	April–July	In rushes, on grass
Black redstart	April–July	Building ledge, crevice, under rafters
Blackbird	March–July	Building ledge, hedgerow, tree
Blue tit	April–May	Tree hole, crevice in wall
Bullfinch	April–July	Thick hedge, bramble, other deep cover
Chaffinch	April–June	Bush, tree, hedgerow
Cirl bunting	May–August	Bush, hedge, tree, ground
Collared dove	March–September	Building ledge, tree
Corn bunting	May–July	Coarse vegetation on or near ground
Fieldfare	April–July	Tree, bush
Goldfinch	May–August	Tree
Great tit	April–May	Hole in tree or wall
Greenfinch	April–August	Bush, tree, hedgerow
Hedge sparrow	March–July	Bush, hedgerow, woodpile
House martin	May–August	Under eaves, under bridge, cave, cliff
House sparrow	April–August	Hole in building or tree, hedgerow

Jackdaw	April–May	Chimney, hole or crevice in tree, cliff, building
Kestrel	April–May	Motorway bridge, cliff ledge, high building, tree hole
Lapwing	March–May	Ground
Little owl	April–May	Tree hole, wall hole, building, quarries, sand pits, cliffs, burrows
Mistle thrush	February–June	Tree
Redwing	May–July	Tree
Robin	March–June	Tree hole, wall, ledge, hollow in bank
Skylark	April–August	Ground
Starling	April–May	Building, tree hole, cliff
Song thrush	March–July	Bush, hedge, tree, ledge
Spotted flycatcher	May–June	Tree, creeper, ledge, cavity
Swallow	May–August	Ledge, rafter
Swift	May–June	Under eaves, in thatch, rock crevice
Tawny owl	March–May	Tree hole, old building, rock crevice, burrow
Tree sparrow	April–July	Tree hole
Wren	April onwards	Bush, creeper, woodpile, hollow in wall
Yellowhammer	April–August	Bank, hedge, ivy, wall, ground

APPENDIX 4
Schedules to the Wildlife and Countryside Act 1981 and European listed species

SCHEDULES TO THE WILDLIFE AND COUNTRYSIDE ACT 1981

Schedule 1   Birds which are specially protected

Part 1   At all times

Avocet	*Recurvirostra avosetta*
Bee-eater	*Merops apiaster*
Bittern	*Botaurus stellaris*
Bittern, Little	*Ixobrychus minutes*
Bluethroat	*Luscinia svecica*
Brambling	*Fringilla montifringilla*
Bunting, Cirl	*Emberiza cirlus*
Bunting, Lapland	*Calcarius lapponicus*
Bunting, Snow	*Plectrophenax nivalis*
Buzzard, Honey	*Pernis apivorus*
Chough	*Pyrrhocorax pyrrhocorax*
Corncrake	*Crex crex*
Crake, Spotted	*Porzana porzana*
Crossbills (all species)	*Loxia*
Curlew, Stone	*Burhinus oedicnemus*
Divers (all species)	*Gavia*
Dotterel	*Charadrius morinellus*
Duck, Long-tailed	*Clangula hyemalis*
Eagle, Golden	*Aquila chrysaetos*
Eagle, White-tailed	*Haliaetus albicilla*
Falcon, Gyr	*Falco rusticolus*

Fieldfare	*Turdus pilaris*
Firecrest	*Regulus ignicapillus*
Garganey	*Anas querquedula*
Godwit, Black-tailed	*Limosa limosa*
Goshawk	*Accipiter gentilis*
Grebe, Black-necked	*Podiceps nigricollis*
Grebe, Slavonian	*Podiceps auritus*
Greenshank	*Tringa nebularia*
Gull, Little	*Larus minutus*
Gull, Mediterranean	*Larus melanocephalus*
Harriers (all species)	*Circus*
Heron, Purple	*Ardea purpurea*
Hobby	*Falco subbuteo*
Hoopoe	*Upupa epops*
Kingfisher	*Alcedo atthis*
Kite, Red	*Milvus milvus*
Merlin	*Falco columbarius*
Oriole, Golden	*Oriolus oriolus*
Osprey	*Pandion haliaetus*
Owl, Barn	*Tyto alba*
Owl, Snowy	*Nyctea scandiaca*
Peregrine	*Falco peregrinus*
Petrel, Leach's	*Oceanodroma leucorhoa*
Phalarope, Red-necked	*Phalaropus lobatus*
Plover, Kentish	*Charadrius alexandrinus*
Plover, Little Ringed	*Charadrius dubius*
Quail, Common	*Coturnix coturnix*
Redstart, Black	*Phoenicurus ochruros*
Redwing	*Turdus iliacus*
Rosefinch, Scarlet	*Carpodacus erythrinus*
Ruff	*Philomachus pugnax*
Sandpiper, Green	*Tringa ochropus*
Sandpiper, Purple	*Calidris maritima*
Sandpiper, Wood	*Tringa glareola*
Scaup	*Aythya marila*
Scoter, Common	*Melanitta nigra*
Scoter, Velvet	*Melan fusca*
Serin	*Serinus serinus*
Shorelark	*Eremophila alpestris*
Shrike, Red-backed	*Lianius collurio*
Spoonbill	*Platalea leucorodia*
Stilt, Black-winged	*Himantopus himantopus*
Stint, Temminck's	*Calidris temminickii*
Swan, Bewick's	*Cygnus bewickii*
Swan, Whooper	*Cygnus cygnus*
Tern, Black	*Chlidonias niger*
Tern, Little	*Sterna albifrons*
Tern, Roseate	*Sterna dougallii*
Tit, Bearded	*Panarus biarmicus*
Tit, Crested	*Parus cristatus*
Treecreeper, Short-toed	*Certhia brachydactyla*

Warbler, Cetti's	*Cettia cetti*
Warbler, Dartford	*Sylvia undata*
Warbler, Marsh	*Acrocephalus palustris*
Warbler, Savi's	*Locustella luscinioides*
Whimbrel	*Numenius phaeopus*
Woodlark	*Lullula arborea*
Wryneck	*Jynx torquilla*

Part 2   Birds which are specially protected during the close season

Goldeneye	*Bucephala clangula*
Goose, Greylag[1]	*Anser anser*
Pintail	*Anas acuta*

Note to Schedule 1
1. Applies in the Outer Hebrides, Caithness, Sutherland and Wester Ross only.

Schedule 2 Birds which may be killed or taken

Part 1 Outside the close season

Capercaillie	*Tetrao urogallus*
Coot	*Fulica atra*
Duck, Tufted	*Aythya fuligula*
Gadwall	*Anas strepera*
Goldeneye	*Bucephala clangula*
Goose, Canada	*Branta canadensis*
Goose, Greylag	*Anser anser*
Goose, Pink-footed	*Anser brachyrhynchus*
Goose, White-fronted[1]	*Anser albifrons*
Mallard	*Anas platyrhynchos*
Moorhen	*Gallinula chloropus*
Pintail	*Anas acuta*
Plover, Golden	*Pluvialis apricaria*
Pochard	*Aythya ferina*
Shoveler	*Anas clypeata*
Snipe, Common	*Gallinago gallinago*
Teal	*Anas crecca*
Wigeon	*Anas penelope*
Woodcock	*Scolopax rusticola*

Part 2   Birds which may be killed or taken by authorised persons at all times. No birds currently listed.

The 13 species listed below were removed by SI 1992/3010.

Crow	*Corvus corone*
Dove, Collared	*Streptopelia decaocto*

Gull, Great Black-backed	*Larus marinus*
Gull, Herring	*Larus argentatus*
Gull, Lesser Black-backed	*Larus fuscus*
Jackdaw	*Corvus monedula*
Jay	*Garrulus glandarius*
Magpie	*Pica pica*
Pigeon, Feral	*Columba livia*
Rook	*Corvus frugilegus*
Sparrow, House	*Passer domesticus*
Starling	*Sturnus vulgaris*
Woodpigeon	*Columba palumbus*

Note to Schedule 2
1. Applies in England and Wales only.

Schedule 3   Birds which may be sold

Part 1   Alive at all times if ringed and bred in captivity

Blackbird	*Turdus merula*
Brambling	*Fringilla montifringilla*
Bullfinch	*Pyrrhula pyrrhula*
Bunting, Reed	*Emberiza schoeniclus*
Chaffinch	*Fringilla coelebs*
Dunnock	*Prunella modularis*
Goldfinch	*Carduelis carduelis*
Greenfinch	*Carduelis chloris*
Jackdaw	*Corvus monedula*
Jay	*Garrulus glandarius*
Linnet	*Carduelis cannabina*
Magpie	*Pica pica*
Owl, Barn	*Tyto alba*
Redpoll	*Carduelis flammea*
Siskin	*Carduelis spinus*
Starling	*Sturnus vulgaris*
Thrush, Song	*Turdus philomelos*
Twite	*Carduelis flavirostris*
Yellowhammer	*Emberiza citrinella*

Part 2   Dead at all times

Woodpigeon	*Columba palumbus*

Part 3   Birds which may be sold dead from 1 September to 28 February

Capercaillie	*Tetrao urogallus*
Coot	*Fulica atra*
Duck, Tufted	*Aythya fuligula*

Mallard	*Anas platyrhynchos*
Pintail	*Anas acuta*
Plover, Golden	*Pluvialis apricaria*
Pochard	*Aythya ferina*
Shoveler	*Anas clypeata*
Snipe, Common	*Gallinago gallinago*
Teal	*Anas crecca*
Wigeon	*Anas penelope*
Woodcock	*Scolopax rusticola*

Schedule 5   Animals which are protected

Adder[1]	*Vipera berus*
Anemone, Ivell's Sea	*Edwardsia ivelli*
Anemone, Starlet Sea	*Nematosella vectensis*
Apus, Tadpole Shrimp	*Triops cancriformis*
Bats, Horseshoe (all species)	Rhinolophidae
Bats, Typical (all species)	Vespertilionidae
Beetle	*Graphoderus zonatus*
Beetle	*Hypebaeus flavipes*
Beetle	*Paracymus aeneus*
Beetle, Leeser Silver Water	*Hydrochara caraboides*
Beetle, Mire Pill[2]	*Curimopsis nigrita*
Beetle, Rainbow Leaf	*Chrysolina cerealis*
Beetle, Stag[3]	*Lucanus cervus*
Beetle, Violet Click	*Limoniscus violaceus*
Burbot	*Lota lota*
Butterfly, Adonis Blue[3]	*Lysandra bellargus*
Butterfly, Black Hairstreak[3]	*Strymonidia pruni*
Butterfly, Brown Hairstreak[3]	*Thecla betulae*
Butterfly, Chalkhill blue[3]	*Lysandra coridon*
Butterfly, Chequered Skipper[3]	*Carterocephalus palaemon*
Butterfly, Duke of Burgundy Fritillary[3]	*Hamearis lucina*
Butterfly, Glanville Fritillary[3]	*Melitaea cinxia*
Butterfly, Heath Fritillary[3]	*Mellicta athalia*
Butterfly, High Brown Fritillary[3]	*Argynnis adippe*
Butterfly, Large Blue	*Maculinea arion*
Butterfly, Large Copper	*Lycaena dispar*
Butterfly, Large Heath[3]	*Coenonympha tullia*
Butterfly, Large Tortoiseshell[3]	*Nymphalis polychloros*
Butterfly, Lulworth Skipper[3]	*Thymelicus acteon*
Butterfly, Marsh Fritillary	*Eurodryas aurinia*
Butterfly, Mountain Ringlet[3]	*Erebia epiphron*
Butterfly, Northern Brown Argus[3]	*Aricia artaxerxes*
Butterfly, Pearl-bordered Fritillary[3]	*Boloria euphrosyne*
Butterfly, Purple Emperor[3]	*Apatura iris*
Butterfly, Silver-spotted Skipper[3]	*Hesperia comm*a
Butterfly, Silver-studded Blue[3]	*Plebejus argus*
Butterfly, Small Blue[3]	*Cupido minimus*
Butterfly, Swallowtail[3]	*Papilio machaon*
Butterfly, White Letter Hairstreak[3]	*Stymonida w-album*

Butterfly, Wood White[3]	*Leptidea sinapis*
Cat, Wild	*Felis silvestris*
Cicada, New Forest	*Cicadetta montana*
Crayfish, Atlantic Stream	*Austropotamobius pallipes*
Cricket, Field	*Gryllus campestris*
Cricket, Mole	*Gryllotalpa gryllotalpa*
Damselfly, Southern	*Coenagrion mercuriale*
Dolphins (all species)	*Cetacea*
Dormouse	*Muscardinus avellanarius*
Dragonfly, Norfolk Aeshna	*Aeshna isosceles*
Frog, Common[3]	*Rana temporaria*
Goby, Couch's	*Gobius couchii*
Goby, Giant	*Gobius cobitis*
Grasshopper, Wart-biter	*Decticus verrucivorus*
Hatchet shell, Northern	*Thyasira gouldi*
Hydroid, Marine	*Clavopsella navis*
Lagoon Snail	*Paludinella littorina*
Lagoon Snail, De Folin's	*Caecum armoricum*
Lagoon Worm, Tentacled	*Alkmaria romijini*
Leech, Medicinal	*Hirudo medicinalis*
Lizard, Sand	*Lacerta agilis*
Lizard, Viviparous[1]	*Lacerta vivipara*
Marten, Pine	*Martes martes*
Moth, Barberry Carpet	*Pareulype berberata*
Moth, Black-veined	*Siona lineata*
Moth, Essex Emerald	*Thetidia smaragdaria*
Moth, Fiery Clearwing	*Bembecia chrysidiformis*
Moth, Fisher's Estuarine	*Gortyna borelii*
Moth, New Forest Burnet	*Zygaena viciae*
Moth, Reddish Buff	*Acosmetia caliginosa*
Moth, Sussex Emerald	*Thalera fimbrialis*
Mussel, Fan[4]	*Atrina fragilis*
Mussel, Freshwater Pearl	*Margaritifera margaritifera*
Newt, Great Crested (or Warty Newt)	*Triturus cristatus*
Newt, Palmate[3]	*Titurus helveticus*
Newt, Smooth[3]	*Triturus vulgaris*
Otter, Common	*Lutra lutra*
Porpoises (all species)	*Cetacea*
Sandworm, Lagoon	*Armandia cirrhosa*
Sea fan, Pink[4]	*Eunicella verrucosa*
Sea mat, Trembling	*Victorella pavida*
Sea slug, Lagoon	*Tenellia adspersa*
Shad, Alis[5]	*Alosa alosa*
Shad, Twaite[2]	*Alosa fallax*
Shark, Basking	*Cetorhinus maximus*
Shrimp, Fairy	*Chirocephalus diaphanus*
Shrimp, Lagoon Sand	*Gammarus insensibilis*
Slow Worm[1]	*Anguis fragilis*
Snail, Glutinous	*Myxas glutinosa*
Snail, Sandbowl	*Catinella arenaria*
Snake, Grass[1]	*Natrix helvetica* (or *N. natrix*)

Snake, Smooth	*Coronella austriaca*
Spider, Fen Raft	*Dolomedes plantarius*
Spider, Ladybird	*Eresus niger*
Squirrel, Red	*Sciurus vulgaris*
Sturgeon	*Acipenser sturio*
Toad, Common[3]	*Bufo bufo*
Toad, Natterjack	*Bufo calamita*
Turtles, Marine (all species)	*Dermochelyidae* and *Cheloriidae*
Vendance	*Coregorus albula*
Vole, Water[6]	*Arvicola terrestris*
Walrus	*Odobenus rosmarus*
Whale (all species)	*Cetacea*
Whitefish	*Coregonus lavaretus*

Notes to Schedule 5
1. In respect of s. 9(1) in so far as it relates to killing or injuring, and s. 9(5) only.
2. In respect of s. 9(4)(a) only, i.e. destroying, damaging or obstructing access to shelters, but not disturbance.
3. In respect of s. 9(5) only.
4. In respect of ss. 9(1), 9(2) and 9(5) only.
5. In respect of ss. 9(1) and 9(4)(a) only.
6. In respect of s. 9(4) only.

Schedule 6   Animals which may not be killed or taken by certain methods

Badger	*Meles meles*
Bats, Horseshoe (all species)	Rhinolophidae
Bats, typical (all species)	Vespertilionidae
Cat, Wild	*Felis silvestris*
Dolphin, Bottle-nosed	*Tursiops truncatus*
Dolphin, Common	*Delphinus delphis*
Dormice (all species)	Gliridae
Hedgehog	*Erinaceus europaeus*
Marten, Pine	*Martes martes*
Otter, Common	*Lutra lutra*
Polecat	*Mustela putorius*
Porpoise, Harbour, or Common	*Phocaena phocaena*
Shrews (all species)	Soricidae
Squirrel, Red	*Sciurus vulgaris*

Schedule 8   Plants which are protected

Adders-tongue, Least	*Ophioglossum lusitanicum*
Alison, Small	*Alyssum alyssoides*
Anomodon, LongLeaved	*Anomodon longifolius*
Beech-lichen, New Forest	*Enterographa elaborata*
Blackwort	*Southbya nigrella*

Bluebell[1]	*Hyacinthoides non-scripta*
Broomrape, Bedstraw	*Orobanche caryophyllacea*
Broomrape, Oxtongue	*Orobanche loricata*
Broomrape, Thistle	*Orobanche reticulata*
Cabbage, Lundy	*Rhynchosinapis wrightii*
Calamint, Wood	*Calamintha sylvatica*
Caloplaca, Snow	*Caloplaca nivalis*
Catapyrenium, Tree	*Catapyrenium psoromoides*
Catchfly, Alpine	*Lychnis alpina*
Catillaria, Laurer's	*Catillaria laureri*
Centaury, Slender	*Centaurium tenuiflorum*
Cinquefoil, Rock	*Potentilla rupestris*
Cladonia, Convoluted	*Caladonia convoluta*
Cladonia, Upright Mountain	*Cladonia stricta*
Clary, Meadow	*Salvia pratensis*
Club-rush, Triangular	*Scirpus triquetrus*
Colt's-foot, Purple	*Homogyne alpina*
Cotoneaster, Wild	*Cotoneaster integerrimus*
Cottongrass, Slender	*Eriophorum gracile*
Cow-wheat, Field	*Melampyrum arvense*
Crocus, Sand	*Romulea columnae*
Crystalwort, Lizard	*Riccia bifurca*
Cudweed, Broad-leaved	*Filago pyramidata*
Cudweed, Jersey	*Gnaphalium luteoalbum*
Cudweed, Red-tipped	*Filago lutescens*
Cut Grass	*Leersia oryzoides*
Deptford Pink[2]	*Dianthus armeria*
Diapensia	*Diapensia lapponica*
Dock, Shore	*Rumex rupestris*
Earwort, Marsh	*Jamesoniella undulifolia*
Eryngo, Field	*Eryngium campestre*
Feather-moss, Polar	*Hygrohypnum polare*
Fern, Dickie's Bladder	*Cystopteris dickieana*
Fern, Killarney	*Trichomanes speciosum*
Flapwort, Norfolk	*Leiocolea rutheana*
Fleabane, Alpine	*Erigeron borealis*
Fleabane, Small	*Pulicaria vulgaris*
Fleawort, South Stack	*Tephroseris integrifolia* ssp. *maritima*
Frostwort, Pointed	*Gymnomitrion apiculatum*
Fungus, Hedgehog	*Hericium erinaceum*
Fungus, Oak Polypore	*Buglossoporus pulvinus*
Fungus, Royal Bolete	*Boletus regius*
Galingale, Brown	*Cyperus fuscus*
Gentian, Alpine	*Gentiana nivalis*
Gentian, Dune	*Gentianella uliginosa*
Gentian, Early	*Gentianella anglica*
Gentian, Fringed	*Gentianella ciliata*
Gentian, Spring	*Gentiana verna*
Germander, Cut-leaved	*Teucrium botrys*
Germander, Water	*Teucrium scordium*

Gladiolus, Wild	*Gladiolus illyricus*
Goblin Lights	*Catolechia wahlenbergii*
Goosefoot, Stinking	*Chenopodium vulvaria*
Grass-poly	*Lythrum hyssopifolia*
Grimmia, Blunt-leaved	*Grimmia unicolor*
Gyalecta, Elm	*Gyalecta ulmi*
Hare's Ear, Sickle-leaved	*Bupleurum falcatum*
Hare's Ear, Small	*Bupleurum baldense*
Hawk's-beard, Stinking	*Crepis foetida*
Hawkweed, Northroe	*Hieracium northroense*
Hawkweed, Shetland	*Hieracium zetlandicum*
Hawkweed, Weak-leaved	*Hieracium attenautifolium*
Heath, Blue	*Phyllodoce caerulea*
Helleborine, Red	*Cephalanthera rubra*
Helleborine, Young's	*Epipactis youngiana*
Horsetail, Branched	*Equisetum ramosissimum*
Hound's Tongue, Green	*Cynoglossum germanicum*
Knawel, Perennial	*Scleranthus perennis*
Knotgrass, Sea	*Polygonum maritimum*
Lady's-slipper	*Cypripedium calceolus*
Lecanactis, Churchyard	*Lecanactis hemisphaerica*
Lecanora, Tarn	*Lecanora archariana*
Lecidea, Copper	*Lecidea inops*
Leek, Round-headed	*Allium sphaerocephalon*
Lettuce, Least	*Lactuca saligna*
Lichen, Arctic Kidney	*Nephroma articum*
Lichen, Ciliate Strap	*Heterodermia leucomelos*
Lichen, Coralloid Rosette	*Heterodermia propagulifera*
Lichen, Ear-lobed Dog	*Peltigera lepidophora*
Lichen, Forked Hair	*Bryroria furcellata*
Lichen, Golden Hair	*Teloschistes flavicans*
Lichen, Orange Fruited Elm	*Caloplaca luteoalba*
Lichen, River Jelly	*Collema dichotomum*
Lichen, Scaly Breck	*Squamarina lentigera*
Lichen, Starry Breck	*Buellia asterella*
Lily, Snowdon	*Lloydia serotina*
Liverwort	*Petallophyllum ralfsi*
Liverwort, Lindenberg's Leafy	*Adelanthus lindenbergianus*
Marsh-mallow, Rough	*Althaea hirsute*
Marshwort, Creeping	*Apium repens*
Milk-parsley, Cambridge	*Selinum carvifolia*
Moss	*Drepanocladius vernicosus*
Moss, Alpine Copper	*Mielichoferia mielichoferi*
Moss, Baltic Bog	*Sphagnum balticum*
Moss, Blue Dew	*Saelania glaucescens*
Moss, Blunt-leaved Bristle	*Orthotrichum obtusifolium*
Moss, Bright Green Cave	*Cyclodictyon laetevirens*
Moss, Cordate Beard	*Barbula cordata*
Moss, Cornish Path	*Ditrichum cornubicum*
Moss, Derbyshire Feather	*Thamnobryum angustifolium*
Moss, Dune Thread	*Bryum mamillatum*

Moss, Flamingo	*Desmatodon cernuus*
Moss, Glaucous Beard	*Barbula glauca*
Moss, Green Shield	*Buxbaumia viridis*
Moss, Hair Silk	*Plagiothecium piliferum*
Moss, Knothole	*Zygodon forsteri*
Moss, Large Yellow Feather	*Scorpidium turgescens*
Moss, Millimetre	*Micromitrium tenerum*
Moss, Multifruited River	*Cryphaea lamyana*
Moss, Nowell's Limestone	*Zygodon gracilis*
Moss, Rigid Apple	*Bartamia stricta*
Moss, Round-leaved Feather	*Rhyncostegium rotundifolium*
Moss, Schleicher's Thread	*Bryum schleicheri*
Moss, Triangular Pygmy	*Acaulon triquetrum*
Moss, Vaucher's Feather	*Hypnum vaucheri*
Mudwort, Welsh	*Limosella australis*
Naiad, Holly-leaved	*Najas marina*
Naiad, Slender	*Najas flexilis*
Orache, Stalked	*Halimione pedunculata*
Orchid, Early Spider	*Ophrys sphegodes*
Orchid, Fen	*Liparis loeselii*
Orchid, Ghost	*Epipogium aphyllum*
Orchid, Lapland Marsh	*Dactylorhiza lapponica*
Orchid, Late Spider	*Ophrys fuciflora*
Orchid, Lizard	*Himantoglossum hircinum*
Orchid, Military	*Orchis militaris*
Orchid, Monkey	*Orchis simia*
Pannaria, Caledonia	*Pannaria ignobilis*
Parmelia, New Forest	*Parmelia minarum*
Parmentaria, Oil Stain	*Parmentaria chilensis*
Pear, Plymouth	*Pyrus cordata*
Penny-cress, Perfoliate	*Thlaspi perfoliatum*
Pennyroyal	*Mentha pulegium*
Pertusaria, Alpine Moss	*Pertusaria bryontha*
Physcia, Southern Grey	*Physcia tribacioides*
Pigmyweed	*Crassula aquatica*
Pine, Ground	*Ajuga chamaepitys*
Pink, Cheddar	*Dianthus gratianopolitanus*
Pink, Childling	*Petroraghia nanteuilii*
Plantain, Floating-leaved Water	*Luronium natans*
Pseudocyphellaria, Ragged	*Pseudocyphellaria lacerata*
Psora, Rusty Alpine	*Psora rubiformis*
Puffball, Sandy Stilt	*Battarraea phalloides*
Ragwort, Fen	*Senecio paludosus*
Ramping-fumitory, Martin's	*Fumaria martinii*
Rampion, Spiked	*Phyteuma spicatum*
Restharrow, Small	*Ononis reclinata*
Rock-cress, Alpine	*Arabis alpina*
Rock-cress, Bristol	*Arabis stricta*
Rustworth, Western	*Marsupella profunda*
Sandwort, Norwegian	*Arenaria norvegica*
Sandwort, Teesdale	*Minuartia stricta*

Saxifrage, Drooping	*Saxifraga cernua*
Saxifrage, Marsh	*Saxifraga hirulus*
Saxifrage, Tufted	*Saxifraga cespitosa*
Solenopsora, Serpentine	*Solenopsora liparina*
Solomon's-seal, Whorled	*Polygonatum verticillatum*
Sow-thistle, Alpine	*Cicerbita alpina*
Spearwort, Adder's Tongue	*Ranunculus ophioglossifolius*
Speedwell, Fingered	*Veronica triphyllos*
Speedwell, Spiked	*Veronica spicata*
Spike-rush, Dwarf	*Eleocharis parvula*
Starfruit	*Damasonium alisma*
Star-of-Bethlehem, Early	*Gagea bohemica*
Stonewort, Bearded	*Chara canescens*
Stonewort, Foxtail	*Lamprothamnium papulosum*
Strapwort	*Corrigiola litoralis*
Sulphur-tresses, Alpine	*Alectoria ochroleuca*
Threadmoss, Long-leaved	*Bryum neodamense*
Turpswort	*Geocalyx graveolens*
Violet, Fen	*Viola persicifolia*
Viper's-grass	*Scorzonera humilis*
Water-plantain, Ribbon-leaved	*Alisma gramineum*
Wood-sedge, Starved	*Carex depauperata*
Woodsia, Alpine	*Woodsia alpina*
Woodsia, Oblong	*Woodsia ilvensis*
Wormwood, Field	*Artemisia campestris*
Woundwort, Downy	*Stachys germanica*
Woundwort, Limestone	*Stachys alpina*
Yellow-rattle, Greater	*Rhinanthus serotinus*

Notes to Schedule 8
1. In respect of s. 13(2) only, which prohibits the sale, advertising the selling or buying of plants etc.
2. Applies to England and Wales only.

Schedule 9 Animals and plants to which section 14 applies

Species which may not be released into, planted or caused to be grown in the wild.

Part 1 Animals which are established in the wild

Bass, Large-mouthed Black	*Micropterus salmoides*
Bass, Rock	*Ambloplites rupestris*
Bitterling	*Rhodeus sericeus*
Budgerigar	*Melopsittacus undulatus*
Capercaillie	*Tetrao urogallus*
Coypu	*Myocastor coypus*
Crayfish, Noble	*Astacus astacus*
Crayfish, Signal	*Pacifastacus leniusculus*
Crayfish, Turkish	*Astacus leptocdactylus*

Deer, *Cervus*[1]	*Cervus* spp.
Deer, *Cervus* (hybrids)[2]	*Cervus* hybrids
Deer, Sika	*Cervus nippon*
Deer, Sika (hybrids)[3]	*Cervus nippon* hybrids
Deer, Muntjac	*Muntiacus reevesi*
Dormouse, Fat	*Glis glis*
Duck, Carolina Wood	*Aix sponsa*
Duck, Mandarin	*Aix galericulata*
Duck, Ruddy	*Oxyura jamaicensis*
Eagle, White-tailed	*Haliaetus albicilla*
Flatworm, New Zealand	*Artiposthia triangulata*
Frog, Edible	*Rana esculenta*
Frog, European Tree (or Common Tree Frog)	*Hyla arborea*
Frog, Marsh	*Rana ridibunda*
Gerbil, Mongolian	*Meriones unguiculatus*
Goose, Canada	*Branta canadensis*
Goose, Egyptian	*Alopochen aegyptiacus*
Heron, Night	*Nycticorax nycticorax*
Lizard, Common Wall	*Podarcis muralis*
Marmot, Prairie (or Prairie Dog)	*Cynomys*
Mink, American	*Mustela vison*
Newt, Alpine	*Triturus alpestris*
Newt, Italian Crested	*Triturus carnifex*
Owl, Barn	*Tyto alba*
Parakeet, Ring-necked	*Psittacula krameri*
Partridge, Chukar	*Alectoris chukar*
Partridge, Rock	*Alectoris graeca*
Pheasant, Golden	*Chrysolophus pictus*
Pheasant, Lady Amherst's	*Chrysolophus amherstiae*
Pheasant, Reeves'	*Syrmaticus reevesii*
Pheasant, Silver	*Lophura nycthemera*
Porcupine, Crested	*Hystrix cristata*
Porcupine, Himalayan	*Hystrix hodgsonii*
Pumpkinseed (or Sun-fish, or Pond-perch)	*Lepomis gibbosus*
Quail, Bobwhite	*Colinus virginianus*
Rat, Black	*Rattus rattus*
Snake, Aesculapian	*Elephe longissima*
Squirrel, Grey	*Sciurus carolinensis*
Terrapin, European Pond	*Emys orbicularis*
Toad, African Clawed	*Xenopus laevis*
Toad, Midwife	*Alytes obstetricans*
Toad, Yellow-bellied	*Bombina variegata*
Wallaby, Red-necked	*Macropus rufogriseus*
Wels (or European catfish)	*Silurus glanis*
Zander	*Stizostedion lucioperca*

Part 2   Plants

Hogweed, Giant	*Heracleum mantegazzianum*
Kelp, Giant	*Macrocystis pyrifera*
Kelp, Giant	*Macrocystis angustifolia*

Kelp, Giant	*Macrocystis integrifolia*
Kelp, Giant	*Macrocystis laevis*
Kelp, Japanese	*Macrocystis japonica*
Knotweed, Japanese	*Polygonum cuspidatum*
Seafingers, Green	*Codium fragile tomentosoides*
Seaweed, California Red	*Pikea californica*
Seaweed, Hooked Asparagus	*Asparagopsis armata*
Seaweed, Japanese	*Sargassum muticum*
Seaweeds, Laver (except native species)	*Porphyra* spp., except –
	P. amethystea
	P. leucosticta
	P. linearis
	P. miniata
	P. purpurea
	P. umbilicalis
Wakame	*Undaria pinnatifida*

Notes to Schedule 9

1. All species of *Cervus* with respect to the Outer Hebrides and the islands of Arran, Islay, Jura and Rhum.
2. Any hybrid one of whose parents or other linear ancestor was a species of *Cervus* Deer.
3. Any hybrid one of whose parents or other lineal ancestor was a sika deer.

THE SCHEDULES TO THE CONSERVATION (NATURAL HABITATS ETC.) REGULATIONS 1994

Schedule 2   European protected species of animals (Habitats Directive Annex IV (a))

Bats, Horseshoe (all species)	Rhinolophidae
Bats, Typical	Vespertilionidae
Butterfly, Large Blue	*Maculinea arion*
Cat, Wild	*Felis silvestris*
Dolphins, Porpoises and Whales (all species)	Cetacea
Dormouse	*Muscardinus avellanarius*
Lizard, Sand	*Lacerta agilis*
Newt, Great Crested	*Triturus cristatus*
Otter, Common	*Lutra lutra*
Snake, Smooth	*Coronella austriaca*
Sturgeon	*Acipenser sturio*
Toad, Natterjack	*Bufo calamita*
Turtles, Marine	*Caretta caretta*
	Chelonia mydas
	Lepidochelys imbricata
	Eretmochelys imbricata
	Dermochelys coriacea

Schedule 3   Animals which may not be killed or taken in certain ways (Habitats Directive Annex V (a))

Barbel	*Barbus barbus*
Grayling	*Thymallus thymallus*
Hare, Mountain	*Lepus timidus*
Lamprey, River	*Lampetra fluviatilis*
Marten, Pine	*Martes martes*
Polecat	*Mustela putorius*
Salmon, Atlantic[1]	*Salmo salar*
Seal, Bearded	*Erignathus barbatus*
Seal, Common	*Phoca vitulina*
Seal, Grey	*Halichoerus grypus*
Seal, Harp	*Phoca groenlandica*
Seal, Hooded	*Cystophora cristata*
Seal, Ringed	*Phoca hispida*
Shad, Allis	*Alosa alosa*
Shad, Twaite	*Alosa fallax*
Vendace	*Coregonus albula*
Whitefish	*Coregonus lavaretus*

Note to Schedule 3
1. Only in freshwater.

Schedule 4   European protected species of plants (Habitats Directive Annex IV (b))

Dock, Shore	*Rumex rupestris*
Fern, Killarney	*Trichomanes speciosum*
Gentian, Early	*Gentianella anglica*
Lady's-slipper	*Cypripedium calceolus*
Marshwort, Creeping	*Apium repens*
Naiad, Slender	*Najas flexilis*
Orchid, Fen	*Liparis loeselii*
Plantain, Floating-leaved Water	*Luronium natans*
Saxifrage, Yellow Marsh	*Saxifraga hirculus*

APPENDIX 5

Annex I projects, EIA Directive

A summary of Annex I projects under the Directive (subject to Art. 4(1))/Schedule 1 projects under the Town and Country Planning (Environmental Impact Assessment) (England and Wales) Regulations 1999.

1. Crude oil refineries, coal or shale gasification or liquefaction installations.
2. Thermal power stations, nuclear power stations, other nuclear reactors etc.
3. Installations for the processing, reprocessing, final disposal or storage of irradiated nuclear fuel, or the production or enrichment of nuclear fuel.
4. Integrated works for the initial smelting of cast-iron and steel.
5. Installations for the extraction, processing and transformation of asbestos.
6. Integrated chemical installations for the industrial scale manufacture of basic organic and inorganic chemicals, fertilisers, plant health products and biocides, pharmaceuticals, and explosives.
7. Construction of long-distance railway lines, airport runways of 2100 metres or more. Construction of motorways and express roads. New roads of four or more lanes and roads which have been improved so as to convert two lanes or less to four lanes or more, where such a road would be 10 kilometres or more in continuous length.
8. Inland waterways and ports for inland-waterway traffic, trading ports and piers.
9. Waste disposal installations for the incineration, chemical treatment or landfill of hazardous waste.
10. Waste disposal installations for the incineration, chemical treatment or landfill of non-hazardous waste.
11. Groundwater abstraction or artificial groundwater recharge schemes.
12. Water transfer schemes between river basins.
13. Waste water treatment plants.
14. Commercial extraction of petroleum and natural gas.
15. Dams and water storage installations.
16. Gas, oil or chemical pipe-lines.
17. Installations for the intensive rearing of poultry or pigs.
18. Paper, board and pulp production plants.
19. Quarries, open-cast mining and peat extraction.

20. Construction of overhead electrical power lines.
21. Installations for the storage of petroleum, petrochemical or chemical products.

Note
Some of the types of projects listed above are covered by Annex I and the Regulations only if they meet certain threshold levels or other criteria.

Annex II projects, EIA Directive

A summary of Annex II projects under the Directive (subject to Art. 4(2))/Schedule 2 projects under the Town and Country Planning (Environmental Impact Assessment) (England and Wales) Regulations 1999.

1. Agriculture, silviculture* and aquaculture
    Restructuring of rural land holdings; use of uncultivated land or semi-natural areas for intensive agriculture; water management projects for agriculture; initial afforestation* and deforestation* for the purpose of conversion to a different land use; intensive livestock installations (projects not included in Annex I); intensive fish farming; reclamation of land from the sea.
2. Extractive industry
    Quarries, open-cast mining, peat extraction (projects not included in Annex I); underground mining; dredging; deep drilling; surface installations for coal, gas, ore and shale extraction.
3. Energy industry
    Installations for production of electricity, steam and hot water and for carrying gas, steam and hot water, and transmission of electricity by overhead cables* (projects not included in Annex I); surface storage of natural gas and fossil fuels; underground storage of combustible gases; briquetting of coal and lignite; installations for processing and storage of radioactive waste (unless included in Annex I); hydroelectric and wind power installations.
4. Production and processing of metals
    Installations for the production of pig iron or steel; processing of ferrous metals; ferrous metal foundries; installations for smelting metals and surface treatment of metals and plastic materials; assembly and manufacture of motor vehicles and motor-vehicle engines; shipyards; installations for construction and repair of aircraft; manufacture of railway equipment; swaging by explosives; and installations for the roasting and sintering of metallic ores.
5. Metal industry
    Coke ovens; installations for the manufacture of glass, cement, asbestos and asbestos products (projects not covered by Annex I); smelting mineral substances; manufacture of ceramic products by burning.

6. Chemical industry (projects not included in Annex I)
   Treatment of intermediate products and production of chemicals; production of pesticides, pharmaceuticals, paint, varnishes, elastomers and peroxides; storage facilities for petroleum, petrochemical products and chemical products.
7. Food industry
   Manufacture of oils, fats, dairy products, confectionery, syrup, industrial starch; packing and canning; brewing and malting; sugar, fish-meal and fish-oil factories; and installations for the slaughter of animals.
8. Textile, leather, wood and paper industries
   Industrial plants for paper and board production (projects not included in Annex I); pre-treatment plants; tanning plants; cellulose-processing and production installations.
9. Rubber industry
   Manufacture and treatment of elastomer-based products.
10. Infrastructure projects (not included in Annex I)
    Industrial estates; urban development projects (including shopping centres and car parks); railways and transhipment facilities; airfields, roads, harbours, ports, inland-waterways; dams and water storage facilities; tramways, elevated and underground passenger railways etc.; oil and gas pipe-lines; long-distance aqueducts; coastal and sea defence works; groundwater abstraction and artificial groundwater recharge schemes; water transfer schemes between river basins; motorway service areas.
11. Other projects
    Permanent motor racing and test tracks; waste disposal projects and waste water treatment plants (projects not included in Annex I); sludge-disposal sites; storage of scrap iron (including scrap vehicles); test benches for engines etc.; installations for the manufacture of artificial mineral fibres and the recovery or destruction of explosives; knackers' yards.
12. Tourism and leisure
    Ski-runs, ski-lifts, cable cars etc.; marinas; holiday villages and hotel complexes outside urban areas; permanent camp sites and caravan sites; theme parks and golf courses.
13. Any change or extension of projects listed in Annex I or Annex II, already authorised, executed or in the process of being executed, which may have adverse environmental effects
    Projects in Annex I, undertaken exclusively or mainly for the development and testing of new methods or products and not used for more than two years.

Note
Some of the types of projects listed above are covered by Annex II and the Regulations only if they meet certain threshold levels or other critieria.
* A project which is included in Annex II of the Directive, but excluded from Schedule 2 of the Regulations (SI 1999/293).

APPENDIX 7

Habitats Directive, Annex I

Natural habitat types of Community Interest whose conservation requires the designation of Special Areas of Conservation

1. *Coastal and halophytic habitats*
   Sandbanks which are slightly covered by seawater all the time
   Estuaries
   Mudflats and sandflats not covered by seawater at low tide
   *Lagoons
   Large shallow inlets and bays
   Reefs
   Annual vegetation of drift lines
   Perennial vegetation of stony banks
   Vegetated sea cliffs of the Atlantic and Baltic coasts
   Salicornia and other annuals colonising mud and sand
   Spartina swards (*Spartinion*)
   Atlantic salt meadows (*Glauco-Puccinellietalia*)
   *Continental salt meadows (*Puccinellietalia distantis*)
   Mediterranean salt meadows (*Juncetalia maritimi*)
   Mediterranean and thermo-Atlantic halophilous scrubs (*Arthrocnemeta fructicosae*)

2. *Coastal sand dunes and continental dunes*
   Embryonic shifting dunes
   Shifting dunes along the shoreline with *Ammophila arenaria* (white dunes)
   *Fixed dunes with herbaceous vegetation (grey dunes)
   *Decalcified fixed dunes with *Empetrum nigrum*
   *Eu-Atlantic decalcified fixed dunes (*Calluno-Ulicetea*)
   Dunes with *Salix arenaria*
   Humid dune slacks
   Machair
   *Dune juniper thickets (*Juniperus* spp.)
   Open grassland with *Corynephorus* and *Agrostis* of continental dunes

3. *Freshwater habitats*
   Oligotrophic waters containing very few minerals of Atlantic sandy plains with amphibious vegetation: *Lobelia, Littorella* and *Isoetes*

Oligotrophic waters in medio-European and perialpine areas with amphibious vegetation: *Littorella* and *Isoetes* or annual vegetation on exposed banks (*Nanocyperetalia*)

Hard oligo-mesotrophic waters with benthic vegetation of *Chara* formations

Natural eutrophic lakes with *Magnopotamion* or *Hydrocharition*-type vegetation

Dystrophic lakes

*Mediterranean temporary ponds

Floating vegetation of *Ranunculus* of plain and sub-mountainous rivers

4. *Temperate heath and scrub*
   North Atlantic wet heaths with *Erica tetralix*
   *Southern Atlantic wet heaths with *Erica ciliaris* and *Erica tetralix*
   Dry heaths (all subtypes)
   *Dry coastal heaths with *Erica vagans* and *Ulex maritimus*
   Alpine and subalpine heaths
   Sub-Arctic willow scrub

5. *Sclerophyllous scrub (Matorral)*
   Stable *Buxus sempervirens* formations on calcareous rock slopes (*Berberidion*)
   Juniperus communis formations on heaths and calcareous grasslands

6. *Natural and semi-natural grassland formations*
   Calaminariun grasslands
   Siliceous alpine and boreal grassland
   Alpine calcareous grasslands
   Semi-natural dry grasslands and scrubland facies on calcareous substrates (*Festuco-Brometalia*)
   Semi-natural dry grasslands and scrubland facies on calcareous substrates (*Festuco-Brometalia*) (*important orchid sites)
   *Species-rich *Nardus* grasslands, on siliceous substrates in mountain areas (and submountain areas, in continental Europe)
   Molinia meadows on chalk and clay (*Eu-Molinion*)
   Eutrophic tall herbs
   Lowland hay meadows (*Alopecurus pratensis, Sanguisorba officinalis*)
   Mountain hay meadows (British types with *Geranium sylvaticum*)

7. *Raised bogs and mires and fens*
   *Active raised bogs
   Degraded raised bogs (still capable of natural regeneration)
   Blanket bogs (*active only)
   Transition mires and quaking bogs
   Depressions on peat substrates (*Rhynchosporion*)
   *Calcareous fens (with *Cladium mariscus* and *Carex davalliana*)
   *Petrifying springs with tufa formation (*Cratoneurion*)
   Alkaline fens
   *Alpine pioneer formations of *Caricion bicoloris-atrofuscae*

8. *Rocky habitats and caves*
   Siliceous scree
   Eutric scree
   Chasmophytic vegetation on rocky slopes – Cacareous subtypes
   Chasmophytic vegetation on rocky slopes – Silicicolous subtypes
   *Limestone pavements
   Submerged or partly submerged sea caves

9. *Forests*
   Beech forests with *Ilex* and *Taxus*, rich in epiphytes (*Ilici-Fagion*)
   Asperulo-Fagetum beech forests
   Stellario-Carpinetum oak-hornbeam forests
   Tilio-Acerion ravine forests
   Old acidophilous oak woods with *Quercus robur* on sandy plains
   Old oak woods with *Ilex* and *Blechnum* in the British Isles
   *Caledonian forest
   *Bog woodland
   *Residual alluvial forest (*Alnion glutinoso-incanae*)
   Taxus baccata woods

This list contains only those habitat types that occur in the UK.
*Priority habitat type.

APPENDIX 8
Vegetation classification systems

A number of systems exist for the classification of natural vegetation. Unfortunately, different systems use different terminology for more-or-less equivalent vegetation types. In the UK several systems have become standards for ecological surveys, mapping, and the implementation of conservation legislation:

- Phase 1 Habitat Survey – developed in the 1980s for mapping terrestrial and freshwater habitats within SSSIs and nature reserves, and for larger scale strategic surveys.
- National Vegetation Classification (NVC) – used for the selection of biological SSSIs for terrestrial habitats, and for the interpretation of Annex I of the Habitats Directive (see Appendix 7).
- Marine Biotopes Classification (BioMar) – adopted by the European Environment Agency for use in European waters and the North Atlantic.
- BAP Broad Habitat Classification – used to categorise habitats in the UK Biodiversity Action Plan.

The Phase 1 and NVC systems are recommended by the Institute of Environmental Assessors as standards for use in environmental statements prepared under the Town and Country Planning (Environmental Impact Assessment) (England and Wales) Regulations 1999.

Phase 1 Habitat Survey

In this system each habitat type is assigned a unique code and colour (solid, hatched or cross-hatched). Over 100 different habitat types are recognised ranging from specific woodland types to buildings. Dominant plant species are indicated with standard codes. The major groups of habitats used in Phase 1 surveys are:

- Woodland and scrub
- Grassland and marsh

- Tall herb and fern
- Heathland
- Mire (including bogs, peat, springs and fens etc.)
- Swamp, marginal and inundation
- Open water (standing and running waters)
- Coastland (saltmarsh, boulders, dunes, cliffs, sand and mudflats etc.)
- Rock exposure and waste (mines, quarries, caves and spoil etc.)
- Miscellaneous (fences, walls, dry ditches, bare ground and buildings etc.)

REFERENCE

JNCC (1993) *Handbook for Phase 1 habitat survey: a technique for environmental audit.* Joint Nature Conservation Committee, Peterborough.

National Vegetation Classification (NVC)

The NVC provides descriptions of British plant communities devised by Lancaster University, based on records from over 33 000 localities in Britain. The system covers all natural and semi-natural habitats and major artificial habitats. It includes:

- Descriptions of around 400 named plant communities
- Characteristic species composition and structure
- Relationships with habitat factors
- Descriptions of common landscape patterns
- Distinctive mosaics and zonations of plant communities
- Successional changes
- National distribution data, including maps

The NVC is used by the statutory nature conservation agencies, DEFRA, the Forestry Commission, the National Trust, the RSPB, the Wildlife Trusts and large corporations such as BNFL, British Gas and British Coal Opencast.

REFERENCE

Rodwell, J. S. (2000) *British Plant Communities.* Volume 1: Woodlands and Scrub; Volume 2: Mires and Heaths; Volume 3: Grasslands and Montane Communities; Volume 4: Aquatic Communities, Swamps and Tall-herb Fens; Volume 5: Maritime Communities and Vegetation of Open Habitats. Cambridge University Press, Cambridge.

Glossary and acronyms

LEGAL TERMS

Appeal	An appeal against a decision of a court in the light of new evidence or as a result of a misinterpretation of the law.
Criminal offence	An act or omission considered harmful of which the state disapproves.
Defendant	An individual who defends an action in a court.
European Directive	A legislative instrument produced by the EC with which members states must comply by producing their own legislation.
European Regulation	A legislative instrument produced by the EC which is law in the member states to which it applies.
Ex parte	On behalf of. Appears in the title of some legal cases.
Indictable offence	A more serious offence which is committed by an adult and triable on indictment; c.f. summary offence.
Indictment	A written or printed accusation of a crime.
Inter alia	Among other things.
Judicial review	A review of the process used to make a legal decision.
Land charges register	Register held by a local authority which contains details of TPOs, SSSIs and other restrictions against individual pieces of land or properties.
Locus standi	The right to bring an action to the courts.
Magistrates' Court	A court in which lay members of the public act as both judge and jury in cases involving less serious (summary) offences.
Plaintiff	An individual who brings an action against another in a court.
Planning Inspectorate	Agency concerned with planning and enforcement appeals, inquiries into local development plans and planning applications called in by the Secretary of State.
Queen's Counsel	A senior barrister. Also known as a 'silk'.
Strict liability offence	The offender is liable for the offence whether or not he intended to break the law.

Summary offence	A minor offence dealt with by the Magistrates' Court; c.f. indictable offence.
Ultra vires	Beyond their powers, e.g. when a Secretary of State makes a decision which he is not empowered to make under the law.

ECOLOGICAL TERMS

Carrying capacity	The capacity of the environment to support a particular species of organism.
Climax community	The assemblage of organisms which represents the end point of an ecological succession.
Community	The assemblage of organisms which live together in a particular habitat, e.g. a sand dune community, a woodland community.
Decomposers	Bacteria and fungi that are responsible for decomposition.
Energy flow	The passage of energy from the sun through the food chain.
Ecosystem	A biological community together with its physical environment. A convenient unit into which the natural world may be divided for the purpose of examining the interrelationships between organisms.
Ecotype	A group of plants (or animals) within a species which is genetically adapted to a particular habitat but able to interbreed freely with other ecotypes of the same species.
Environment	All of the factors that affect the survival of an organism. Sometimes divided into the physical and the biological environment.
Food chain	A linear sequence of feeding relationships between organisms.
Food web	The complex of feeding relationships within a community.
Habitat	The environment of an organism, comprising the vegetation, soil and climatic factors which affect its survival.
Homeotherm	An animal that is capable of controlling its body temperature by physiological means, independent of the environmental temperature ('warm-blooded').
Niche	The role an organism plays in the ecosystem.
Nutrient cycle	Route taken by a nutrient as it passes in a cycle between the biological and physical components of the environment, e.g. nitrogen cycle, phosphorus cycle.
Organism	A general term for any living thing.
Photosynthesis	The production of sugars by green plants using carbon dioxide, water and light energy.
Poikilotherm	An animal whose body temperature is determined by the external environment ('cold-blooded').

Population	A group of organisms of the same species which occupies a particular place at a particular time, e.g. the population of deer in a particular woodland.
Seral stage	A stage in an ecological succession.
Species	A group of organisms of the same type which can interbreed to produce viable offspring.
Succession	The sequence with which animals and plants colonise a piece of ground, each community being replaced by the last, until a climax community is established.
Taxon	A group of genetically related organisms, e.g. a species, a family or a phylum.

SOME ACRONYMS USED IN THE TEXT

AONB	Area of Outstanding Natural Beauty
ASSI	Area of Special Scientific Interest
BAP	Biodiversity Action Plan
CCW	Countryside Council for Wales
CITES	Convention on International Trade in Endangered Species of Wild Fauna and Flora
CRoW 2000	Countryside and Rights of Way Act 2000
DEFRA	Department for Environment, Food and Rural Affairs
DETR	Department of the Environment, Transport and the Regions
EC	European Community
EIA	Environmental Impact Assessment
EN	English Nature
ERDP	England Rural Development Programme
ESA	Environmentally Sensitive Area
EU	European Union
HAP	Habitat Action Plan
IUCN	International Union for the Conservation of Nature and Natural Resources (World Conservation Union)
JNCC	Joint Nature Conservation Committee
LAP	Local Action Plan
MAFF	Ministry of Agriculture, Fisheries and Food
MPG	Minerals Policy Guidance (Notes)
MNR	Marine Nature Reserve
NCC	Nature Conservancy Council*
NNR	National Nature Reserve
NP	National Park
NSA	Nitrate Sensitive Area
NWCI	National Wildlife Crime Intelligence Unit
PAW	Partnership for Action Against Wildlife Crime
PPG	Planning Policy Guidance (Notes)
RSPB	Royal Society for the Protection of Birds
RSPCA	Royal Society for the Prevention of Cruelty to Animals
SAC	Special Area of Conservation
SAP	Species Action Plan
SEPA	Scottish Environment Protection Agency

SI	Statutory Instrument
SNH	Scottish Natural Heritage
SPA	Special Protection Area
SSPCA	Scottish Society for the Prevention of Cruelty to Animals
SSSI	Site of Special Scientific Interest
TCPA 1990	Town and Country Planning Act 1990
TPO	Tree Preservation Order
TRAFFIC	Trades Records Analysis of Flora and Fauna in Commerce
UN	United Nations
UNEP	United Nations Environment Programme
UNESCO	United Nations Educational, Scientific and Cultural Organisation
WCA 1981	Wildlife and Countryside Act 1981
WWF	World Wide Fund for Nature
WWT	Wildfowl and Wetlands Trust

Note
* NCC is used as an abbreviation for any or all of the following organisations, depending upon the context: English Nature (formerly the Nature Conservancy Council for England), Countryside Council for Wales and Scottish Natural Heritage. This follows the convention established by the Wildlife and Countryside Act 1981 s. 27A.

Information sources

Organisation	Address	Website
Arboricultural Association	Ampfield House, Ampfield, Romsey, Hants, SO5 9PA	www.trees.org.uk
Barn Owl Trust	Waterleat, Ashburton, Devon, TQ13 7HU	www.barnowltrust.org.uk
Bat Conservation Trust	15 Cloisters House, 8 Battersea Park Road, London, SW8 4BG	www.bats.org.uk
Biological Records Centre	Monks Wood, Abbots Ripon, Huntingdon, Cambs, PE28 2LS	www.brc.ac.uk
Botanical Society of the British Isles		www.bsbi.org.uk
British Butterfly Conservation Society Ltd	Head Office, PO Box 222, Dedham, Colchester, Essex, CO7 6EY	www.butterfly-conservation.org.uk
British Geological Survey	See NERC	www.bgs.ac.uk
British Herpetological Society	c/o Zoological Society of London, Regents Park, London, NW1 4RY	www.herplit.com
British Lichen Society		www.argonet.co.uk/users/jmgray
British Trust for Conservation Volunteers	36 St Mary's Street, Wallingford, Oxfordshire, OX10 0EU	www.btcv.org
British Trust for Ornithology (BTO)	The Nunnery, Nunnery Place, Thetford, Norfolk	www.bto.org
British Waterways	Willow Grange, Church Road, Watford, Herts, WD1 3QA	www.british-waterways.org
Broads Authority	Thomas Harvey House, 18 Colegate, Norwich, NR3 1BQ	www.broads-authority.gov.uk

Cadw (Welsh Historic Monuments)	Crown Building, Cathays Park, Cardiff, CF1 3NQ	*www.cadw.wales.gov.uk*
Cambridgeshire's Biodiversity Partnership	Cambridgeshire County Council, Box ET 1001, Castle Court, Shire Hall, Cambridge, CB3 0AP	*www.camcnty.gov.uk/sub/ cntryside/biodiv/partnership*
Centre for Ecology and Hydrology	CEH Directorate, Monks Wood, Abbots Ripton, Huntingdon, PE17 2LS	*www.ceh.ac.uk*
Centre for Urban Ecology	Birmingham Settlement, 318 Summer Lane, Birmingham, B19 3BL	
Civic Trees	Forestry House PO Box 23, Tring, Herts, HP23 4AE	*www.civictrees.co.uk*
Council for the Protection of Rural England (CPRE)	Warwick House, 25 Buckingham Palace Road, London, SW1W 0PP	*www.cpre.org.uk*
Countryside Agency	John Dower House, Crescent Place, Cheltenham, Glos., GL50 3RA	*www.countryside.gov.uk*
Countryside Council for Wales/Cyngor Cefn Gwlad Cymru	Plas Penrhos, Ffordd Penrhos, Bangor, Gwynedd, LL57 2LQ	*www.ccw.gov.uk*
Deer Commission for Scotland	Knowsley, 82 Fairfield Road, Inverness, IV3 5LH	*www.dcs.gov.uk*
Department for Environment, Food and Rural Affairs (DEFRA)	Nobel House, 17 Smith Square, London, SW1P 3JR	*www.defra.gov.uk*
Department for Transport	Eland House, Bressenden Place, London, SW1E 5DU	*www.dft.gov.uk*
Department of the Environment for Northern Ireland	Clarence Court, 10–18 Adelaide Street, Belfast, BT2 8GB	*www.doeni.gov.uk*
English Heritage	Customer Services Department, PO Box 569, Swindon, SN2 2YP	*www.english-heritage.org.uk*
English Nature	Northminster House, Peterborough, PE1 1UA	*www.english-nature.org.uk*
Environment Agency	25th Floor, Millbank Tower, 21–24 Millbank, London, SW1P 4XL	*www.environment-agency.gov.uk*

Environment and Heritage Service (Northern Ireland)	Commonwealth House, Castle Street, Belfast, BT1 1GU	*www.ehsni.gov.uk*
Fauna and Flora International	c/o Zoological Society of London, Regents Park, London, NW1 4RY	*www.fauna-flora.org*
Flora Locale,	36 Kingfisher Court, Hambridge Road, Newbury, RG14 5SJ	*www.floralocale.org*
Forestry Commission/ Forest Enterprise/ Forest Research	231 Corstorphine Road, Edinburgh, EH12 7AT	*www.forestry.gov.uk*
Forest Service (Northern Ireland)	Dundonald House, Upper Newtownards Road, Belfast, BT4 3SB	*www.forestserviceni.gov.uk*
Friends of the Earth	26/28 Underwood Street, London, N1 7JQ	*www.foe.co.uk*
Froglife	Mansion House, 27–28 Market Place, Halesworth, Suffolk, IP19 8AY	*www.froglife.org*
Greenpeace	Greenpeace, Canonbury Villas, London, N1 2PN	*www.greenpeace.org.uk*
Hawk and Owl Trust	c/o Zoological Society of London, Regents Park, London, NW1 4RY	*www.hawkandowl.co.uk*
Her Majesty's Stationery Office	St Clements House, 2–16 Colegate, Norwich, NR3 1BQ	*www.hmso.gov.uk*
Highways Agency	Traffic Safety and Environment Division, St Christopher House, Southwark Street, London, SE1 0TE	*www.highways.gov.uk*
Historic Scotland	Longmore House, Salisbury Place, Edinburgh, EH9 1SH	*www.historic-scotland.gov.uk*
Insect Identification Services	Department of Entomology, Natural History Museum, Cromwell Road, London, SW7 5BD	*www.nhm.ac.uk/entomology/insident/ index.html*
International Union for the Conservation of Nature and Natural Resources (IUCN)	219c Huntingdon Road, Cambridge, CB3 0DL	*www.iucn.org*

Joint Nature Conservation Committee	Monkstone House, City Road, Peterborough PE1 1JY	*www.jncc.gov.uk*
Mammal Society	15 Cloisters Business Centre, 8 Battersea Park Road, London, SW8 4BG	*www.abdn.ac.uk/mammal*
Marine Conservation Society	9 Gloucester Road, Ross on Wye, Herefordshire, HR9 5BU	*www.mcsuk.org*
Maritime and Coastguard Agency	Spring Place, 105 Commercial Road, Southampton, SO15 1EG	*www.mcagency.org.uk*
Metropolitan Police Service, Wildlife Crime Unit	New Scotland Yard, Broadway, London, SW1H 0BG	*www.met.police.uk/police/mps/ wildlife*
Ministry of Defence Conservation Office	Blandford House, Farnborough Road, Aldershot, Hampshire, GU11 2HA	*www.mod.uk*
National Biodiversity Network Trust	NBN Secretariat, c/o The Wildlife Trusts (see below)	*www.nbn.org.uk*
National Federation of Badger Groups	15 Cloisters Business Centre, 8 Battersea Park Road, London, SW8 4BG	*www.badgers.org.uk/nfbg/*
National Hedgelaying Society	Secretary, Mrs J. Hallam, 16 Narcot Lane, Chalfont St Giles, Bucks, HP8 4DA	*http://members.tripod.co.uk/ hedgelaying/*
National Land Use Database	NLUD Helpdesk, Office of the Deputy Prime Minister, Zone 3/K10, Eland House, Bressenden Place, London, SW1E 5DU	*www.nlud.org.uk*
National Trust	PO Box 39, Bromley, Kent, BR1 3XL	www.nationaltrust.org.uk
National Wildflower Centre	National Wildflower Centre, Court Hey Park, Knowsley, Merseyside	*www.nwc.org.uk*
National Wildlife Crime Intelligence Unit	National Criminal Intelligence Service, PO Box 8000, London, SE11 5EN	*www.ncis.co.uk*
Natural Environment Research Council	Polaris House, North Star Avenue, Swindon, Wiltshire, SN2 1EU	*www.nerc.ac.uk*

Natural History Museum	Cromwell Road, London, SW7 5BD	*www.nhm.ac.uk*
Ordnance Survey	Romsey Road, Maybush, Southampton, SO16 4GU	*www.ordnancesurvey.gov.uk*
Ordnance Survey of Northern Ireland	Colby House, Stranmillis Court, Belfast, BT9 5BJ	*www.osni.gov.uk*
Otter Trust	Earsham, nr Bungay, Suffolk, NR35 2AF	*www.ottertrust.org.uk*
Partnership for Action Against Wildlife Crime, DEFRA	PAW Secretariat, DEFRA, Nobel House, 17 Smith Square, London SWIP 3JR	*www.defra.gov.uk/paw*
Planning Inspectorate	Tollgate House, Houlton Street, Bristol, BS2 9DJ; Crown Building, Cathays Park, Cardiff, CF1 3NQ	*www.open.gov.uk/pi*
Planning Service (Northern Ireland)	Clarence House, 10–18 Adelaide Street, Belfast, BT2 8GB	*www.doeni.gov.uk/planning/*
Plantlife	21 Elizabeth Street, London, SW1W 9RP	*www.plantlife.org.uk*
Rivers Agency (Northern Ireland)	4 Hospital Road, Belfast, BT8 8JP	*www.dardni.gov.uk*
Royal Botanic Gardens, Kew	Richmond, Surrey, TW9 3AB	*www.rbgkew.org.uk*
Royal Society for the Prevention of Cruelty to Animals (RSPCA)	Causeway, Horsham, West Sussex, RH12 1HG	*www.rspca.org.uk*
Royal Society for the Protection of Birds	The Lodge, Sandy, Bedfordshire	*www.rspb.org.uk*
Scottish Environment Protection Agency	Erskine Court, The Castle Business Park, Stirling, FK9 4TR	*www.sepa.org.uk*
Scottish Fisheries Protection Agency	Pentland House, 47 Robb's Loan, Edinburgh, EH14 1TY	
Scottish Natural Heritage	12 Hope Terrace, Edinburgh, EH9 2AS	*www.snh.org.uk*
Tree Register	The Tree Register, 77a Hall End, Wootton, Bedford, MK43 9HP	*www.tree-register.org*
Trust for Urban Ecology	PO Box 514, London, SE16 1AS	

Whale and Dolphin Conservation Society	19a James Street West, Bath, Avon, BA1 2BT	*www.wdcs.org*
Wildfowl and Wetlands Trust	Slimbridge, Gloucester, GL2 7BT	*www.wwt.org.uk*
Wildlife Inspectorate	Room 810, Tollgate House, Houlton Street, Bristol, BS2 9DJ	*www.defra.gov.uk/wildlife-countryside/wildcrime/index.htm*
Wildlife Trusts	The Kiln, Waterside, Mather Road, Newark, Notts, NG24 1WT	www.wildlifetrusts.org
Woodland Trust	Autumn Park, Dysart Road, Grantham, Lincs, NG31 6LL	*www.woodland-trust.org.uk*
World Wide Fund for Nature	Panda House, Weyside Park, Godalming, Surrey, GU7 1XR	*www.panda.org*

Vertebrate wildlife management problems

Inquiries and applications for licences should be made to:

England

Wildlife Administration Unit, DEFRA, Burghill Road, Westbury-on-Trym, Bristol. *www.defra.gov.uk/wildlife-countryside/vertebrates*

Wales

National Assembly for Wales, Agriculture Department, Food Farming Development Division, Yr Hen Ysgol Gymrag, Ffordd Alexandra, Aberystwyth, Ceredigion SY23 1LD

Scotland

Environment and Rural Affairs Department (ERAD), PEP2 Division, Branch 2, Agricultrual Pollution, Pesticides and Pest Control, Pentland House, 47 Robb's Loan, Edinburgh EH14 1TY

Northern Ireland

Environment and Heritage Service, 35 Castle Street, Belfast BT1 1GU

Table of laws

International law

Agreement on the Conservation of African–
    Eurasian Migratory Waterbirds
    (AEWA) 1995   87, 286
Agreement on the Conservation of Bats in
    Europe (EUROBATS) 1991   87, 286
Agreement on the Conservation of
    Cetaceans in the Black Sea,
    Mediterranean Sea and Contiguous
    Atlantic Area (ACCOBAMS) 1996
    87, 286
Agreement on the Conservation of Small
    Cetaceans in the Baltic and North
    Seas (ASCOBANS) 1991   87, 286
Convention Concerning the Protection of
    the World Cultural and Natural
    Heritage 1972 (World Heritage
    Convention)   87, 282–4
  Art. 1   282
  Art. 2   282
  Art. 5   282
  Art. 5(a)   284
  Art. 5(b)   284
  Art. 5(c)   284
  Art. 5(d)   284
  Art. 11   282
Convention for the Protection of Human
    Rights and Fundamental Freedoms
    1950
  Art. 6(1)   359
Convention on Biological Diversity 1992
    (Biodiversity Convention)   85, 87,
    169, 272, 293–8
  Art. 1   295
  Art. 2   295
  Art. 3   295
  Art. 6(a)   295
  Art. 6(b)   295
  Art. 7   296
  Art. 8   296
  Art. 8(d)   169
  Art. 8(f)   169, 297–8
  Art. 8(h)   297
  Art. 9(c)   169

  Annex I   296
Convention on International Trade in
    Endangered Species of Wild Fauna
    and Flora 1973 (CITES) (Washington
    Convention)   22, 24, 29, 87, 289–93,
    299
  Art. I   292
  Art. III   291
  Art. III(3)(a)   291
  Art. III(3)(b)   291
  Art. III(3)(c)   291
  Appendix I   291, 293
  Appendix II   291, 293
  Appendix III   291
Convention on the Conservation of
    European Wildlife and Natural
    Habitats 1979 (Berne Convention)
    87, 286–8, 289, 299, 326
  Art. 1–3   286
  Art. 2   286
  Art. 3(1)   286
  Art. 3(2)   286
  Art. 4   287
  Art. 4(1)   287
  Art. 4(2)   287
  Art. 4(3)   287
  Art. 5   287
  Art. 5–9   287
  Art. 6   287
  Art. 6(a)   288
  Art. 6(b)   288
  Art. 6(c)   288
  Art. 6(d)   288
  Art. 6(e)   288
  Art. 7   287
  Art. 9   287
  Art. 10   288
  Art. 11(2)   169
  Art. 11(2)(a)   288
  Art. 11(2)(b)   288
  Appendix I   286–7
  Appendix II   286–7
  Appendix III   286–7, 300

Convention on the Conservation of
    Migratory Species of Wild Animals
    1979 (Bonn Convention)   87, 284–6,
    289, 299
  Art. I(1)(a)   284
  Art. II(3)(b)   285
  Art. II(3)(c)   285
  Art. III(1)   284–5
  Art. III(4)(a)   285
  Art. III(4)(b)   285
  Art. III(4)(c)   285
  Art. III(5)   285
  Art. IV(1)   285
  Appendix I   284–5
  Appendix II   285
Convention on Wetlands of International
    Importance Especially as Waterfowl
    Habitat 1971 (the Rasmar

    Convention)   87, 184, 277
  Art. 1   277
  Art. 1(1)   278
  Art. 1(2)   278
  Art. 2(1)   277
  Art. 2(2)   27
  Art. 8   277
International Convention for the Regulation
    of Whaling 1946 (Whaling
    Convention)   87, 288–9
  Art. I   288
  Art. III   288
  Art. V(1)   288–9
Kyoto Protocol to the UN Framework
    Convention on Climate Change 1997
    (Kyoto Protocol)   235, 289–90
  Art. 3(3)   289
  Annex I   289

European Community law

Treaties establishing the European
Community
Single European Act 1986   17
Treaty Establishing the European Atomic
    Energy Authority (Euratom) Treaty
    1957   17
Treaty Establishing the European Coal and
    Steel Community Treaty 1951   17
Treaty Establishing the European Economic
    Community (Rome) 1957 (EC
    Treaty)   17
  Art. 130r   17–18
  Art. 174   17–18
  Art. 174(2)   84
  Art. 189   18
  Art. 249   18
Treaty of Amsterdam 1997   17, 84
Treaty on European Union (Maastricht)
    1992   17, 84

Directives (chronological list)
76/160/EEC Standards for Water used for
    Bathing Purposes Directive   273–4
  Art. 3(1)   274
78/659/EEC Quality of Freshwater to
    Support Fish Life Directive   84,
    220, 272
  Art. 1(3)   273
79/409/EEC Wild Birds Directive   83–4,
    87, 184, 194, 255, 260, 263–4, 267, 299,
    332, 342, 358
  Art. 1   258

  Art. 2   256
  Art. 3(2)   256
  Art. 4   256
  Art. 4(1)   257
  Art. 4(2)   256
  Art. 4(3)   256
  Art. 4(4)   256–7, 259, 263
  Art. 5   258
  Art. 6   258
  Art. 7   258, 263
  Art. 8   258
  Art. 9   258
  Art. 11   258
  Art. 14   258
  Annex I   256–7
  Annex II   258
79/923/EEC Quality of Water for Shellfish
    Directive   274
83/129/EEC Importation of Seal Pup Skins
    Directive   85, 292
85/337/EEC Environmental Impact
    Assessment Directive   86–7, 255,
    275, 337–49
  Art. 1(1)   338
  Art. 1(2)   338
  Art. 1(4)   342
  Art. 1(5)   342
  Art. 2(1)   338
  Art. 2(3)   341
  Art. 3   345
  Art. 4(1)   382
  Art. 7   341
  Annex I   341, 382–3

Annex II   341, 384–5
Annex IV   345
85/374/EEC Product Liability
   Directive   20
91/271/EEC Urban Waste Water Treatment
   Directive   273
   Art. 4(1)   273
92/43/EEC Habitats Directive   29, 83, 85,
   87, 164, 194, 255, 259, 266, 271–2, 299,
   302, 326, 332, 342, 358
   Art. 2   259, 264–5
   Art. 2(1)   260
   Art. 2(2)   260
   Art. 2(3)   260, 341
   Art. 3   259, 264
   Art. 4(1)   261
   Art. 4(2)   261, 266–7
   Art. 4(4)   261, 341
   Art. 5   261, 264
   Art. 5(1)   261, 267
   Art. 5(2)   261, 266
   Art. 5(3)   267
   Art. 6(1)   262
   Art. 6(2)   262–3
   Art. 6(3)   263
   Art. 6(4)   259, 263
   Art. 7   263, 341
   Art. 10   263
   Art. 11   264
   Art. 12   264
   Art. 13   264
   Art. 14   264
   Art. 19   264
   Art. 22(a)   264
   Art. 22(b)   265
   Annex I   260, 341, 386–8
   Annex II   260, 341
   Annex III   260
   Annex IV   264, 269, 289, 380–1
   Annex V   264, 381
   Annex VI   264
97/11/EC Environmental Impact
   (Amendment) Directive   337

Regulations (chronological list)
348/81/EEC Whale and other Cetacean
   Products Importation Regulation
   85, 289, 292

UK law

Acts of Parliament
Acquisition of Land Act 1981
   s. 19   330

797/85/EEC Improving the Efficiency of
   Agriculture Regulation
   Art. 19   224
2328/91/EEC Improving the Efficiency of
   Agriculture Regulation
   Art. 21   224
3254/91/EEC Leghold Traps
   Regulation   292
2078/92/EEC Agri-environment
   Regulation   224
338/97/EC Trade in Endangered Species
   Regulation   20, 85, 275, 289, 292
2087/2001/EC Introduced Species
   Regulation   85

European Community case law
Commission v. United Kingdom (Case C56/
   90) [1993] ECR I-4109   274
Commission v. Germany (Case C-57/89)
   [1991] ECR I-883   259
Commission v. Netherlands (Case C-3/96)
   [1999] Env LR 147   259
Commission v. Spain (Case C-355/90) [1993]
   ECR I-4221   258
Francovich and Others v. *Italian Republic*
   (Cases C-6/90 and C-9/90) [1991]
   ECR I-5357   21
Marleasing SA v. *La Comercial Internacional de
   Alimentación SA* (Case C-106/89)
   [1990] ECR I-4135   21
R v. *Secretary of State for the Environment, ex
   parte Royal Society for the Protection of
   Birds* (Case C-44/95) [1997] QB
   206   258
R v. *Secretary of State for the Environment,
   Transport and the Regions, ex parte First
   Corporate Shipping Limited* (*World
   Wide Fund for Nature UK and Avon
   Wildlife Trust, interveners*) (Case
   C-371/98) *The Times*, 16 November
   2000, Times 2, p. 29   271
Van Gend en Loos v. *Nederlandse Administratie
   der Belastingen* (Case 26/62) [1963]
   ECR 1   20
Von Colson and Kamann v. *Land Nordrhein-
   Westfalen* (Case 14/83) [1984] ECR
   1891   22

Agriculture (Scotland) Act 1948   173
   s. 39   126, 138
Agriculture Act 1947   173
   s. 98   126, 138, 155

Agriculture Act 1986   224
Ancient Monuments and Archaeological
    Areas Act 1979   66, 231
  s. 1   147
  s. 1(11)   177, 229
Animal Health Act 1981   127, 138, 226
Animal Protection Act 1822   83
Channel Tunnel Rail Link Act 1996   342–3
Commons Registration Act 1965   248
  s. 22   203–4
Conservation of Seals Act 1970   11, 157–
    60, 223
  s. 1(1)(a)   159
  s. 1(1)(b)   159
  s. 1(2)   159
  s. 2   160
  s. 2(1)   159
  s. 2(2)   159
  s. 3   160
  s. 3(1)   160
  s. 3(2)   160
  s. 6   159
  s. 8(1)   159
  s. 8(2)   159
  s. 9(1)(a)   160
  s. 9(1)(b)   160
  s. 9(1)(c)   160
  s. 9(2)   160
  s. 10   160
  s. 10(1)(a)   160
  s. 10(3)   160
  s. 10(4)   160
  s. 11   160
  s. 13   160
Consumer Protection Act 1987   20
Countryside Act 1968   200, 211
  s. 15   197, 199
  s. 15A   200
Countryside and Rights of Way Act
    2000   5, 11–12, 62, 83, 88–90, 103–4,
    113, 127, 140, 177, 198, 203–4, 227,
    357, 359
  s. 2(1)   177
  s. 21(5)   177, 229
  s. 26   176, 229–30
  s. 26(1)   177
  s. 26(3)   229
  s. 26(3)(a)   177
  s. 26(3)(b)   177
  s. 26(4)   229
  s. 27(1)   230
  s. 40   176
  s. 73   63
  s. 74   169

  s. 74(1)   298
  s. 74(2)   298
  s. 74(3)   298
  s. 74(7)   298
  s. 75(3)   200
  s. 75(4)   200
  s. 77   278
  s. 80   202
  s. 82   187
  s. 84(4)   187
  s. 86   187
  Sch. 9   195, 202
Cruelty to Animals Act 1876   82
Customs and Excise Management Act 1979
  s. 170(2)   293
Deer Act 1991   154, 173
  s. 1   156
  s. 1(1)   154
  s. 1(2)(a)   154
  s. 1(2)(b)   154
  s. 1(2)(c)   154
  s. 1(3)(a)   154
  s. 1(4)   154
  s. 2(1)   155
  s. 2(3)   155
  s. 3   155
  s. 4   155
  s. 4(1)(a)   155
  s. 4(1)(b)   155
  s. 4(5)   155
  s. 5   155
  s. 6   155
  s. 7   155
  s. 9(2)   155
  s. 10   155–6
  s. 11   155–6
  s. 13(1)   155
  s. 14   156
  s. 16   154
  Sch. 1   156
  Sch. 2   155, 157
  Sch. 3   157
Deer (Scotland) Act 1996   154, 173
Environment Act 1995   64–5, 67
  s. 61   186
  s. 67   187
  s. 69(1)   215
  s. 97   248, 250
Environmental Audit Bill 2002   358
Environmental Protection Act 1990   62
  s. 128   63
  Sch. 9   199
European Communities Act 1972
  s. 2   20

s. 2(2)   20
Flamborough Enclosure Act 1765   253
Forestry Act 1986   240
Game (Scotland) Act 1772   82, 173
Game (Scotland) Act 1832   173
Game Act 1831   173
Game Licences Act 1860   173
Game Preservation Act (Northern Ireland)
    1928   173
Government of Wales Act 1998   3
Ground Game (Amendment) Act
    1906   173
Ground Game Act 1880   173
Harbours Act 1964
    s. 14   330
Hares Act 1848   173
Hares (Scotland) Act 1848   173
Hares Preservation Act 1892   173
Highways Act 1980   245
    s. 41   245
    s. 79   245
    s. 142   245
    s. 154   245
Housing, Town Planning etc. Act 1909   83
Human Rights Act 1998   204, 359
Interpretation Act 1978   xvi
Land Drainage Act 1991
    s. 61C(1)   279
Local Government Act 1972
    s. 101   209
    s. 111   211
    s. 120(1)(b)   211
Magna Carta 1215   81
Marine Wildlife Conservation Bill
    2001   223, 356
National Heritage Act 1983   66
National Parks and Access to the Country-
    side Act 1949   83, 185, 187, 192, 206
    s. 5(1)(a)   186–7
    s. 5(1)(b)   186–7
    s. 11A(2)   187
    s. 15(a)   209
    s. 15(b)   209
    s. 16   197, 206
    s. 17   200, 206
    s. 20   206, 268
    s. 21   209
    s. 21(1)   209
    s. 23   194
    s. 89   241
    s. 114   241
National Parks (Scotland) Act 2000   186
Natural Heritage (Scotland) Act 1991   64,
    189

Night Poaching Act 1828   173
Night Poaching Act 1844   173
Northern Ireland (Elections) Act 1998   3
Pests Act 1954   173
Planning (Listed Buildings and
    Conservation Areas) Act 1990   232
Poaching Prevention Act 1862   173
Protection of Animals Act 1911   173, 178,
    359
    s. 1(1)(a)   179
Protection of Animals (Scotland) Act 1912
    173
Protection of Badgers Act 1992   93, 144,
    175
    s. 1(1)   145
    s. 1(2)   145
    s. 1(3)   145, 145
    s. 1(4)(a)   145
    s. 1(4)(b)   145
    s. 2(1)(a)   145
    s. 2(1)(b)   145
    s. 2(1)(c)   145
    s. 2(1)(d)   145
    s. 2(2)   145
    s. 3   146
    s. 3(a)   146
    s. 3(b)   146
    s. 3(c)   146
    s. 3(d)   146
    s. 3(e)   146
    s. 4   146
    s. 5   147
    s. 6(a)   147
    s. 7(1)   145
    s. 7(2)(a)   145
    s. 8(2)   146
    s. 8(3)   146
    s. 8(4)   148
    s. 8(4–9)   146, 176
    s. 9   146
    s. 10   146–7
    s. 10(1)   147
    s. 10(2)   147
    s. 10(3)   148
    s. 12(1)   148
    s. 12(2)   148
    s. 12(4)   148
    s. 13(1)   148
    s. 14   146
Protection of Wild Mammals (Scotland) Act
    2002   175–6, 179
    s. 1(1)   175
    s. 1(2)   175
    s. 1(3)   175

s. 2   176
s. 3   176
s. 4   176
s. 5(1)(a)   176
s. 5(1)(b)   176
s. 5(1)(c)   176
s. 5(2)(c)   176
Protection of Wrecks Act 1970   232
Public Order Act 1994
  s. 68   230
Salmon Act 1986   174
Salmon and Freshwater Fisheries Act 1975
   174, 220
  s. 4(1)   220
Salmon and Freshwater Fisheries
   (Protection) (Scotland) Act 1951
   174
Salmon Conservation (Scotland) Act 2001
   174
Salmon Fishery Act 1861   82
Scotland Act 1998   3
Sea Fisheries Regulation Act 1966
  s. 20(1)   26
Theft Act 1968   164
Town and Country Planning (Scotland) Act
   1997
  s. 26   147
Town and Country Planning Act 1990
   239, 243
  s. 55(1)   147
  s. 197   241
  s. 197(a)   241
  s. 197(b)   241
  s. 198   241
  s. 211   245
  s. 212   243
  s. 248   330
Transport and Works Act 1992   344
  s. 1   330
  s. 5   330
Water Industry Act 1991
  s. 3(2)(a)   74
  s. 4(1)   279
Water Resources Act 1991
  s. 85(1)   219–20
  s. 90   219
  s. 93   225
  s. 94   225
  s. 104   219
Wild Mammals (Protection) Act 1996   7–8,
   83, 125, 144, 154, 173, 175, 177
  s. 1   178
  s. 2(a)   178
  s. 2(b)   175

s. 2(d)   178
s. 5(2)   179
s. 6   178
Wildlife and Countryside Act 1981   4, 5,
   26, 29, 83, 89–90, 93–4, 113, 135, 144,
   161, 167, 174–5, 184, 194, 223, 238,
   250, 258, 265, 269, 299
s. 1   104–5
s. 1(1)(a)   103
s. 1(1)(b)   103
s. 1(1)(c)   103
s. 1(2)(a)   103
s. 1(2)(b)   103–4
s. 1(3)   104
s. 1(5)(a)   104
s. 1(5)(b)   104
s. 2   104
s. 2(5)   104
s. 3   104–5
s. 3(1)   104
s. 4   104
s. 4(2)(c)   105
s. 5   103, 105
s. 5(1)   105
s. 5(1)(d)   109
s. 5(1)(f)   105
s. 5(5)   105
s. 6   105
s. 7   105
s. 8(1)   106
s. 8(3)(a)   106
s. 8(3)(b)   106
s. 9   126–7, 138
s. 9(1)   125–6, 137
s. 9(2)   125–6, 136–7
s. 9(3)(a)   126
s. 9(3)(b)   126
s. 9(4)   140
s. 9(4)(a)   126
s. 9(4)(b)   126
s. 9(4A)(a)   127
s. 9(4A)(b)   127
s. 9(5)   126
s. 9(5)(a)   126–7, 137
s. 9(5)(b)   127
s. 9(6)   126
s. 9–11   136
s. 10(1)(a)   129, 138
s. 10(2)   126, 137
s. 10(3)(a)   126–7, 137
s. 10(3)(b)   126–7, 137
s. 10(3)(c)   127, 137
s. 10(4)   126, 137
s. 10(5)   125, 137

s. 10(6)   126, 137
s. 11   128
s. 11(1)(d)   130
s. 11(2)(f)   130
s. 13(1)   164
s. 13(1)(a)   163
s. 13(1)(b)   163, 238
s. 13(2)   164
s. 13(2)(a)   163–4
s. 13(2)(b)   163
s. 13(3)   164
s. 13(4)   164
s. 14   168
s. 14(1)   171
s. 14(1)(a)   167
s. 14(1)(b)   167
s. 14(2)   167
s. 14(3)   168
s. 16   103, 106, 109
s. 16(1)–(11)   130
s. 16(4)(c)   168
s. 16(9)   130
s. 18   107, 109
s. 19   107
s. 27(1)   26, 103, 125–6, 163, 238
s. 27(3)   167
s. 27A   62, 394
s. 28   194–5, 197, 203, 267, 359
s. 28(1)(a)   195
s. 28(1)(b)   195
s. 28(1)(c)   195
s. 28(2)   195
s. 28(3)   195
s. 28(4)   198
s. 28(4)(a)   195
s. 28(4)(b)   195
s. 28(5)   195
s. 28(5)(b)   196
s. 28(6)   197
s. 28(6)(b)   195
s. 28A   195
s. 28B   196
s. 28C   196
s. 28D   196
s. 28D(2)   196
s. 28D(3)   196
s. 28D(5)   196
s. 28E   197
s. 28E(1)   200
s. 28E(4)   197
s. 28F   197
s. 28G   197, 201
s. 28G(2)   198
s. 28H   197–8, 201

s. 28H(5)   198
s. 28I   198, 201
s. 28J   197, 200
s. 28J(1)   199
s. 28J(3)   199
s. 28J(6)   199
s. 28J(7)   199
s. 28J(8)   199
s. 28J(9)(b)   199
s. 28J(11)   199
s. 28K   197, 199
s. 28L   199
s. 28M(2)   199
s. 28N   200
s. 28P   200–1, 204
s. 28P(1)   197, 200, 202
s. 28P(2)   202
s. 28P(3)   202
s. 28P(4)   201
s. 28P(5)   201
s. 28P(6)   202
s. 28P(6)(a)   201
s. 28P(6)(b)   201
s. 28P(7)   201
s. 28P(8)   201
s. 28P(10)   201
s. 28Q   201
s. 28R   202
s. 29   200
s. 31   200, 202
s. 34   214
s. 34(1)   214–5
s. 34(2)   215
s. 34(4)(a)   215
s. 34(4)(b)   215
s. 34(5)   215
s. 34(6)   214–5
s. 36   222
s. 36(1)(a)   222
s. 36(1)(b)   222
s. 37   222
s. 37A   278
s. 37A(2)   279
s. 43   187
s. 52(2C)   203
s. 69   204
Sch. 1   103–4, 108, 111, 250, 368–70
Sch. 2   103–4, 108, 370–1
Sch. 3   103, 105–6, 108, 371–2
Sch. 4   65–6, 103, 105–6, 108
Sch. 5   124–7, 135, 140, 174, 250, 289, 300, 372–4
Sch. 6   300, 374
Sch. 8   164, 238, 250, 374–8

Sch. 9   108, 167–8, 378–80
Wildlife and Countryside (Amendment)
    Act 1985   11

Statutory Instruments
Bathing Waters (Classification) Regulations
    1991 (SI 1991/1597)   274
Conservation (Natural Habitats etc.)
    Regulations 1994 (SI 1994/2716)
    83, 93, 130, 136, 164, 215, 258, 265–271
  reg. 2(1)   267
  reg. 5   269
  reg. 7   266–7
  reg. 8(1)   266
  reg. 8(2)   266
  reg. 8(2)(a)   266
  reg. 8(2)(b)   266
  reg. 9   266
  reg. 10(1)(c)   267
  reg. 11   267
  reg. 12   267
  reg. 14   267
  reg. 16(1)   267
  reg. 18   267
  reg. 19   268
  reg. 20(1)   267
  reg. 20(2)   267
  reg. 22   268
  reg. 23   268
  reg. 24   267
  reg. 25   268
  reg. 26(5)   268
  reg. 26(6)   268
  reg. 28   268
  reg. 32   268
  reg. 33   268
  reg. 34   269
  reg. 36   269
  reg. 37   269
  reg. 39(1)(d)   140
  reg. 44   269
  reg. 48   267, 269–70, 330
  reg. 48(1)(a)   270
  reg. 48(1)(b)   270
  reg. 48(2)   269
  reg. 48(3)   269
  reg. 48(4)   269
  reg. 48(5)   269, 270
  reg. 49   270
  reg. 54–85   270
  Sch. 2   269, 289, 380
  Sch. 3   381
  Sch. 4   269, 381
Conservation (Natural Habitats etc.)
    (Amendment) (England)
    Regulations 2000 (SI 2000/192)   267
Control of Trade in Endangered Species
    Enforcement Regulations 1985 (SI
    1985/1155)
  reg. 8(2)   293
Controlled Waters (Lakes and Ponds) Order
    1989 (SI 1989/1149)   219
Electricity Works (Environmental Impact
    Assessment) (England and Wales)
    Regulations 2000 (SI 2000/1927)
    344
Environmental Impact Assessment (Fish
    Farming in Marine Waters)
    Regulations 1999 (SI 1999/367)   344
Environmental Impact Assessment (Fish
    Farming in Marine Waters)
    Regulations (Northern Ireland) 1999
    (SR 1999/415)   345
Environmental Impact Assessment
    (Forestry) Regulations (Northern
    Ireland) 2000 (SR 2000/84)   345
Environmental Impact Assessment
    (Forestry) (England and Wales)
    Regulations 1999 (SI 1999/2228)
    344
Environmental Impact Assessment
    (Forestry) (Scotland) Regulations
    1999 (SSI 1999/43)   344
Environmental Impact Assessment (Land
    Drainage Improvement Works)
    Regulations 1999 (SI 1999/1783)
    344
Environmental Impact Assessment
    (Scotland) Regulations 1999 (SSI
    1999/1)   344
Fawley Branch Line Improvements Order
    (2000)   330
Game Birds Preservation Order (Northern
    Ireland) 1999 (SR 1999/328)   173
Harbour Works (Environmental Impact
    Assessment) Regulations 1999 (SI
    1999/3445)   344
Hedgerows Regulations 1997 (SI 1997/
    1160)   248, 249
  reg. 3(1)   248
  reg. 3(3)   248
  reg. 3(4)   248
  reg. 3(5)   248
  reg. 4   250
  reg. 5(1)(b)(i)   252
  reg. 5(5)   250
  reg. 6   251
  reg. 7(1)   252

reg. 7(2)   252
reg. 7(4)(a)–(g)   251
reg. 8   252
reg. 10   252
reg. 12   252
reg. 13   252
reg. 14   252
Sch. 1   250
Sch. 3   251
Sch. 4   248
Highways (Assessment of Environmental
      Effects) Regulations 1999 (SI 1999/
      369)   343
Nature Conservation and Amenity Lands
      (Northern Ireland) Order 1985 (SR
      1985/170 (NI 1))   299
Nuclear Reactors (Environmental Impact
      Assessment for Decommissioning)
      Regulations 1999 (SI 1999/2892)
      344
Offshore Petroleum Production and Pipe-
      lines (Assessment of Environmental
      Effects) Regulations 1999 (SI 1999/
      360)   344
Pipe-line Works (Environmental Impact
      Assessment) Regulations 2000 (SI
      2000/1928)   343
Planning (Environmental Impact
      Assessment) Regulations (Northern
      Ireland) 1999 (SR 1999/73)   344
Port of Southampton (Dibden Terminal)
      Harbour Revision Order (2000)
      330
Public Gas Transporter Pipe-line Works
      (Environmental Impact Assessment)
      Regulations 1999 (SI 1999/1672)
      343
Roads (Environmental Impact Assessment)
      Regulations (Northern Ireland) 1999
      (SR 1999/89)   345
Spring Traps (Approval) Order 1995 (SI
      1995/2427)   173
Spring Traps (Scotland) (Approval) Order
      1996 (SI 1996/2202 (S 178))   173
Surface Waters (Shellfish) (Classification)
      Regulations 1997 (1997/1332)   274
Town and Country Planning
      (Environmental Impact Assessment)
      (England and Wales) Regulations
      1999 (SI 1999/293)   326–7, 337–42,
      346–9, 389
   reg. 2(1)   338–9, 346
   reg. 3(2)   338–9
   reg. 4(4)   341

reg. 5   339
reg. 6   339
reg. 6(7)   339
reg. 10   339
reg. 11   339
reg. 13   340
reg. 15   339
reg. 16   340
reg. 17   339
reg. 19   339
reg. 20   340
reg. 21   340
reg. 23   341
reg. 24   341
reg. 25   341
reg. 27   341
reg. 28   341
Sch. 1   338, 341–2, 351, 382–3
Sch. 2   338, 341–2, 384–5
Sch. 3   341, 342
Sch. 4   346–7
Town and Country Planning
      (Environmental Impact Assessment)
      (England and Wales) (Amendment)
      Regulations 2000 (SI 2000/2867)
      337
Town and Country Planning (General
      Development Procedure) Order 1995
      (SI 1995/419)   326
Town and Country Planning (Trees)
      Regulations 1999 (SI 1999/1892)
      238
   reg. 2(1)(b)   238
   reg. 2(2)   239
   reg. 2(3)   239
   Schedule
      Art. 5   239
Transport and Works (Applications and
      Objections Procedure) (England and
      Wales) Rules 2000 (SI 2000/2190)
      344
Urban Waste Water Treatment (England
      and Wales) Regulations 1994 (SI
      1994/2841)   273
Urban Waste Water Treatment (Scotland)
      Regulations 1994 (SI 1994/2842)
      273
Welfare of Animals (Transport) Order 1997
      (1997/1480)   173
Wildlife (Northern Ireland) Order 1985 (SR
      1985/171 (NI 2))   103, 144, 154, 174,
      258, 265, 289
   Sch. 5   300
   Sch. 6   300

Wildlife and Countryside (Registration and Ringing of Captive Birds) Regulations 1982 (SI 1982/1221) 105
  reg. 3(1)   105
  reg. 5(1)   106

UK case law
Bullock v. *Secretary of State for the Environment* (1980) 40 P&CR 246 241
Caygill v. *Thwaite* (1885) 49 JP 614   26
Delaware Mansions Ltd and Another v. *Westminster City Council* (2001) House of Lords *The Times*, 26 October 2001, Times 2, p. 22   240
Duke v. *GEC Reliance Ltd* [1988] AC 618   21
David Green, Ian Peter Reynolds, David Rowbotham and Martin George Trench v. *Stipendary Magistrate for the County of Lincolnshire* [2001] Env LR 295 148
Holbeck Hall Hotel v. *Scarborough Borough Council,* (2000) *The Times*, 23 February 2000, p. 4   191
Lafarge Redland Aggregates Ltd v. *The Scottish Ministers* [2001] Court of Session, Outer House Env LR 507   311
Marshall v. *Southampton & South West Hampshire Area Health Authority (Teaching)* [1986] QB 401   21
R v. *English Nature and Secretary of State for the Environment, Food and Rural Affairs, ex parte Aggregate Industries UK Ltd* [2002] EWHC 908   359
R v. *Jeans* (1844) 1 Car & Kir 539   109
R v. *Poole Borough Council, ex parte Beebee* [1991] JPL 643   16, 204
R. v. *Secretary of State for Scotland & Others, ex parte World Wildlife Fund UK Ltd, and Royal Society for the Protection of Birds*

(1998) Court of Session, Outer House LTL 20 November 1998 Case Law C7600057   271
R v. *Secretary of State for Transport, ex parte Berkshire, Buckinghamshire and Oxfordshire Naturalists Trust* [1997] Env LR 80   271
R v. *Secretary of State for Trade and Industry, ex parte Greenpeace (No. 2)* [2000] Env LR 221   272
R. v. *Sefton Metropolitan Borough Council, ex parte British Association of Shooting & Conservation Ltd* [2001] Queen's Bench Division Env LR 182   211
Raymond Holden v. *Lancaster Justices* (1998) COD 429   109
Royal Society for the Protection of Birds and the Wildfowl & Wetlands Trust Ltd (Petitioners) v. *The Secretary of State for Scotland (Respondent)* [2001] Court of Session, Inner House Env LR 356   109
Royal Society for the Prevention of Cruelty to Animals v. *Craig Cundey* (2001) LTL 27 November 2001 Case Law C0102240   109
Seymour v. *Flamborough Parish Council* (1997) *The Times*, 3 January 1997   253
Southern Water Authority v. *NCC* [1992] 3 All ER 481   203
Tilly v. *DPP, DPP* v. *Tilly and Others* (2001) Queen's Bench Division *The Times*, 27 November, *Law Supplement*, p. 10 230
Stuart Bandeira and Darren Brannigan v. *Royal Society for the Prevention of Cruelty to Animals* (2000) LTL 28 February 2000 Case Law C9300302   179
Sweet v. *Secretary of State and Nature Conservancy Council* [1989] JEL 245 202

Index

access to land
  conservation purposes, for, 176–7
  hedgerows, for protection of, 252
  restriction of, 177, 227, 229
  SSSIs and, 202
Acts of Parliament see law sources
  Government Bill, 5
  Private Member's Bill, 5, 83, 356
  Royal Assent, 5
  structure of, 5–8
afforestation, 289
Agenda 21, 294, 298, 301, 304, 306–7
agriculture
  disease control, 225–8
  wildlife and, 223–30
agri-environment schemes, 224–5
air pollution see pollution
airport
  Heathrow, 357
    CITES Enforcement Team, 292
  Manchester, 140–1, 149, 351–3
amphibians see toad, wild animals,
  classification of, 120
  conservation see Froglife
  cruelty to, 179
  natural history, 121
  protection during development, 326, 353–4
  SSSI and, 16, 204
  toad crossings see roads
ancient monuments, 229, 231
  caves as see caves
Ancient Monuments Board for Scotland, 66

Ancient Monuments Board for Wales, 67
animals see wild animals
  classification of, 30
archaeological sites see World Heritage Sites
  access to, 229
  EIA and, 348
  hedgerows and, 250
  protection during development, 353, 355
  scheduling of, 231
Areas of Archaeological Importance (AAIs), 231
Areas of Outstanding Natural Beauty (AONBs), 83, 89, 111, 187–9
  conservation boards, 187
  locations of, 188
  management plans, 187
Areas of Special Scientific Interest (ASSIs), 194
Asiantaeth yr Amgylchedd (Welsh Environment Agency), 67
Associated British Ports see Dibden Container Terminal Inquiry
Attorney-General, 13–4, 16

badger groups, 75, 77, 141, 150–1
badgers, 141–51
  agriculture, protection of, 146
  baiting of, 141, 151
  bovine TB and, 141–2
  breeding season, 144
  culling of, 141
  digging, 141, 145, 148–51
  disabled, 146–7
  fences, 149, 316

gates, 149,
  humane killing of, 147
  legal protection, 144–8, 150
  marking of, 147
  natural history, 141–4
  property, protection of, 146
  road deaths, 141, 319, 321–2
  sale, illegal, 146
  setts, 143–4, 179
    ancient monuments and, 147
    artificial, 149–50, 343, 353
    capping of, 148–9
    fox hunting and, 146, 148, 150
    legal definition, 146, 148
    planning permission and, 146
    size of, 143
    signs, 144
    tunnels, 315, 317
barn owl
  barn conversions and, 324
  captive breeding of, 78
  decline of, 95
  licences to release, 65
  nest boxes for, 110
barristers, 13–4
bathing waters, 273–4
bat boxes, 138–9, 326
bat bricks, 138–9
Bat Conservation Trust, 76–7
bats, 132–41
  agriculture, protection of, 137–8
  attracting, 138–9
  barns, specially constructed for, 353
  breeding, 135
  classification of, 133

decline in numbers,
    reasons for, 132–3
disabled, 137
hibernation, 134–5
humane killing of, 137
international law and, 286
legal protection of, 135–8,
    140
monitoring programme,
    national, 76
natural history, 133–5
new species, 29, 133
property, protection of,
    137
protection during
    development, 137
reckless disturbance of,
    140
removal from dwelling,
    137
roosting places in
    buildings, 132–3, 140,
    324, 334
wood treatments, 140
battlefield sites, 232
beaches *see* sand dunes,
    bathing waters
beavers, re–introduction of,
    170–1
binomial system of
    nomenclature, 26–9
biodiversity
    ecosystems and, 33, 36–8
    legal definition of, 294–5
    moths, 37
    recording of, 296
    sustainable use of, 295
    UK, 37
    woodland, 39
Biodiversity Action Plans,
    68, 89, 169, 294, 298–9,
    314, 324
    Habitat Action Plans, 299,
        301–6, 325
    Local Action Plan, Bolton,
        305–6
    otter, 301
    red squirrel, 300–1
    sand dunes, 302–4
    secretariat, 61
    Species Action Plans,
        299–301, 305, 325
    trunk roads in Scotland,
        319–20

Biodiversity Group, UK, 301
Biodiversity Partnership,
    Cambridgeshire's,
    324–8
Biogenetic Reserves, 277
biological oxygen demand
    (BOD), 225
Biological Records Centre, 48
Biosphere Reserves, 275–6
    examples of
        Braunton Burrows, 276
        Caerlaverock, 276
    function of, 275–6
birds, *see* barn owl, black
    redstart, Canada
    goose, red kite,
    sparrow, starling,
    93–113
    agriculture, protection of,
        104, 106, 109
    areas of special protection,
        104
    buildings, exclusion from,
        98, 334–6
    cages and, 106–7
    classification of, 98–100
    conservation
        organisations *see*
        BTO, RSPB, WWT
    decline in numbers, 95–8
    disabled, 104
    eggs
        clutch size, 101
        collectors of, 107, 113
        deduction in shell
            thickness, 50
        laying times, 366–7
        legal protection of,
            103–109, 255–9, 288
        'Operation Easter' 113
    humane killing of, 104
    legal protection of, 103–13,
        255–9
    history of, 82
    scheduled species
        (WCA 1981), 368–372
    maiming of, 109
    monitoring of population
        changes, 95–6
    natural history of, 100–2
    nests
        disturbance of, 104
        legal protection of,
            103–6, 256, 258

nest boxes, 109–10, 131
nesting behaviour, 101–2
nesting times, 366–7
of prey
    monitoring of, 176–7,
        202
    nest boxes and nesting
        baskets for, 109–10
    poisoning of, 50, 94
    re–introduction of *see*
        red kite
    shooting of, 107–8
    trade in, 290
    trapping of, 112
road deaths of, 318
registration of captive, 105
ringing of, 105
sale of, 105–6
trade in endangered
    species, 290, 292–3
waterfowl *see* Ramsar
    sites, Wildfowl and
    Wetlands Trust
black redstart, 110–11
Botanical Society of the
    British Isles, 48
British Geological Survey, 71
British Herpetological
    Society, 16
British Rail, 164–5
British Trust for
    Ornithology, 76
British Waterways, 141, 221
Broads Authority, 186
brownfield development, 88,
    313–4
building
    design, effect on wildlife,
        334–6
    exclusion of birds *see* birds
    house longhorn, damage
        caused by, 335
    listed, 232
    regulations, 334–5
bulb rustlers, 161
butterflies, 124, 130, 290
butterfly conservation, 77–8

Cadw, 67
Cairngorms, 186, 271
Canada goose, 167–8
canals
    as controlled waters, 219
    wildlife and, 221

case law *see* law sources
caves
  as ancient monuments,
    216–7
  conservation plans for, 216
  register of, 216
  sea, 265–6
Centre for Ecology and
    Hydrology, 70
cetaceans *see* whaling, 266,
    286, 288–9
  reckless disturbance of,
    127
  strandings of, 127, 129
Channel Tunnel Rail Link,
    342–3, 357
Civic Trees, 246–7
civil offences, 4, 15
classification of living
    things, 26–31
cliffs, 190–1
coastal erosion, 189–91
coastlines *see* bathing waters,
    Heritage Coasts,
Common Agricultural
    Policy, 60, 86, 358
Common Fisheries Policy,
    60, 86
conservation *see* grassland
    management
  access to land for *see*
    access to land
  justification for, 48–50
  management, 53–6
conservation areas, 231–2,
    245
conservation boards *see*
    AONBs
controlled waters
  definition of, 219
  pollution of, 219–20, 225
Council for the Protection of
    Rural England
    (CPRE), 80, 357
Council of Europe, 277
Council of the European
    Union, 17–9
counter pollution and
    salvage officers, 69
country parks, 211
Countryside Agency, 60, 62,
    186, 189, 193, 208, 214,
    230, 234
  functions, 63

Countryside Commission,
    63, 186
Countryside Council for
    Wales (CWW), 62,
    186, 189
  EIA and, 339
  functions, 63–4
County Recorder, 48
courts, 4, 8–9, 14–5
criminal offences, 4, 14–5
cruelty to animals, 82–3,
    177–9, 359
Customs and Excise, HM *see*
    trade in endangered
    species, 75
Cyngor Cefn Gwlad Cymru
    see Countryside
    Council for Wales

DDT in food chains, 50
deer, 80–1, 152–7
  agriculture, protection of,
    155
  classification of, 151
  close seasons, 155–6
  distribution of, 152–3
  fencing, 316
  humane killing of, 155
  injured, 155
  legal definition of, 154
  legal protection of, 154–7
  natural history of, 152–4
  offences committed by
    corporate bodies, 156
  prohibition of certain
    methods of killing,
    155
  property, protection of,
    155
  red, 152–4, 156
  rutting season of, 154
  venison, illegal sale of,
    155–6
Deer Commission for
    Scotland, 71, 157
Department for
    Environment, Food
    and Rural Affairs
    (DEFRA), 59–62, 67,
    75, 113, 223, 226
Department for Transport,
    59, 69
Department for Transport,
    Local Government

and the Regions
    (DTLR), 59
Department of Agriculture
    and Rural
    Development for
    Northern Ireland, 71,
    208
Department of
    Environment,
    Transport and the
    Regions (DETR), 59
Department of the
    Environment for
    Northern Ireland, 64
  functions of, 64
Deptford pink, 164–5
development control *see*
    planning
devolution of legal powers, 3
Dibden Container Terminal
    Inquiry, 329–34
Director of Public
    Prosecutions, 13–4,
    201
disease
  bovine tuberculosis, 141–2
  BSE, 34
  CJD, 34
  control of *see* agriculture
  food chains and, 34
  foot-and-mouth, 9, 141,
    223, 226–8, 358
    effect on biodiversity,
    228
  rabies, 173
dolphins *see* cetaceans
dwellings, wild animals
    within *see* bats,
    buildings

ecological networks *see*
    Natura 2000, Econet,
    Life
ecological surveys *see*
    National Vegetation
    Classification, habitat
    survey, 44–8
  mark–recapture
    techniques, 45–6
  transects, 44–5
  traps, 46–7
    pitfall, 46–7
    Longworth small
    mammal, 46

quadrat, 44–5
vegetation classification
    see National
    Vegetation
    Classification, habitat
    survey
ecology *see* ecosystem
    definition of, 33
Econet, *Life*, 328–9
ecosystem, 33–43
    carrying capacity, 38–9, 41
    characteristics of, 33
    climax community, 41–2
    energy flow in, 33–4
    food chains, 33–4
    food web, 34
    management *see*
        conservation
    nutrient cycles in, 35–6
    succession, 40–2
    trophic levels, 33–4
        consumers, 33–4
        decomposers, 33, 35
        primary producers, 33–4
        pyramid of numbers, 34
eggs *see* birds
England Rural Development
    Programme (ERDP),
    89, 223–4
English China Clay
    International, 323
English Heritage
    functions of, 66
English Nature, 60, 62, 191–2
    EIA and, 339
    functions, 63
    SSSIs and, 16
Environment Agency, 60, 71,
    218–20, 238, 269, 274
    EIA and, 339
    nature conservation role
        of, 68
    functions of, 67–8
    SSSIs and, 196
Environment Action
    Programme, EU
    Fifth, 84
    Sixth, 85–7
Environment and Heritage
    Service (Northern
    Ireland), 62, 67, 211
    functions of, 64
environmental audit, 358
environmental impact

assessment, 275,
    337–55
consultation with
    environmental
    bodies, 339
developer, definition of,
    338
development consent,
    definition of, 338
enterprise zone orders
    and, 341
environmental statement,
    339–40, 353
    impacts which must be
        assessed, 345–9
    mitigating measures,
        349, 351,
    public assess to, 339
member states, effects of
    project on, 341
nature conservation and,
    349–351
planning register,
    documents to be
        placed on, 340
project, definition of, 338
projects
    exemptions from EIA,
        341–2
    not approved under the
        planning system,
        342–4
    requiring EIA, 338, 341,
        382–5
    SACs, 263, 269–70
    Scotland and Northern
        Ireland, 344–5
scoping
    directions, 339–40
    opinions, 339–40
screening
    directions, 339–40
    opinions, 339–40
simplified planning zones
    and, 341
Environmentally Sensitive
    Areas (ESAs), 224
European Commission,
    17–20
European Court of Justice *see*
    European law
European Environment
    Agency, 84
European law

decisions, 20
development of, 83–7
direct effect, 20–1
indirect effect, 20–2
European Directive,
    18–22, 84–87
European Regulation,
    18–20, 85
European Court of Justice,
    17, 19–22
Francovich liability, 21
nature conservation and,
    255–75
supremacy of, 20–2
European Marine Site, 267
byelaws, 269
definition of, 267
management schemes for,
    269
markers, 268
European Parliament, 17–19
European protected species
    see wild animals, wild
    plants
European site, 258, 266
byelaws, 268, 271
compulsory purchase of,
    268
damaging operation
    within, 268, 270
definition of, 266–7
management agreement
    for, 267–8
planning controls, 269–70
registers of, 267
restoration order, 268
special nature
    conservation order
    for, 268
European Union, 16–23
history of, 16–7
institutions of, 17–9
exotic species *see* introduced
    species

farming *see* agriculture
fences *see* badgers, deer
feral cats, 335
fish *see* wild animals
    classification of, 121, 123
    historical protection of,
        81–2
    ladders, 123
    legislation, 173–4, 272–3

natural history of, 123
pollution of waters, 220
fisheries, damage by birds, 113
Flora Locale, 254
Forest Authority, 71
Forest, Community, 208, 234
Forest Enterprise, 70, 207
Forest, National, 208, 234
Forest Nature Reserves, 71, 207–8
Forest Research, 71
Forest Reserves, Caledonian, 207
Forest Service (Northern Ireland), 71
forests, as carbon sinks *see* trees
Forestry Commission, 70–1, 207–8, 234, 245
foxes *see* badgers
hunting of, 175–6, 178, 359
urban, 174–5
freshwater habitats, 218–20
legal protection of, 219–220, 272–3
Friends of the Earth, 81, 357
Froglife, 77
fungi *see* mushrooms, 161
Fungus Conservation Forum, 77

game, 81–2, 173
gamekeepers, unlawful actions of, 107–8, 141, 150
genetically modified organisms, 230, 296
Geological Sites, Regionally Important (RIGS), 215–216
geological structures, protection of, 212–17
global warming, 222, 289
Global Wildlife Division of DEFRA, 61, 65
government policy *see* Green Papers, White Papers, 87–90, 313, 364–5
grassland management, 320–1
Green Papers, 4, 13, 364–5
Greenpeace, 16, 81
legal cases involving, 272

Habitat Action Plans *see* Biodiversity Action Plans
Habitat Survey, Phase 1, 389–90
Habitats of Community Interest (Habitats Directive, Annex I), 386–8
Hadrian's Wall, 282–3
Hansard, 5, 13
Hanson Quarry Products, 205
Hawk and Owl Trust, 109
heathland, damaged by arson, 206–7
Hebridean Mink Project, 172
hedgerows, 247–54
aging of, 247
cutting of, 248
gaps in, 248–9
habitats, 247–8
'important', 250–2
definition of, 249–51
permitted work, 251
loss of, 247
native plant species, 253–4
natural history of, 247–8
offences, 252
planting new, 253–4
protection using Enclosure Acts, 252–3
Red Data Book species in, 251
records held by local authority, 252
removal notice, 248, 250
retention notice, 74, 250
species associated with, 247–8
Her Majesty's Inspectorate of Pollution (HMIP), 68
Heritage Coasts, 189–90
heritage conservation *see* archaeological sites, Areas of Archaeological Importance, Cadw, English Heritage, Environment and Heritage Service (Northern Ireland), Historic Scotland, World Heritage Sites

highways *see* highways Agency, roads
Highways Agency, 69, 245, 314–5
Environmental Committee of, 314
heritage features and, 315
Historic Buildings Council for Northern Ireland, 67
for Scotland, 66–7
for Wales, 67
Historic Monuments Council for Northern Ireland, 67
Historic Scotland, 66
history of nature conservation law, 81–3, 361–3
Hudson, W.H., 95
human rights, 204, 359
hunting *see* game, deer
fox *see* fox hunting
with dogs, 175–6, 178

identification of species, 29–32,
guides to, 58
services, 73
insects, 124–5
attracting, 130
classification of, 124–5
international law, 22–24
Conference of the Parties, 23
EU and, 22–3, 87
nature conservation and, 275–307
principles of, 23–4
treaties, 87
introduced species, 40, 112–3, 115, 165–8
law and, 167–8, 258, 285, 288, 296
scheduled species (WCA 1981), 378–80
invertebrates *see* snails, insects, butterflies
legal protection of *see* wild animals
soil, 43, 46
International Union for the Conservation of Nature and natural Resources (IUCN), 277, 290

Joint Nature Conservation
    Committee (JNCC),
    61, 79
  functions of, 64–5
judicial review, 9, 204, 271–2

Karst landscapes, protection
    of *see* Limestone
    Pavement Orders,
    216–7
lakes *see* freshwater habitats
Lancashire Biodiversity
    Action Partnership,
    48–9
landscape *see* Areas of
    Outstanding Natural
    Beauty, Econet,
    Heritage Coasts,
    National Parks,
    Natural Areas, 217,
  historic, 232
  management
    highways, 319
    in SACs, 263, 269
Lappel Bank, 258
law reports, 4, 8–10
  judgement, 8–9
  obiter dicta, 8
  precedent, 4, 8–9
  ratio decidendi, 8–9
law sources
  citation of, 10–1
  European
    Directives, 22
    case law, 22
    Regulations, 22
  international
    treaties, 24
  UK
    Acts of Parliament, 4–6,
      10–13
    case law, 4, 9–10
    Northern Ireland
      Orders in Council, 5
    Statutory Instruments,
      4–6, 12
legal personality, 15–6
legal profession, 13–4
Leybucht Dykes, 259
licences, 93, 174, 400
  animals, 130
  badgers, 147–8
  bats, 137, 140
  birds, 105–6, 109, 130

general licences, 106–7
disease control, 227
endangered species
    (CITES), for trade in,
    291, 293
European protected
    species, 269
introduced species, 168
oil exploration, 272
plants, 130,164,
seals, 160
tree felling, 239–40
  exemptions, 239–40
vegetation in the highway,
    245
Wildlife Inspectorate,
    powers, 65–6
lichens, 38, 40
  as air pollution monitors,
    51–2
Life programme, EU *see*
    Econet, 84, 86, 259, 302
limestone pavements, 213–5
  legal definition of, 214
  Limestone Pavement
    Order, 74, 214–5
  planning permission
    and, 214
Lincoln Index, 45–6
Linnaeus, Carl, 26
Local Action Plan *see*
    Biodiversity Action
    Plans
local authorities, 88, 189
  nature conservation
    responsibilities of, 74,
    209–11
Local Land Charges Register,
    74, 245, 267–8
locus standi, 15–16
  Attorney–General, 16
  British Herpetological
    Society, 16
  Greenpeace, 16
  WWF, 16
Lord Chancellor's Panel, 15

Mammal Society, 48, 76
mammals *see* badgers, bats,
    beavers, cetaceans,
    deer, feral cats, mink,
    otter, seals, squirrel,
    voles, wild animals,
    wildcat, wolves

classification of, 115–6
conservation
    organisations *see*
    Mammal Society
  cruelty to *see* cruelty
  extinct species, 115
  natural history of, 117–8
  Scotland, legal protection
    in, 175–6
man and the Biosphere
    Programme, 275
marine animals *see*
    cetaceans, seals and
    Marine SACs
  intentional or reckless
    disturbance of, 127
  protection under Habitats
    Directive *see* Marine
    Special Areas of
    Conservation 272
Marine Conservation
    Society, 79
Marine Sites of Special
    Interest (MSSIs),
    356–7
Marine Special Areas of
    Conservation
    (Habitats Directive),
    265–6
Marine Stewardship Report,
    358
Marismas de Santoña, 258
Maritime and Coastguard
    Agency, 69
Medway Estuary and
    Marshes, 258
microorganisms
  classification of, 31
  role in decomposition, 35
migratory species, 55, 284–6,
    287
Millennium Dome, 110–11
mineral extraction, 205
Mineral Policy Guidance
    Notes, 313
Ministry of Agriculture,
    Fisheries and Food *see*
    DEFRA, 59, 150, 223
Ministry of Defence (MOD),
    184–5
mink, 171–172
mobile phone mast,
    planning permission
    for, 312

Museum, Natural History,
32, 36, 72–3, 254
museums, 29, 32, 36
mushrooms *see* fungi, 164

National Anti-Vivisection
Society, 82
National Assembly for
Wales, 3
National Biodiversity
Network, 48
National Cave, Karst and
Mine Register, 216–7
National Caving
Association, 216
National Federation of
Badger Groups, 77,
141
National Forest Company,
61
National Geosciences
Information Service
(NGIS), 71
National Inventory of
Woodlands and
Trees, 70
National Land Use Database
(NLUD), 313–4
National Monuments
Record
of Scotland, 67
of Wales, 67
National Parks, 64, 83, 89,
185–7
locations of, 186
maps of, 72, 187
purpose of, 186
National Parks Authorities,
65, 186
as planning authority, 187
duty to prepare maps, 187
National Parks Commission,
186
National Rivers Authority,
67
National Scenic Areas, 188–9
National Trails, 193
National Trust, 79–80
National Vegetation
Classification (NVC),
389–90
National Wildflower Centre,
164
National Wildlife Crime

Intelligence Unit
(NWCI), 75–6, 292
Natura 2000, 61, 86, 259, 266,
272
Natural Areas, 191–2
Urban Mersey Basin, 192
Natural Environment
Research Council, 69–
71, 160–1, 194
Natural Heritage Areas, 189
natural history societies, 32
Nature Conservancy, 83
Nature Conservancy
Councils *see* English
Nature, Countryside
Council for Wales,
and Scottish Natural
Heritage
nature reserves, 77–80
design of, 55–6
legal definition of, 209
management of, *see*
conservation, 183
private, 212
Nature Reserve
Local (LNR), 74, 208–10
Local Authority (LANR),
74, 210
Marine (MNR), 222–3
National (NNR), 206–8
compulsory purchase,
206
byelaws, 206
Gibraltar Point, 53–4
nature reserve
agreement, 206
Rostherne Mere, 280–1
Roadside, 321
nests *see* birds
Newbury Bypass, 271
Nitrate Sensitive Areas
(NSAs), 225
North West Water *see* United
Utilities
Northern Ireland Assembly,
3

Occupier *see* SSSI,
orchid, military, 165
Ordnance Survey, 72
otter, 73, 218
natural history, 297
road deaths, 319, 322
road signs, 318

parks, 56–7, 88
birds in, 57
Historic, 232
Partnership for Action
Against Wildlife
Crime (PAW), 75, 292
pests, 74, 97, 107, 115,154,
166, 218, 334–6
pipe-lines *see* Transco
planning *see* Biodiversity
Partnership,
Cambridgeshire's,
Environmental
Impact assessment,
311–4
authorities, 311
development plans, 311
inquiries *see* Dibden
Container Terminal
Inquiry, 15
permission *see* SSSIs
conditions attached to,
311, 328
Secretary of State, role of,
312
Planning Inspectorate, 312,
330
planning inspectors, 15, 330,
339
Planning Policy Guidance
Notes, 190, 312–3
Plantlife, 77
Reserves, 77
plants *see* wild plants
classification of, 30–31, 162
poisoning of wildlife *see*
birds of prey, 179
police *see* PAW, NWCI,
Wildlife Crime Unit,
74–77, 113, 150–1, 226
Policy Commission on the
Future of Farming
and Food, 358
'polluter pays' principle, 84
pollution
air, 51, 235
acid rain, 52–3
agricultural *see*
insecticides, manure,
224
biological indicators of,
49–53
biomagnification of, 33–4,
94

insecticides, 50
water, 51, 218–20
fish and, 123, 220
manure, 51, 225
sewage, 220, 273
shopping trolleys, 220
ponds, 53, 218–9
population growth, 38–41
population sizes of
bats, 134
carnivores, 117
deer, 153
estimation of, 45–6
insectivores, 114
lagomorphs, 114
natterjack toad, 122
rodents, 114
seals, 157
priority habitats (Habitats
Directive), 259, 267,
306, 386–8
priority species (Habitats
Directive), 259, 267
protected habitats *see*
AONBs, National
Parks, priority
habitats, Limestone
Pavement Orders,
MNRs, NNRs,
Ramsar sites, SACs,
SPAs, SSSIs
protected species *see* birds,
wild animals, wild
plants

quarries, 205, 210, 313
quarry species, 173
Queen's Council (QC), 13–4

rabbit warren, 217–8
railways *see* British Rail
funicular, 271
Ramsar sites, 277–81
examples of
Chesil Beach and the
Fleet, 281
Claish Moss, 281
Rostherne Mere, 280–1
Rutland Water, 280
Solent and Southampton
Water, 281
List of Wetlands of
International
Importance, 277

notification of, 278–9
selection criteria for, 279
sites, number in UK, 280
waterfowl, legal definition
of, 278
wetlands, legal definition
of, 278
red kite, 170
re-introduction of species,
169–73, 264, 287–8
reptiles *see* turtles, wild
animals
classification of, 118–9
natural history of, 119–20
restoration
of derelict land
powers of local
authorities, 241
of habitats, 53–4, 164,
169–70, 212, 256, 260,
267–8, 285, 296–7, 355
effect of CRoW 2000 on,
199, 230, 298
of species *see* re-
introduction of
species
rhododendron, 166–7
rivers *see* freshwater habitats
diversion, 353
pollution of *see* pollution,
water
Rivers Agency (Northern
Ireland), 218
roads *see* Highways Agency
*Design Manual for Roads
and Bridges*, 319
roadkills, prevention of
315–9
bridges, 317, 322
culverts, 322
fences, 316
reflecting devices, 317–8
signs, 315, 318–9
toads crossings, 318, 322
tunnels, 316–7
trees and *see* trees
trunk roads, approval of,
315–6
roadside verges,
management of, 320–1
Royal Commission
on Environmental
Pollution, 60
on the Ancient and

Historical Buildings
of Scotland, 66
on the Ancient and
Historical
Monuments of Wales,
67
on the Historical
Monuments of
England, 66
Royal Society for the
Prevention of Cruelty
to Animals (RSPCA),
77, 80–82, 106, 179
legal cases involving, 109,
150, 293, 293
Royal Society for the
Protection of Birds
(RSPB), 75, 78–9, 82,
113
legal cases involving, 108,
150, 258, 271
Rural Development
Commission, 63

salmon, 81, 123, 225, 272
sand dunes *see* Biodiversity
Action Plans
Scheduled Ancient
Monuments (SAMs),
231
scientific names, 25–9
Scott, Sir Peter, 95
Scottish Environment
Protection Agency
(SEPA), 67, 218, 225
Scottish Executive, 3
Scottish Fisheries Protection
Agency, 69
Scottish Natural Heritage
(SNH), 62, 173, 186
functions, 64
sea bed, damage or
disturbance of, 222
seals, 157–161
close seasons, 159
fisheries, protection of, 160
legal protection of, 158–61,
265–6
natural history of, 157–8
products, importation of,
292
prohibited methods of
killing, 159–60
Severn Estuary, 272

shark, basking, 127–8
  reckless disturbance of,
    127
shellfish, 274–5
Sites and Monuments
    Record (SMR), 231
Sites of Special Scientific
    Interest (SSSIs), 74,
    78, 83, 87–8, 193–206
  Brock's Farm, 323
  byelaws, 202
  compulsory purchase of
    land, 199–200
  damaging operations
    affecting, 195, 197–8,
    202–3
  denotification, 196
  designation of, 194–7
  history of, 193–4
  human rights challenges
    to, 359
  Huxley Committee, 193
  intentional or reckless
    damage to, 201
  judicial review, 204
  local land charge, 195
  low water mark and
    designation of, 196
  management notice, 199,
    201
  management schemes and
    agreements, 198–200
  Nob End, 194
  notification of, 194–7
  number of sites, 194
  occupiers of, 4, 203–4
  offences, 200–2
  planning permission, 201
  political interference, 204
  powers of entry onto land,
    202
  public bodies, duties of,
    197–8
  restoration order, 200, 202
  review team, 61
  statutory conservation
    agencies and, 195–205
snails, 271
soil *see* translocation, 35–6,
    42–3
  compaction, 43
  nitrogen-fixing bacteria,
    35–6
  organisms, 35–6, 43

seed bank, 42
  structure, 43
solicitors, 13–14
sparrow, decline of, 97–8
Special Areas of
    Conservation (SACs)
    (Habitats Directive),
    206, 259–272
  case law, 271–2
  boundaries, discretion
    in drawing of, 271
  economic
    considerations, 271–2
  frustration of future
    decision to designate,
    271
  territorial waters,
    application beyond,
    272
  development projects,
    assessment of, 263
  draft lists, notification to
    Commission, 261,
    264, 267
  marine *see* Marine Special
    Areas of Conservation
  numbers of cSACs in UK,
    260
Special Protection Areas
    (SPAs) (Birds
    Directive), 256–9, 267
  case law, 109, 258–9
  duty to designate, 258
  economic
    considerations, 258–9
  sufficient sites, 259
  Natura 2000 and, 259
  numbers in UK, 257
  pollution, protection from,
    256–7
  presumption against
    development, 258–9,
    263
  reduction in area of, 259
  wetlands and, 256
Species Action Plans *see*
    Biodiversity Action
    Plans
species diversity *see*
    biodiversity
squirrel
  grey, 40, 166–7
  red, 40, 73, 166
    re-introduction of, 170

Species Action Plan for
    see Biodiversity
    Action Plans
starling, 39, 97–9
Statutory Instrument *see* law
    sources
sustainability indicators, 307
sustainable development,
    296, 301

Tarmac Quarry Products,
    246, 253
Tesco, 220
toads *see* roads
  natterjack, 53, 79, 122, 299
trade in endangered species,
    358
  Berne Convention and,
    288
  bushmeat, sale of, 293
  cetaceans, 292
  EU and, 275, 292
  extent of, 290
  HM Customs and Excise
    seizures, 290, 292–3
  'Operation Retort', 293
  specimen, legal definition
    of, 292
TRAFFIC, 290
Transco, 331, 353–5
translocation,
  ancient woodland, 323
  buildings, 353
  failure of, 323
  soil, 323
  species, 221, 271, 353
  trees, 246–7, 323
tree felling licences *see*
    licences
tree officer, 245, 326
Tree Preservation Order
    (TPO), 74, 238–41,
    243, 246, 354
  felling licences and,
    239–40
  offences, 239
  planning permission and,
    239
  protection of tree stumps
    by, 238, 240
  protection of woodlands
    by, 238
  replacement trees, 239
Tree Register, 245

tree spades, 246
trees, 234–47
  carbon sinks, 235, 289–90
  conservation areas and,
    243
  dangerous, 239
  diseased, 239
  building sites and, 243–5
  habitats for other species,
    237
  highways and, 245
    dangers to users, 245
  legal definition of, 240–1
  legal protection of *see*
    TPO, tree felling
    licences, 238–43
  natural history, 235–8
  nuisance caused by,
    239–40
  oak, pedunculate, 236
  planning permission and,
    239, 241–3
  powers and duties of local
    authorities, 241–2
  public register of trees
    felled and planted,
    245
  replacement of, 239–41
  roadside tree database,
    245
  roots
    damage to, 235–6
    precautionary area,
      244–5
  translocation of *see*
    translocation
trespass, aggravated, 230
turtles, 119

Ultra vires, 9
UNESCO, 275, 282
United Utilities, 54, 73–4,
  184, 212
urban ecology, 56–7, 97–8
Urban Summit, 88

vegetation classification *see*
  Habitat Survey,
  National Vegetation
  Classification
voles, 322, 343

waste disposal authorities,
  68

waste regulation authorities,
  67
waste water treatment, 273
water companies *see* United
  Utilities, 73–4, 203
water pollution *see* pollution
water protection zones, 225
Welsh Historic Monuments
  see Cadw
wetlands *see* Ramsar sites
whales *see* cetaceans
whaling, 288–9
  EC law and, 289
  International Whaling
    Commission, 288
White Papers, 4–5, 13, 364
  Rural White Paper, 61,
    88–9, 223, 313
  Urban White Paper, 88,
    313
wild animals *see*
    amphibians,
    invertebrates, fish,
    mammals, reptiles
  agriculture, protection of,
    126
  destruction of shelters,
    126
  disabled, 127
  disturbance of, 126
  humane killing of, 127
  illegal sale of, 126–7
  legal definition of, 26, 125
  legal protection of,
    125–130
    European protected
      species (Habitats
      Directive), 380–1
    scheduled species
      (WCA 1981), 372–4
  prohibited methods of
    taking, 128
  property, protection of,
    126
wild bird *see* birds
  legal definition of, 103
wild plant *see* bulb rustling,
    Deptford pink, orchid
  classification of, 162
  conservation organisation
    see Plantlife
  legal definition of, 163
  legal protection of, 163–4
    European protected

    species (Habitats
      Directive), 381
    scheduled species
      (WCA 1981), 374–8
  illegal sale of, 163–4
  natural history of, 162–3
  picking and uprooting,
    163, 287
  theft of, 164
Wildcat, 262
Wildfowl *see* Ramsar sites,
  286
Wildfowl and Wetlands
  Trust (WWT), 79
wildlife
  agriculture and *see*
    agriculture
  culling of, 225–7
  feeders, 131–2
  shelters, 131
Wildlife and Countryside
  Directorate, 60–2
Wildlife Crime Unit, 75
Wildlife Enhancement
  Scheme, 199–200
Wildlife Groups, Urban, 210
Wildlife Incident
  Investigation Scheme,
  179
Wildlife Inspectorate, 65–6
Wildlife Liaison Officer, 75,
  292
Wildlife Sites, Local, 210–1
Wildlife Trusts, 80, 210
wind turbines, 111–2, 311
wolves, 172–3
woodland *see* trees
  ancient, 80, 237
  secondary, 237
Woodland Trust, 80
World Heritage Sites, 282–4
  natural heritage,
    definition of, 282
  sites in UK, 283
  World Heritage List, 282
Worldwide Fund for Nature
  (WWF), 81, 290
  legal cases involving, 16,
    271
wrecks of historic interest,
  232

zoos, 29, 61, 78, 166, 221, 293,
  343, 353